NUREG-0713
Vol. 32

# Occupational Radiation Exposure at Commercial Nuclear Power Reactors and Other Facilities 2010

## Forty-Third Annual Report

Office of Nuclear Regulatory Research

# AVAILABILITY OF REFERENCE MATERIALS
## IN NRC PUBLICATIONS

**NRC Reference Material**

As of November 1999, you may electronically access NUREG-series publications and other NRC records at NRC's Public Electronic Reading Room at http://www.nrc.gov/reading-rm.html.
Publicly released records include, to name a few, NUREG-series publications; *Federal Register* notices; applicant, licensee, and vendor documents and correspondence; NRC correspondence and internal memoranda; bulletins and information notices; inspection and investigative reports; licensee event reports; and Commission papers and their attachments.

NRC publications in the NUREG series, NRC regulations, and *Title 10, Energy*, in the Code of *Federal Regulations* may also be purchased from one of these two sources.
1. The Superintendent of Documents
   U.S. Government Printing Office
   Mail Stop SSOP
   Washington, DC 20402–0001
   Internet: bookstore.gpo.gov
   Telephone: 202-512-1800
   Fax: 202-512-2250
2. The National Technical Information Service
   Springfield, VA 22161–0002
   www.ntis.gov
   1–800–553–6847 or, locally, 703–605–6000

A single copy of each NRC draft report for comment is available free, to the extent of supply, upon written request as follows:
Address:  U.S. Nuclear Regulatory Commission
          Office of Administration
          Publications Branch
          Washington, DC 20555-0001
E-mail:    DISTRIBUTION.SERVICES@NRC.GOV
Facsimile: 301–415–2289

Some publications in the NUREG series that are posted at NRC's Web site address http://www.nrc.gov/reading-rm/doc-collections/nuregs are updated periodically and may differ from the last printed version. Although references to material found on a Web site bear the date the material was accessed, the material available on the date cited may subsequently be removed from the site.

**Non-NRC Reference Material**

Documents available from public and special technical libraries include all open literature items, such as books, journal articles, and transactions, *Federal Register* notices, Federal and State legislation, and congressional reports. Such documents as theses, dissertations, foreign reports and translations, and non-NRC conference proceedings may be purchased from their sponsoring organization.

Copies of industry codes and standards used in a substantive manner in the NRC regulatory process are maintained at—
      The NRC Technical Library
      Two White Flint North
      11545 Rockville Pike
      Rockville, MD 20852–2738

These standards are available in the library for reference use by the public. Codes and standards are usually copyrighted and may be purchased from the originating organization or, if they are American National Standards, from—
      American National Standards Institute
      11 West 42$^{nd}$ Street
      New York, NY  10036–8002
      www.ansi.org
      212–642–4900

Legally binding regulatory requirements are stated only in laws; NRC regulations; licenses, including technical specifications; or orders, not in NUREG-series publications. The views expressed in contractor-prepared publications in this series are not necessarily those of the NRC.

The NUREG series comprises (1) technical and administrative reports and books prepared by the staff (NUREG–XXXX) or agency contractors (NUREG/CR–XXXX), (2) proceedings of conferences (NUREG/CP–XXXX), (3) reports resulting from international agreements (NUREG/IA–XXXX), (4) brochures (NUREG/BR–XXXX), and (5) compilations of legal decisions and orders of the Commission and Atomic and Safety Licensing Boards and of Directors' decisions under Section 2.206 of NRC's regulations (NUREG–0750).

United States Nuclear Regulatory Commission

*Protecting People and the Environment*

NUREG-0713
Vol. 32

# Occupational Radiation Exposure at Commercial Nuclear Power Reactors and Other Facilities 2010

# Forty-Third Annual Report

Manuscript Completed: March 2012
Date Published: May 2012

Prepared by
D.E. Lewis
D.A. Hagemeyer*
Y.U. McCormick*

*Oak Ridge Associated Universities
100 ORAU Way
Oak Ridge, TN 37830

Office of Nuclear Regulatory Research

# PREVIOUS REPORTS IN THIS SERIES

WASH-1311    A Compilation of Occupational Radiation Exposure from Light Water Cooled Nuclear Power Plants, 1969–1973, U.S. Atomic Energy Commission, May 1974.

NUREG-75/032    Occupational Radiation Exposure at Light Water Cooled Power Reactors, 1969–1974, U.S. Nuclear Regulatory Commission, June 1975.

NUREG-0109    Occupational Radiation Exposure at Light Water Cooled Power Reactors, 1969–1975, U.S. Nuclear Regulatory Commission, August 1976.

NUREG-0323    Occupational Radiation Exposure at Light Water Cooled Power Reactors, 1969–1976, U.S. Nuclear Regulatory Commission, March 1978.

NUREG-0482    Occupational Radiation Exposure at Light Water Cooled Power Reactors, 1977, U.S. Nuclear Regulatory Commission, May 1979.

NUREG-0594    Occupational Radiation Exposure at Commercial Nuclear Power Reactors, 1978, U.S. Nuclear Regulatory Commission, November 1979.

NUREG-0713    Occupational Radiation Exposure at Commercial Nuclear Power Reactors, 1979, Vol. 1, U.S. Nuclear Regulatory Commission, March 1981.

NUREG-0713    Occupational Radiation Exposure at Commercial Nuclear Power Reactors, 1980, Vol. 2, U.S. Nuclear Regulatory Commission, December 1981.

NUREG-0713    Occupational Radiation Exposure at Commercial Nuclear Power Reactors, 1981, Vol. 3, U.S. Nuclear Regulatory Commission, November 1982.

NUREG-0713    Occupational Radiation Exposure at Commercial Nuclear Power Reactors, 1982, Vol. 4, U.S. Nuclear Regulatory Commission, December 1983.

NUREG-0713    Occupational Radiation Exposure at Commercial Nuclear Power Reactors, 1983, Vol. 5, U.S. Nuclear Regulatory Commission, March 1985.

NUREG-0713    Occupational Radiation Exposure at Commercial Nuclear Power Reactors and Other Facilities, 1984, Vol. 6, U.S. Nuclear Regulatory Commission, October 1986.

NUREG-0713    Occupational Radiation Exposure at Commercial Nuclear Power Reactors and Other Facilities, 1985, Vol. 7, U.S. Nuclear Regulatory Commission, April 1988.

NUREG-0713    Occupational Radiation Exposure at Commercial Nuclear Power Reactors and Other Facilities, 1986, Vol. 8, U.S. Nuclear Regulatory Commission, August 1989.

NUREG-0713    Occupational Radiation Exposure at Commercial Nuclear Power Reactors and Other Facilities, 1987, Vol. 9, U.S. Nuclear Regulatory Commission, November 1990.

NUREG-0713    Occupational Radiation Exposure at Commercial Nuclear Power Reactors and Other Facilities, 1988, Vol. 10, U.S. Nuclear Regulatory Commission, July 1991.

NUREG-0713    Occupational Radiation Exposure at Commercial Nuclear Power Reactors and Other Facilities, 1989, Vol. 11, U.S. Nuclear Regulatory Commission, April 1992.

NUREG-0713    Occupational Radiation Exposure at Commercial Nuclear Power Reactors and Other Facilities, 1990, Vol. 12, U.S. Nuclear Regulatory Commission, January 1993.

NUREG-0713    Occupational Radiation Exposure at Commercial Nuclear Power Reactors and Other Facilities, 1991, Vol. 13, U.S. Nuclear Regulatory Commission, July 1993.

NUREG-0713    Occupational Radiation Exposure at Commercial Nuclear Power Reactors and Other Facilities, 1992, Vol. 14, U.S. Nuclear Regulatory Commission, December 1993.

NUREG-0713    Occupational Radiation Exposure at Commercial Nuclear Power Reactors and Other Facilities, 1993, Vol. 15, U.S. Nuclear Regulatory Commission, January 1995.

NUREG-0713    Occupational Radiation Exposure at Commercial Nuclear Power Reactors and Other Facilities, 1994, Vol. 16, U.S. Nuclear Regulatory Commission, January 1996.

NUREG-0713    Occupational Radiation Exposure at Commercial Nuclear Power Reactors and Other Facilities, 1995, Vol. 17, U.S. Nuclear Regulatory Commission, January 1997.

NUREG-0713    Occupational Radiation Exposure at Commercial Nuclear Power Reactors and Other Facilities, 1996, Vol. 18, U.S. Nuclear Regulatory Commission, February 1998.

NUREG-0713    Occupational Radiation Exposure at Commercial Nuclear Power Reactors and Other Facilities, 1997, Vol. 19, U.S. Nuclear Regulatory Commission, November 1998.

NUREG-0713    Occupational Radiation Exposure at Commercial Nuclear Power Reactors and Other Facilities, 1998, Vol. 20, U.S. Nuclear Regulatory Commission, November 1999.

NUREG-0713    Occupational Radiation Exposure at Commercial Nuclear Power Reactors and Other Facilities, 1999, Vol. 21, U.S. Nuclear Regulatory Commission, October 2000.

NUREG-0713    Occupational Radiation Exposure at Commercial Nuclear Power Reactors and Other Facilities, 2000, Vol. 22, U.S. Nuclear Regulatory Commission, September 2001.

NUREG-0713    Occupational Radiation Exposure at Commercial Nuclear Power Reactors and Other Facilities, 2001, Vol. 23, U.S. Nuclear Regulatory Commission, September 2002.

NUREG-0713    Occupational Radiation Exposure at Commercial Nuclear Power Reactors and Other Facilities, 2002, Vol. 24, U.S. Nuclear Regulatory Commission, October 2003.

NUREG-0713    Occupational Radiation Exposure at Commercial Nuclear Power Reactors and Other Facilities, 2003, Vol. 25, U.S. Nuclear Regulatory Commission, October 2004.

NUREG-0713    Occupational Radiation Exposure at Commercial Nuclear Power Reactors and Other Facilities, 2004, Vol. 26, U.S. Nuclear Regulatory Commission, December 2005.

NUREG-0713    Occupational Radiation Exposure at Commercial Nuclear Power Reactors and Other Facilities, 2005, Vol. 27, U.S. Nuclear Regulatory Commission, December 2006.

NUREG-0713    Occupational Radiation Exposure at Commercial Nuclear Power Reactors and Other Facilities, 2006, Vol. 28, U.S. Nuclear Regulatory Commission, November 2007.

NUREG-0713    Occupational Radiation Exposure at Commercial Nuclear Power Reactors and Other Facilities, 2007, Vol. 29, U.S. Nuclear Regulatory Commission, December 2008.

NUREG-0713    Occupational Radiation Exposure at Commercial Nuclear Power Reactors and Other Facilities, 2008, Vol. 30, U.S. Nuclear Regulatory Commission, December 2009.

NUREG-0713    Occupational Radiation Exposure at Commercial Nuclear Power Reactors and Other Facilities, 2009, Vol. 31, U.S. Nuclear Regulatory Commission, April 2011.

Previous reports in the NUREG-0714 series, which are now combined with NUREG-0713, are as follows:

WASH-1350-R1 through WASH-1350 R6    First through Sixth Annual Reports of the Operation of the U.S. AEC's Centralized Ionizing Radiation Exposure Records and Reporting System, U.S. Atomic Energy Commission.

NUREG-75/108    Seventh Annual Occupational Radiation Exposure Report for Certain NRC Licensees, 1974, U.S. Nuclear Regulatory Commission, October 1975.

NUREG-0119    Eighth Annual Occupational Radiation Exposure Report for 1975, U.S. Nuclear Regulatory Commission, October 1976.

NUREG-0322    Ninth Annual Occupational Radiation Exposure Report for 1976, U.S. Nuclear Regulatory Commission, October 1977.

NUREG-0463    Tenth Annual Occupational Radiation Exposure Report for 1977, U.S. Nuclear Regulatory Commission, October 1978.

NUREG-0593    Eleventh Annual Occupational Radiation Exposure Report for 1978, U.S. Nuclear Regulatory Commission, January 1981.

NUREG-0714    Twelfth Annual Occupational Radiation Exposure Report for 1979, Vol. 1, U.S. Nuclear Regulatory Commission, August 1982.

NUREG-0714    Occupational Radiation Exposure, Thirteenth and Fourteenth Annual Reports, 1980 and 1981, Vols. 2 and 3, U.S. Nuclear Regulatory Commission, October 1983.

NUREG-0714    Occupational Radiation Exposure, Fifteenth and Sixteenth Annual Reports, 1982 and 1983, Vols. 4 and 5, U.S. Nuclear Regulatory Commission, October 1985.

# ABSTRACT

This report summarizes the occupational exposure data that are maintained in the U.S. Nuclear Regulatory Commission's (NRC) Radiation Exposure Information and Reporting System (REIRS). The bulk of the information contained in the report was compiled from the 2010 annual reports submitted by five of the seven categories[1] of NRC licensees subject to the reporting requirements of 10 CFR 20.2206. Because there are no geologic repositories for high-level waste currently licensed and no NRC-licensed low-level waste disposal facilities currently in operation, only five categories will be considered in this report. The annual reports submitted by these licensees consist of radiation exposure records for each monitored individual. These records are analyzed for trends and presented in this report in terms of collective dose and the distribution of dose among the monitored individuals.

Annual reports for 2010 were received from a total of **190** NRC licensees. The summation of reports submitted by the **190** licensees indicated that **192,424** individuals were monitored, **81,961** of whom received a measurable dose (Table 3.1).[2] When adjusted for transient workers who worked at more than one licensee during the year, there were actually **142,471** monitored individuals and **62,782** who received a measurable dose (See Section 5).

The collective dose incurred by these individuals was **10,617** person-rem, which represents a **12% decrease** from the 2009 value. This decrease was primarily due to the decrease in collective dose at commercial nuclear power reactors, as well as a decrease in the collective dose for most of the other categories of NRC licensees. The number of individuals receiving a measurable dose also decreased, resulting in an average measurable dose of 0.13 rem for 2010. The average measurable dose is defined as the total effective dose equivalent (TEDE) divided by the number of individuals receiving a measurable dose.

In calendar year 2010, the average annual collective dose per reactor for light water reactor (LWR) licensees was **83** person-rem. This represents a **14% decrease** from the value reported for 2009 (96 person-rem). The decrease in collective dose for commercial nuclear power reactors was due to an 11% decrease in total outage hours in 2010. During outages, activities involving increased radiation exposure such as refueling and maintenance are performed while the reactor is not in operation. The average annual collective dose per reactor for boiling water reactors (BWRs) was **137** person-rem for **35** BWRs, and **55** person-rem for **69** pressurized water reactors (PWRs).

Analyses of transient individual data indicate that **29,333** individuals completed work assignments at two or more licensees during the monitoring year. The dose distributions are adjusted each year to account for the duplicate reporting of transient individuals by multiple licensees. The adjustment to account for transient individuals has been specifically noted in footnotes in the figures and tables for commercial nuclear power reactors. In 2010, the average measurable dose per individual for all licensees calculated from reported data was **0.13** rem. Although the average measurable dose per individual from data submitted by licensees was 0.13 rem, a corrected dose distribution resulted in an average measurable dose per individual of **0.17** rem.

---

[1] Commercial nuclear power reactors and test reactor facilities, industrial radiographers; fuel processors (including uranium enrichment facilities), fabricators, and reprocessors; manufacturing and distribution of byproduct material; independent spent fuel storage installations; facilities for land disposal of low-level waste; and geologic repositories for high-level waste. There are currently no NRC licensees involved in low-level waste disposal or geologic repositories for high-level waste.

[2] The number of individuals with measurable dose includes any individual with a dose greater than zero rem and does not include doses reported as "not detectable."

# EDITOR'S NOTE

Staff in the Offices of Nuclear Reactor Regulation, Nuclear Material Safety and Safeguards, New Reactors, Federal and State Materials and Environmental Management Programs, and Nuclear Regulatory Research assisted in the preparation of this NUREG, serving as technical reviewers. The NRC welcomes responses from readers.

Comments should be directed to:

> Doris E. Lewis
> REIRS Project Manager
> Office of Nuclear Regulatory Research
> U.S. Nuclear Regulatory Commission
> Washington, DC 20555
> Phone: 301-251-7559
> E-mail Address: Doris.Lewis@nrc.gov

---

**Paperwork Reduction Act Statement**
The information collections contained in this NUREG are subject to the Paperwork Reduction Act of 1995 (44 U.S.C. 3501 et seq.), which were approved by the Office of Management and Budget, approval number 3150-0014.

**Public Protection Notification**
The NRC may not conduct or sponsor, and a person is not required to respond to, a request for information or an information collection requirement unless the requesting document displays a currently valid OMB control number.

# TABLE OF CONTENTS

# TABLE OF CONTENTS (Continued)

# TABLE OF CONTENTS (Continued)

## LIST OF FIGURES

# TABLE OF CONTENTS (Continued)

## LIST OF TABLES

# PREFACE

A number of NRC licensees have inquired as to how the occupational radiation exposure data that are compiled from the individual exposure reports required by 10 CFR 20.2206 are used by the NRC staff. In combination with other sources of information, the principal uses of the data are to provide facts regarding routine occupational exposures to radiation and radioactive material that occur in connection with certain NRC-licensed activities. The data can be used by the NRC staff as indicated below:

1.  The data permit evaluation of trends, both favorable and unfavorable, from the viewpoint of the effectiveness of overall NRC/licensee radiation protection and as low as is reasonably achievable (ALARA) efforts by licensees.

2.  The data assist in the evaluation of the radiological risk associated with certain categories of NRC-licensed activities and are used for comparative analyses of radiation protection performance: U.S./foreign, boiling water reactors/pressurized water reactors (BWRs/PWRs), civilian/military, facility/facility, nuclear industry/other industries, etc.

3.  The data are used as one of the metrics of the NRC Reactor Oversight Program to evaluate the effectiveness of the licensees' ALARA programs and also for inspection planning purposes.

4.  The data permit evaluation of transient individuals who may affect dose distribution statistics through multiple counting.

5.  The data are used in the establishment of priorities for the utilization of NRC health physics resources: research, standards development, and regulatory program development.

6.  The data provide facts for answering Congressional and administration inquiries and for responding to questions raised by the public.

7.  The data are used to provide radiation exposure histories to individuals who were exposed to radiation at NRC-licensed facilities.

8.  The data provide information that may be used to conduct epidemiologic studies.

# FOREWORD

Through this annual report, the NRC supports openness in its regulatory process by providing the public with accurate and timely information about the radiation protection program of NRC's licensees. Toward that end, NUREG-0713, Volume 32, summarizes the 2010 occupational radiation exposure data maintained in the NRC's Radiation Exposure Information and Reporting System (REIRS) database.

Seven categories of NRC licensees are required to report annually on individual exposure in accordance with Title 10 of the Code of Federal Regulations, Section 20.2206 (10 CFR 20.2206, "Reports of Individual Monitoring"). Specifically, these categories include commercial nuclear power reactors; industrial radiographers; fuel processors (including uranium enrichment facilities), fabricators, and reprocessors; manufacturing and distribution of byproduct material; independent spent fuel storage installations; facilities for land disposal of low-level waste; and geologic repositories for high-level waste. Because NRC has not licensed any geologic repositories for high-level waste and no NRC-licensed low-level waste disposal facilities are currently in operation, this report considers only the first five categories of NRC licensees. As such, this report reflects the occupational radiation exposure data that NRC received from 190 licensees.

The data submitted by licensees consist of radiation exposure records for each monitored individual. In 2010, 142,471 individuals were monitored and 62,782 received a measurable dose when adjusted for transient individuals who worked at more than one facility during the year. This report analyzes and presents these records in terms of collective dose and the distribution of dose among the monitored individuals. During 2010, these individuals incurred a collective dose of 10,617 person-rem, which represents a 12% decrease from the 2009 value of 12,056 person-rem. This decrease was primarily due to the decrease in collective dose at commercial nuclear power reactors, as well as a decrease in the collective dose for most of the other categories of NRC licensees. The average measurable dose is the total collective dose divided by the number of individuals receiving a measurable dose. While the collective dose decreased from 2009 to 2010, there was a proportional decrease in the number of individuals receiving a measurable dose, resulting in the average measurable dose decreasing from 0.18 rem in 2009 to 0.17 rem in 2010 when adjusted for transient workers. This value can be compared with the 0.31 rem [Ref. 1] that the average person in the United States receives annually from natural background radiation. Worldwide annual exposures to natural background radiation are generally expected to be in the range of 0.1 rem to 1.3 rem, with 0.24 rem [Ref. 2] being the current average worldwide value.

This annual report is useful in evaluating trends in occupational radiation exposure to assess the effectiveness of licensees' radiation protection programs to maintain exposures as low as is reasonably achievable (ALARA). For example, the NRC staff uses the data presented in this report as one of the metrics of the NRC's Reactor Oversight Program to evaluate the effectiveness of licensees' ALARA programs.

# ABBREVIATIONS

| | |
|---|---|
| AEC | U.S. Atomic Energy Commission |
| ALARA | as low as is reasonably achievable |
| | |
| BWR | boiling water reactor |
| | |
| CDE | committed dose equivalent |
| CEDE | committed effective dose equivalent |
| CFR | Code of Federal Regulations |
| | |
| D&D | decontamination and decommissioning |
| DDE | deep dose equivalent |
| DOE | U.S. Department of Energy |
| | |
| ERDA | Energy Research and Development Administration |
| | |
| FSME | Office of Federal and State Materials and Environmental Management Programs |
| FSSR | final status survey report |
| | |
| ICRP | International Commission on Radiological Protection |
| ISFSI | independent spent fuel storage installation |
| | |
| LDE | lens dose equivalent |
| LES | Louisiana Energy Services |
| LTP | license termination plan |
| LWR | light water reactor |
| | |
| M&D | manufacturing and distribution |
| mSv | millisievert |
| MWe | megawatts electric |
| MW-yr | megawatt-year |
| | |
| ND | not detectable |
| NMSS | Office of Nuclear Material Safety and Safeguards |
| NR | not required to be reported |
| NRC | U.S. Nuclear Regulatory Commission |
| NRR | Office of Nuclear Reactor Regulation |

# ABBREVIATIONS (Continued)

| | |
|---|---|
| PSDAR | Post shut-down decommissioning activities report |
| PSE | planned special exposure |
| PWR | pressurized water reactor |
| | |
| REIRS | Radiation Exposure Information and Reporting System |
| RES | Office of Nuclear Regulatory Research |
| | |
| SDE-ME | shallow dose equivalent maximum extremity |
| SDE-WB | shallow dose equivalent whole body |
| SI | international system of units |
| $SR_E$ | collective dose distribution ratio |
| SSC | safety related structures, systems and components |
| Sv | sieverts |
| | |
| TEDE | total effective dose equivalent |
| TMI | Three Mile Island |
| TODE | total organ dose equivalent |
| | |
| $UF_6$ | uranium hexafluoride |
| USEC | United States Enrichment Corporation, Inc. |

# Section 1
# INTRODUCTION

## 1.1 BACKGROUND

One of the basic purposes of the Atomic Energy Act and the implementing regulations in Title 10, Part 20, of the *Code of Federal Regulations* (10 CFR Part 20), is to protect the health and safety of the public, including the employees of the licensees conducting operations under those regulations. The regulations at 10 CFR 20.1502 specifies conditions that require individual monitoring of external and internal occupational dose. Each licensee is also required, under 10 CFR 20.2106(f), to maintain records of the results of such monitoring until the Commission terminates the license. However, there was no initial provision that these records or any summary of them be transmitted to a central location where the data could be retrieved and analyzed.

On November 4, 1968, the U.S. Atomic Energy Commission (AEC) published an amendment to 10 CFR Part 20 requiring the reporting of certain occupational radiation exposure information to a central repository at AEC Headquarters. At that time, there were only four categories[3] of AEC licensees required to report. These facilities were considered to have the greatest potential for significant occupational doses. A procedure was established whereby the appropriate occupational exposure data were extracted from these reports and entered into the AEC Radiation Exposure Information and Reporting System (REIRS), a computer system that was maintained at the Oak Ridge National Laboratory Computer Technology Center in Oak Ridge, Tennessee, until May 1990.

At that time, the data were transferred to a database management system and are now maintained at the Oak Ridge Institute for Science and Education, which is managed by Oak Ridge Associated Universities. The computerization of these data facilitates their retrieval and analysis. The data maintained in REIRS have been summarized and published in a report every year since 1969. Annual reports for each of the years 1969 through 1973 presented the data reported by both AEC licensees and contractors and were published in six documents designated as WASH-1350-R1 through WASH-1350-R6.

In January 1975, with the separation of AEC into the Energy Research and Development Administration (ERDA) and the U.S. Nuclear Regulatory Commission (NRC), each agency assumed responsibility for collecting and maintaining occupational radiation exposure information reported by the facilities under its jurisdiction. The annual reports published by NRC on occupational exposure for calendar year 1974 and subsequent years do not contain information pertaining to ERDA facilities or contractors. Comparable information for facilities and contractors under ERDA, now the U.S. Department of Energy (DOE), is collected and published by the DOE Office of Corporate Analysis within the Office of Health, Safety and Security, in Germantown, Maryland.

---

[3] Commercial nuclear power reactors; industrial radiographers; fuel processors (including uranium enrichment facilities as of 1997), fabricators, and reprocessors; and manufacturing and distribution of specified quantities of byproduct material.

In 1982 and 1983, 10 CFR 20.408(a) was amended to require three additional categories of NRC licensees to submit annual statistical exposure reports and individual termination exposure reports. The three additional NRC licensee categories were: (1) geologic repositories for high-level radioactive waste, (2) independent spent fuel storage installations, and (3) facilities for the land disposal of low-level radioactive waste. This document presents the exposure information that was reported by NRC licensees representing one of these categories – independent spent fuel storage installations; there are no geologic repositories for high-level waste currently licensed, and there are no low-level land disposal facilities currently in operation that report to the NRC.

In May 1991, 10 CFR Part 20 was revised. The revision redefined the radiation monitoring and reporting requirements of NRC licensees. Instead of submitting summary annual reports (§20.407) and termination reports (§20.408), licensees are now required to submit an annual report of the dose received by each monitored individual (§20.2206). Licensees were required to implement the new requirements no later than January 1994.

This report summarizes information reported for the current year and previous 10 years. More licensee-specific data for the previous 10 years, such as the annual reports submitted by each commercial nuclear power reactor pursuant to 10 CFR 20.407 and 20.2206 (after 1993) and their technical specifications (prior to Volume 20 of this report), may be found in the documents listed on the inside of the front cover of this report for the specific year desired. Additional operating data and statistics for each commercial nuclear power reactor for the years 1973 through 1982 may be found in a series of reports, Nuclear Power Plant Operating Experience [Refs. 3–11]. These documents are available for viewing at all NRC public document rooms, as well as on the NRC public Web site (www.nrc.gov), or they may be purchased from the National Technical Information Service, as shown in the References section.

## 1.2 RADIATION EXPOSURE INFORMATION ON THE INTERNET

In May 1995, NRC began pursuing the dissemination of radiation exposure information via a Web site on the Internet. This site allows interested parties with the appropriate equipment to access the data electronically rather than through the published NUREG-0713 document. A Web site was created for radiation exposure and linked into the main NRC Web page. The Web site contains up-to-date information on radiation exposure, as well as information and guidance on reporting radiation exposure information to NRC. Interested parties may read the documents online or download information to their systems for further analysis. The Radiation Exposure Monitoring and Information Transmittal System, a software application designed to maintain licensee dose records, and REIRView, a software package designed to validate a licensee's annual data submittal, are also available for downloading via the Web site. There are also links to other Web sites dealing with the topics of radiation and health physics. Individuals may submit requests for their dose records contained in REIRS on this Web site. In addition, organizations that have provided documentation to the NRC may also submit requests for dose records contained in REIRS on this website.

NRC intends to continue pursuing the dissemination of radiation exposure information via the Web and will focus more resources on the electronic distribution of information rather than the publication of hard-copy reports.

The main Web address for NRC is

**http://www.nrc.gov**

The NRC radiation exposure information Web URL is

**http://www.reirs.com**

Comments on this report or the NRC's radiation exposure Web page should be directed to

**Doris E. Lewis**
**REIRS Project Manager**
**Office of Nuclear Regulatory Research**
**U.S. Nuclear Regulatory Commission**
**Washington, DC 20555**
**Phone: 301-251-7559**
**E-mail Address: Doris.Lewis@nrc.gov**

# Section 2
# LIMITATIONS OF THE DATA

All of the figures compiled in this report relating to exposures and occupational doses are based on the results and interpretations of the readings of various types of personnel-monitoring devices employed by each licensee. This information, obtained from routine personnel-monitoring programs, is sufficient to characterize the radiation exposure incident to individuals' work and is used in evaluating the radiation protection program.

Monitoring requirements are specified in 10 CFR 20.1502, which requires licensees to monitor individuals who receive or are likely to receive, in one year, a dose in excess of 10 percent of the applicable limits. For occupational individuals, the annual limit for the whole body is 5 rem, so 0.5 rem per year is the level above which monitoring is required. Separate dose limits have been established for minors, declared pregnant women, and members of the public. Monitoring is also required for any individual entering a high or very high radiation area. Depending on the administrative policy of each licensee, persons such as visitors and clerical individuals may also be provided with monitoring devices, even though the probability of their exposure to measurable levels of radiation is extremely small.

Pursuant to 10 CFR 20.2206(b), certain categories of licensees must submit an annual report of the results of individual monitoring carried out by the licensee for each individual for whom monitoring was required by Section 20.1502. In addition to this requirement, many licensees elect to report the doses for every individual for whom they provided monitoring. This practice increases the number of individuals that are monitored for radiation exposure. In an effort to account for this increase, the number of individuals reported as having "no measurable dose"[4] is subtracted from the total number of monitored individuals. This resulting number can then be used to calculate the average measurable dose per individual as well as the average dose per monitored individual.

This report contains information reported by NRC licensees. Since NRC licenses all commercial nuclear power reactors, fuel processors and fabricators, and independent spent fuel storage installations, information shown for these categories reflect all relevant activity in the United States (U.S.). This is not the case, however, for the remaining categories of industrial radiography, manufacturing and distribution of specified quantities of byproduct material, and low-level waste disposal. Many companies that conduct these types of activities are located in Agreement States. More than seven times as many facilities are licensed and regulated by Agreement States than are licensed and regulated by NRC. Agreement States are not required to adopt the reporting requirements in 10 CFR 20.2206. As a result, Agreement State licensees are not required to submit occupational dose reports to NRC.

---

[4] The number of workers with measurable dose includes any individual with a total effective dose equivalent greater than zero rem. Workers reported with zero dose, or no detectable dose, are included in the number of workers with no measurable exposure.

Although some Agreement State licensees voluntarily submit occupational dose reports to NRC, these results are not included in the analyses presented in Sections 3, 5, and 6 of this report. The NRC staff is currently developing the report *Occupational Radiation Exposure at Agreement State-Licensed Materials Facilities*. This report provides information regarding occupational radiation exposures at Agreement State-licensed facilities. This report will be available in Summer 2012 and may be obtained from the website, www.reirs.com. In addition, this report does not include compilations of nonoccupational exposure, such as exposure received by medical patients from X-rays, fluoroscopy, or accelerators.

The average dose per individual, as well as the dose distributions shown for groups of licensees, also can be affected by the multiple reporting of individuals who were monitored by two or more licensees during the year. Licensees are only required to report the doses received by individuals at their licensed facilities. A dose distribution for a single licensee does not consider that some of the individuals may have received doses at other facilities. When the data are summed to determine the total number of individuals monitored by a group of licensees, individuals may be counted more than once if they have worked at more than one facility during the calendar year. These occurrences can also affect the distribution of doses because individuals may be counted multiple times in the lower dose ranges rather than one time in the higher dose range corresponding to the actual accumulated dose for the year (the sum of an individual's dose accrued at all facilities). This source of error

has the greatest potential impact on the data reported by commercial nuclear power reactors since they employ many short-term individuals. Section 5 contains an analysis that corrects for transient individuals being counted more than once.

When examining the annual statistical data, it is important to note that all of the personnel included in the report may not have been monitored throughout the entire year. Many licensees, such as radiography firms and commercial nuclear power reactors, may monitor numerous individuals for periods much less than a year. The average doses calculated from these data, therefore, are less than the average dose that an individual involved in that activity would receive for the full year.

Considerable attention should be given when referencing the collective totals presented in this report. The differences between the totals presented for all licensees that reported versus only those licensees that are required to report should be noted. See Section 1.1 for the categories of licensees that are required to report to REIRS. A number of licensees are not required to report to REIRS but voluntarily report for convenient recordkeeping or because they have reported in the past and have decided to continue to do so. These licensees are listed in Appendix A, Table A2 – Other Facilities Reporting to the NRC.

Likewise, one should distinguish between the doses attributed to the pressurized water reactors (PWRs) and the doses attributed to boiling water reactors (BWRs). The totals may be inclusive or exclusive of those licensees that were in commercial operation for less than one full year. These parameters vary throughout the tables and appendices of this report. The apparent discrepancies among the various tables are a necessary side effect of this endeavor.

The data contained in this report are subject to change because licensees may submit corrections or additions to data for previous years. For the 2010 report, additional data received from a uranium hexafluoride ($UF_6$) production plant were added to the report for the years 2000 - 2010. This provides a more comprehensive and accurate analysis for the fuel cycle licensees.

All dose equivalent values in this report are given in units of rem in accordance with the general provisions for records in 10 CFR 20.2101(a). In order to convert rem into the International System of Units (SI) unit of sieverts (Sv), readers should divide the value in rem by 100. Therefore, 1 rem = 0.01 Sv. In order to convert rem into millisieverts (mSv), readers should multiply the value in rem by 10.

# Section 3

# ANNUAL PERSONNEL MONITORING REPORTS – 10 CFR 20.2206

## 3.1 DEFINITION OF TERMS AND METHODOLOGIES

### 3.1.1 Number of Licensees Reporting

The number of licensees refers to the NRC licenses issued to use radioactive material for certain activities that would place the licensees in one of the seven[5] categories that are required to report pursuant to 10 CFR 20.2206. The third column in Table 3.1 shows the number of licensees that have filed such reports during the past eleven years. All commercial nuclear power reactors, fuel processors and fabricators, and independent spent fuel storage installations are required to report occupational exposure to NRC, whether or not they are in an Agreement State.

Many companies that conduct industrial radiography and manufacturing and distribution activities are located in and regulated by Agreement States and are, therefore, not required to adopt the reporting requirements of 10 CFR 20.2206. However, industrial radiography and manufacturing and distribution licensees that are licensed and regulated by NRC are required to report occupational exposure to NRC. Appendix A, Table A1 lists all non-reactor licensees that reported occupational data to NRC in 2010.

### 3.1.2 Number of Monitored Individuals

The number of monitored individuals refers to the total number of individuals that NRC licensees reported as being monitored for exposure to external and internal radiation during the year. This number includes all individuals for whom monitoring is required under 10 CFR 20.1502. This number also includes visitors, service representatives, contract individuals, clerical individuals, and any other individuals for whom the licensee determines that monitoring devices should be provided, although monitoring was not required.

The total number of individuals was determined from the number of unique personal identification numbers submitted per licensee. Uniqueness is defined by the combination of identification number and identification type [Ref. 12].

### 3.1.3 Number of Individuals with Measurable Dose

The number of individuals with measurable dose includes any individual with a total effective dose equivalent (TEDE) greater than zero rem.

### 3.1.4 Collective Dose

The concept of collective dose is used in this report to denote the summation of the TEDE received by all monitored individuals and is reported in units of person-rem. Since 10 CFR 20.2206 requires that the TEDE be reported, the collective dose is calculated by summing the TEDE for all monitored individuals.

---

[5] These categories are commercial nuclear power reactors; industrial radiographers; fuel processors (including uranium enrichment facilities), fabricators, and reprocessors; manufacturing and distribution of byproduct material; independent spent fuel storage installations; facilities for land disposal of low-level waste; and geologic repositories for high-level waste. There are currently no NRC licensees involved in low-level waste disposal or geologic repositories for high-level waste.

## TABLE 3.1
### Average Annual Exposure Data for Certain Categories of NRC Licensees
### 2000–2010

| NRC License Category * and Program code | Calendar Year | Number of Licensees Reporting | Number of Monitored Individuals | Number of Individuals with Measurable TEDE | Collective TEDE (person-rem) | Average TEDE (rem) | Average Measurable TEDE per Individual (rem) |
|---|---|---|---|---|---|---|---|
| Industrial Radiography<br><br>03310<br>03320 | 2000 | 128 | 3,157 | 2,454 | 1,525.143 | 0.48 | 0.62 |
| | 2001 | 123 | 3,560 | 3,040 | 2,106 213 | 0 59 | 0.69 |
| | 2002 | 100 | 3,420 | 2,842 | 1,729 222 | 0 51 | 0.61 |
| | 2003 | 118 | 3,115 | 2,651 | 1,584 249 | 0 51 | 0.60 |
| | 2004 | 113 | 3,568 | 3,014 | 1,603.591 | 0.45 | 0.53 |
| | 2005 | 90 | 3,009 | 2,623 | 1,504 575 | 0 50 | 0.57 |
| | 2006 | 78 | 2,388 | 1,981 | 1,109 347 | 0.46 | 0.56 |
| | 2007 | 74 | 2,607 | 2,224 | 1,315.171 | 0 50 | 0.59 |
| | 2008 | 61 | 2,967 | 2,587 | 1,460.757 | 0.49 | 0.56 |
| | 2009 | 64 | 2,651 | 2,302 | 1,317.135 | 0 50 | 0.57 |
| | 2010 | 56 | 2,371 | 2,030 | 1,296 291 | 0 55 | 0.64 |
| Manufacturing and Distribution<br><br>02500<br>03211<br>03212<br>03214 | 2000 | 39 | 2,460 | 1,187 | 415.402 | 0.17 | 0.35 |
| | 2001 | 35 | 1,705 | 1,184 | 344.743 | 0 20 | 0.29 |
| | 2002 | 29 | 1,437 | 1,052 | 328 092 | 0 23 | 0.31 |
| | 2003 | 33 | 2,372 | 1,796 | 436 660 | 0.18 | 0.24 |
| | 2004 | 28 | 2,539 | 1,787 | 347.258 | 0.14 | 0.19 |
| | 2005 | 23 | 2,566 | 1,557 | 388.547 | 0.15 | 0.25 |
| | 2006 | 22 | 1,256 | 795 | 273.028 | 0 22 | 0.34 |
| | 2007 | 23 | 2,106 | 1,463 | 291.326 | 0.14 | 0.20 |
| | 2008 | 18 | 1,934 | 1,341 | 222.123 | 0.11 | 0.17 |
| | 2009 | 16 | 1,933 | 1,386 | 179 222 | 0 09 | 0.13 |
| | 2010 | 17 | 970 | 670 | 146 365 | 0.15 | 0.22 |
| Independent Spent Fuel Storage<br><br>23100<br>23200 | 2000 | 2 | 146 | 83 | 5.571 | 0 04 | 0.07 |
| | 2001 | 2 | 154 | 107 | 13 088 | 0 08 | 0.12 |
| | 2002 | 2 | 75 | 67 | 6 013 | 0 08 | 0.09 |
| | 2003 | 2 | 55 | 46 | 2.791 | 0 05 | 0.06 |
| | 2004 | 1 | 37 | 27 | 1.257 | 0 03 | 0.05 |
| | 2005 | 2 | 59 | 30 | 0.769 | 0 01 | 0.03 |
| | 2006 | 2 | 59 | 26 | 2.108 | 0 04 | 0.08 |
| | 2007 | 2 | 57 | 26 | 1 697 | 0 03 | 0.07 |
| | 2008 | 2 | 53 | 21 | 1 248 | 0 02 | 0.06 |
| | 2009 | 2 | 72 | 34 | 1.465 | 0 02 | 0.04 |
| | 2010 | 2 | 73 | 39 | 1 337 | 0 02 | 0.03 |
| Fuel Cycle Licenses - Fabrication Processing and Uranium Enrichment and UF$_6$ Production Plants<br><br>11400<br>21200<br>21210 | 2000 | 10 | 9,681 | 4,954 | 1,409.055 | 0.15 | 0.28 |
| | 2001 | 10 | 8,498 | 4,316 | 1,286 572 | 0.15 | 0.30 |
| | 2002 | 9 | 8,270 | 4,209 | 820.442 | 0.10 | 0.19 |
| | 2003 | 9 | 8,103 | 3,986 | 676 082 | 0 08 | 0.17 |
| | 2004 | 9 | 8,060 | 4,283 | 657.799 | 0 08 | 0.15 |
| | 2005 | 10 | 8,215 | 3,839 | 643 631 | 0 08 | 0.17 |
| | 2006 | 10 | 8,097 | 4,017 | 677 025 | 0 08 | 0.17 |
| | 2007 | 10 | 8,402 | 4,007 | 588.837 | 0 07 | 0.15 |
| | 2008 | 10 | 7,807 | 3,424 | 538.201 | 0 07 | 0.16 |
| | 2009 | 11 | 8,918 | 3,738 | 533.721 | 0 06 | 0.14 |
| | 2010 | 11 | 9,362 | 4,212 | 541.876 | 0 06 | 0.13 |
| Commercial Light Water Reactors (LWRs) **<br><br>41111 | 2000 | 104 | 147,901 | 74,108 | 12,651.682 | 0 09 | 0.17 |
| | 2001 | 104 | 140,776 | 67,570 | 11,108.552 | 0 08 | 0.16 |
| | 2002 | 104 | 149,512 | 73,242 | 12,126.190 | 0 08 | 0.17 |
| | 2003 | 104 | 152,702 | 74,813 | 11,955 570 | 0 08 | 0.16 |
| | 2004 | 104 | 150,322 | 69,849 | 10,367.897 | 0 07 | 0.15 |
| | 2005 | 104 | 160,701 | 78,127 | 11,455.807 | 0 07 | 0.15 |
| | 2006 | 104 | 164,823 | 80,265 | 11,021.186 | 0 07 | 0.14 |
| | 2007 | 104 | 164,081 | 79,530 | 10,120 013 | 0 06 | 0.13 |
| | 2008 | 104 | 169,324 | 79,450 | 9,195 940 | 0 05 | 0.12 |
| | 2009 | 104 | 176,381 | 81,754 | 10,024 804 | 0 06 | 0.12 |
| | 2010 | 104 | 179,648 | 75,010 | 8,631 384 | 0 05 | 0.12 |
| Grand Totals and Averages | 2000 | 283 | 163,345 | 82,786 | 16,006.853 | 0.10 | 0.19 |
| | 2001 | 274 | 154,693 | 76,217 | 14,859.168 | 0.10 | 0.19 |
| | 2002 | 244 | 162,714 | 81,412 | 15,009.959 | 0 09 | 0.18 |
| | 2003 | 266 | 166,347 | 83,292 | 14,655.352 | 0 09 | 0.18 |
| | 2004 | 255 | 164,526 | 78,960 | 12,977.802 | 0 08 | 0.16 |
| | 2005 | 229 | 174,550 | 86,176 | 13,993.329 | 0 08 | 0.16 |
| | 2006 | 216 | 176,623 | 87,084 | 13,082.694 | 0 07 | 0.15 |
| | 2007 | 213 | 177,253 | 87,250 | 12,317.044 | 0 07 | 0.14 |
| | 2008 | 195 | 182,085 | 86,823 | 11,418.269 | 0 06 | 0.13 |
| | 2009 | 197 | 189,955 | 89,214 | 12,056 347 | 0 06 | 0.14 |
| | 2010 | 190 | 192,424 | 81,961 | 10,617 253 | 0 06 | 0.13 |

\* These categories consist only of NRC licensees required to submit an annual report (see Section 2).

\*\* This category includes all LWRs in commercial operation for a full year for each of the years indicated. Reactor data have not been corrected to account for the multiple counting of transient reactor workers (see Section 5).

The phrase "collective dose" is used throughout this report to mean the collective TEDE, unless otherwise specified.

Prior to the implementation of the revised dose reporting requirements of 10 CFR 20.2206 in 1994, the collective dose, in some cases, was calculated from the dose distributions by multiplying the number of individuals reported in each of the dose ranges by the midpoint of the corresponding dose range and then summing the products. This assumed that the midpoint of the range was equal to the arithmetic mean of the individual doses in the range. Experience has shown that the actual mean dose of individuals reported in each dose range is less than the midpoint of the range. For this reason, the resultant calculated collective doses shown in this report for these licensees may be approximately 10% higher than the sum of the actual individual doses. Care should be taken when comparing the actual collective dose calculated for 1994 to 2010 with the collective dose for years prior to 1994 because of this change in methodology.

In addition, prior to 1994, doses only included the external whole-body dose with no internal dose contribution. Although the contribution of internal dose to the TEDE is minimal for most licensees, it should be considered when comparing collective doses for 1994 and later with the collective dose for years prior to 1994. One noted exception is for fuel fabrication licensees, where the committed effective dose equivalent (CEDE), in some cases, contributes the majority of the TEDE (see Section 3.3.5).

### *3.1.5 Average Individual Dose*

The average individual dose is obtained by dividing the collective dose by the total number of monitored individuals. This figure is usually less than the average measurable dose because it includes the number of those individuals who received zero or less than measurable doses.

### *3.1.6 Average Measurable Dose*

The average measurable dose is obtained by dividing the collective TEDE by the number of individuals with a measurable dose. This is the average most commonly used in this and other reports when examining trends and comparing doses received by individuals in various segments of the nuclear industry.

## 3.2 ANNUAL TEDE DOSE DISTRIBUTIONS

Table 3.2 provides a statistical compilation of the occupational dose reports by categories of licensees (see Section 3.3 for a description of each licensee category). The dose distributions are generated by summing the TEDE for each individual and counting the number of individuals in each dose range. In nearly every licensee category, a large number of individuals receive doses that are less than measurable, and only one individual exceeded 4 rem in 2010. Ninety-two percent of the reported individuals with measurable doses (shown in Table 3.2) were monitored by commercial nuclear power reactors in 2010, where they received 81% of the total collective dose.

## TABLE 3.2
### Distribution of Annual Collective TEDE by License Category
### 2010

| License Category (Number of sites reporting) | Number of Individuals with TEDE in the Ranges (rem) * | | | | | | | | | | | | | Total Number Monitored | Number with Meas. Dose | Total Collective Dose (TEDE) (person-rem) |
|---|---|---|---|---|---|---|---|---|---|---|---|---|---|---|---|---|
| | No meas. | Meas. <0.1 | 0.10-0.25 | 0.25-0.50 | 0.50-0.75 | 0.75-1.00 | 1.00-2.00 | 2.00-3.00 | 3.00-4.00 | 4.00-5.00 | 5.00-6.00 | 6.00-12.00 | >12 | | | |
| **INDUSTRIAL RADIOGRAPHY** | | | | | | | | | | | | | | | | |
| Fixed Locations (2) | 71 | 12 | 1 | - | - | - | - | - | - | - | - | - | - | 84 | 13 | 0.496 |
| Temporary Job Sites (54) | 270 | 552 | 285 | 299 | 255 | 153 | 336 | 103 | 34 | - | - | - | - | 2,287 | 2,017 | 1,295.795 |
| Total (56) | 341 | 564 | 286 | 299 | 255 | 153 | 336 | 103 | 34 | - | - | - | - | 2,371 | 2,030 | 1,296.291 |
| **MANUFACTURING AND DISTRIBUTION** | | | | | | | | | | | | | | | | |
| Type "A" Broad (2) | 47 | 141 | 81 | 38 | 17 | 21 | 41 | - | - | - | - | - | - | 386 | 339 | 115.733 |
| Type "B" Broad and Other (3) | 41 | 12 | 3 | - | - | 1 | - | 1 | - | - | - | - | - | 58 | 17 | 4.410 |
| Nuclear Pharmacies (12) | 212 | 257 | 38 | 10 | 2 | 5 | 1 | - | - | 1 | - | - | - | 526 | 314 | 26.222 |
| Total (17) | 300 | 410 | 122 | 48 | 19 | 27 | 42 | 1 | - | 1 | - | - | - | 970 | 670 | 146.365 |
| **INDEPENDENT SPENT FUEL STORAGE** | | | | | | | | | | | | | | | | |
| Total (2) | 34 | 35 | 4 | - | - | - | - | - | - | - | - | - | - | 73 | 39 | 1.337 |
| **FUEL CYCLE **** | | | | | | | | | | | | | | | | |
| Total (11) | 5,150 | 2,704 | 832 | 436 | 158 | 61 | 21 | - | - | - | - | - | - | 9,362 | 4,212 | 541.876 |
| **COMMERCIAL POWER REACTORS **** * | | | | | | | | | | | | | | | | |
| Boiling Water (35) | 31,960 | 23,284 | 8,187 | 4,019 | 1,143 | 385 | 195 | 1 | - | - | - | - | - | 69,174 | 37,214 | 4,807.656 |
| Pressurized Water (69) | 72,678 | 26,287 | 7,855 | 2,637 | 658 | 217 | 138 | 4 | - | - | - | - | - | 110,474 | 37,796 | 3,823.728 |
| Total (104) | 104,638 | 49,571 | 16,042 | 6,656 | 1,801 | 602 | 333 | 5 | - | - | - | - | - | 179,648 | 75,010 | 8,631.384 |
| **GRAND TOTALS** | 110,463 | 53,284 | 17,286 | 7,439 | 2,233 | 843 | 732 | 109 | 34 | 1 | - | - | - | 192,424 | 81,961 | 10,617.253 |

* Dose values exactly equal to the values separating ranges are reported in the next higher range.

** This category includes fabrication, processing, and uranium enrichment plants (see Sec ion 3.3 5).

*** This category includes all reactors in commercial operation for a full year during 2010. Al hough Brown's Ferry 1 was placed on administrative hold in 1985, it remains in the count of opera ing reactors and has resumed operation as of June, 2007. These values have not been adjusted for the multiple counting of transient reactor workers (see Section 5).

## 3.3 SUMMARY OF OCCUPATIONAL DOSE DATA BY LICENSE CATEGORY

### *3.3.1 Industrial Radiography Licenses, Fixed Locations and Temporary Job Sites*

Industrial radiography licenses are issued to allow the use of sealed radioactive materials, usually in exposure devices or "cameras," that primarily emit gamma rays for nondestructive testing of pipeline weld joints, steel structures, boilers, aircraft and ship parts, and other high-stress alloy parts. Some firms are licensed to conduct such activities in one location, usually in a permanent facility designed and shielded for radiography; others perform radiography at temporary job sites in the field. The radioisotopes most commonly used are cobalt-60 and iridium-192. As shown in Table 3.1, annual reports were received for 56 radiography licensees in 2010. Table 3.3 summarizes the reported data for the two types of industrial radiography licenses for 2008, 2009, and 2010 for comparison purposes.

The average measurable dose for individuals performing radiography at a fixed location ranged from 4% to 7% of the average measurable dose of individuals at temporary job sites over the past three years. This is because it is more difficult for individuals to avoid exposure to radiation at temporary job sites in the field, where conditions are not optimal and may change daily.

High exposures in radiography can be directly attributable to the type and location of the radiography field work. For example, locations such as oil drilling platforms and aerial tanks offer the radiographer little available shielding. In these situations, there may not be an opportunity to use distance as a means of reducing exposure. Although these licensed activities usually result in average measurable doses that are higher than those received by other licensees, they involve a relatively small number of exposed individuals.

Figure 3.1 shows the number of individuals with measurable dose, the total collective dose, and

**TABLE 3.3**
Annual Exposure Information for Industrial Radiography Licensees
2008-2010

| Year | Type of License | Number of Licensees | Number of Monitored Individuals | Individuals with Measurable Dose | Collective Dose (person-rem) | Average Measurable Dose (rem) |
|------|-----------------|---------------------|---------------------------------|----------------------------------|------------------------------|-------------------------------|
|      | Fixed Location | 3 | 61 | 26 | 0.509 | 0.02 |
| 2008 | Temporary Job Sites | 58 | 2,906 | 2,561 | 1,460.248 | 0.57 |
|      | **Total** | **61** | **2,967** | **2,587** | **1,460.757** | **0.56** |
|      | Fixed Location | 2 | 80 | 45 | 1.805 | 0.04 |
| 2009 | Temporary Job Sites | 62 | 2,571 | 2,257 | 1,315.330 | 0.58 |
|      | **Total** | **64** | **2,651** | **2,302** | **1,317.135** | **0.57** |
|      | Fixed Location | 2 | 84 | 13 | 0.496 | 0.04 |
| 2010 | Temporary Job Sites | 54 | 2,287 | 2,017 | 1,295.795 | 0.64 |
|      | **Total** | **56** | **2,371** | **2,030** | **1,296.291** | **0.64** |

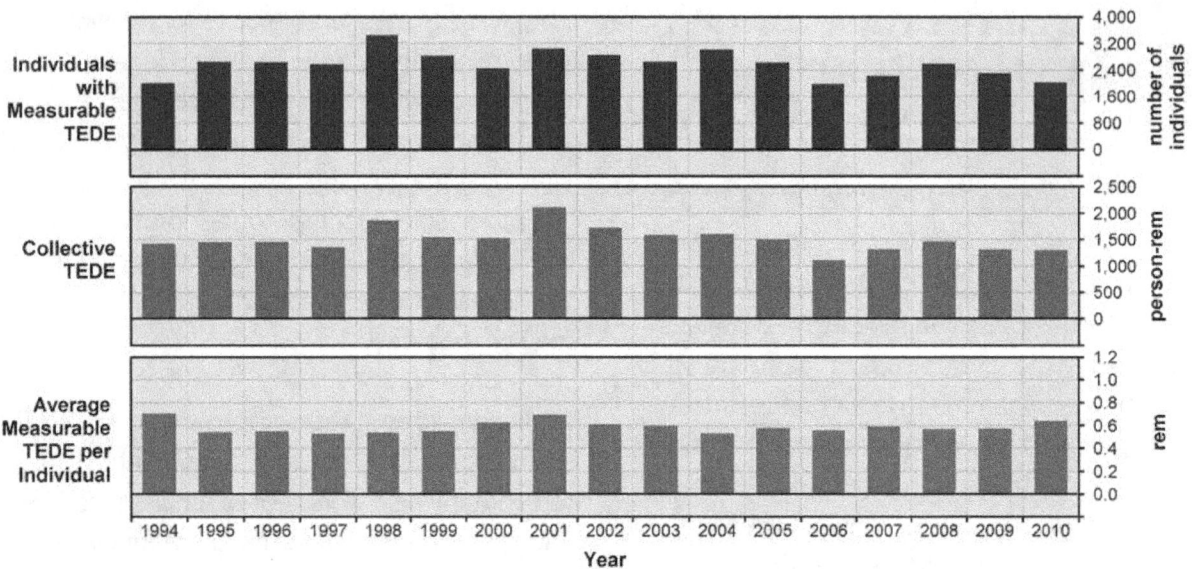

**FIGURE 3.1.** Average Annual Values for Industrial Radiography Licensees
1994–2010

the average measurable dose per individual for both types of industrial radiography licensees from 1994 through 2010. From 2009 to 2010, there was a 12% decrease in the number of individuals with measurable TEDE and a 2% decrease in the collective TEDE. As shown in Table 3.3, eight  fewer temporary job site radiography licensees reported in 2010 affecting a decrease in both the number of individuals with measurable TEDE and the collective TEDE.

### 3.3.2 Manufacturing and Distribution Licenses, Type "A" Broad, Type "B" Broad, Other, and Nuclear Pharmacies

Manufacturing and distribution (M&D) licenses are issued to allow the manufacture and distribution of radionuclides in various forms for a number of diverse purposes. The products are usually distributed to organizations/companies specifically licensed by NRC. Type "A" Broad licenses are issued to larger organizations that may use many different radionuclides in many

different ways and that have a comprehensive radiation protection program. Some Type "A" Broad license firms are medical suppliers that process, package, or distribute such products as diagnostic test kits, radioactive surgical implants, and tagged radiochemicals for use in medical research, diagnosis, and therapy. Type "B" Broad and Other licenses are usually issued to smaller firms requiring a more restrictive license. These firms are suppliers of industrial radionuclides and are involved in the processing, encapsulation, packaging, and distribution of the radionuclides that they have purchased in bulk quantities from production reactors and cyclotrons. Major products include gamma radiography sources, cobalt irradiation sources, well-logging sources, sealed sources for gauges and smoke detectors, and radiochemicals for nonmedical research. Nuclear pharmacies are involved in the compounding and dispensing of radioactive materials for use in nuclear medicine procedures.

Table 3.4 presents the annual data that were reported by the three types of licensees for 2008, 2009, and 2010. Looking at the information shown for manufacturing and distribution licensees, it can be seen that the average measurable dose is generally higher for the Type "A" Broad licensees. These licensees can be authorized to handle larger quantities of radioactive materials which can result in higher average doses during possession and use. Only two Type "A" Broad licensees reported in 2010.

Table 3.4 and Figure 3.2 show the number of individuals with measurable dose, the total collective dose, and the average measurable dose per individual for Type "A" Broad, Type "B" Broad, Other, and Nuclear Pharmacy licensees. The number of individuals with measurable dose decreased by 52% because fewer Type "A" Broad licensees submitted

2010 annual data and the nuclear pharmacies submitted fewer individuals with measurable dose. The collective TEDE decreased nearly 18% in 2010. The average measurable dose increased by 69% from 0.13 rem to 0.22 rem due to the moderate decrease in collective TEDE and the significant drop in the number of individuals with measurable dose. The values for Type "A" Broad licensees are attributed to Covidien-Mallinckrodt, Inc. and International Isotopes Idaho, Inc., which accounted for 79% of the collective dose in 2010 for this licensee category.

For Type "B" Broad, Other, and Nuclear Pharmacy licensees, the decrease in values for 2008 through 2010 has been due to one licensee (Cardinal Health) decreasing its collective TEDE and number of individuals with measurable dose by 66% and 65%, respectively, from the 2009 values.

**TABLE 3.4**
Annual Exposure Information for Manufacturing and Distribution Licensees
2008–2010

| Year | Type of License | Number of Licensees | Number of Monitored Individuals | Individuals with Measurable Dose | Collective Dose (person-rem) | Average Measurable Dose (rem) |
|------|-----------------|---------------------|---------------------------------|----------------------------------|------------------------------|-------------------------------|
| 2008 | M & D - Type "A" Broad | 2 | 465 | 312 | 95.790 | 0.31 |
|  | M & D - Type "B" Broad and Other | 4 | 205 | 114 | 8.421 | 0.07 |
|  | M & D - Nuclear Pharmacies | 12 | 1,264 | 915 | 117.912 | 0.13 |
|  | Total | 18 | 1,934 | 1,341 | 222.123 | 0.17 |
| 2009 | M & D - Type "A" Broad | 3 | 738 | 525 | 103.094 | 0.20 |
|  | M & D - Type "B" Broad and Other | 3 | 88 | 44 | 3.785 | 0.09 |
|  | M & D - Nuclear Pharmacies | 10 | 1,107 | 817 | 72.343 | 0.09 |
|  | Total | 16 | 1,933 | 1,386 | 179.222 | 0.13 |
| 2010 | M & D - Type "A" Broad | 2 | 386 | 339 | 115.733 | 0.34 |
|  | M & D - Type "B" Broad and Other | 3 | 58 | 17 | 4.410 | 0.26 |
|  | M & D - Nuclear Pharmacies | 12 | 526 | 314 | 26.222 | 0.08 |
|  | Total | 17 | 970 | 670 | 146.365 | 0.22 |

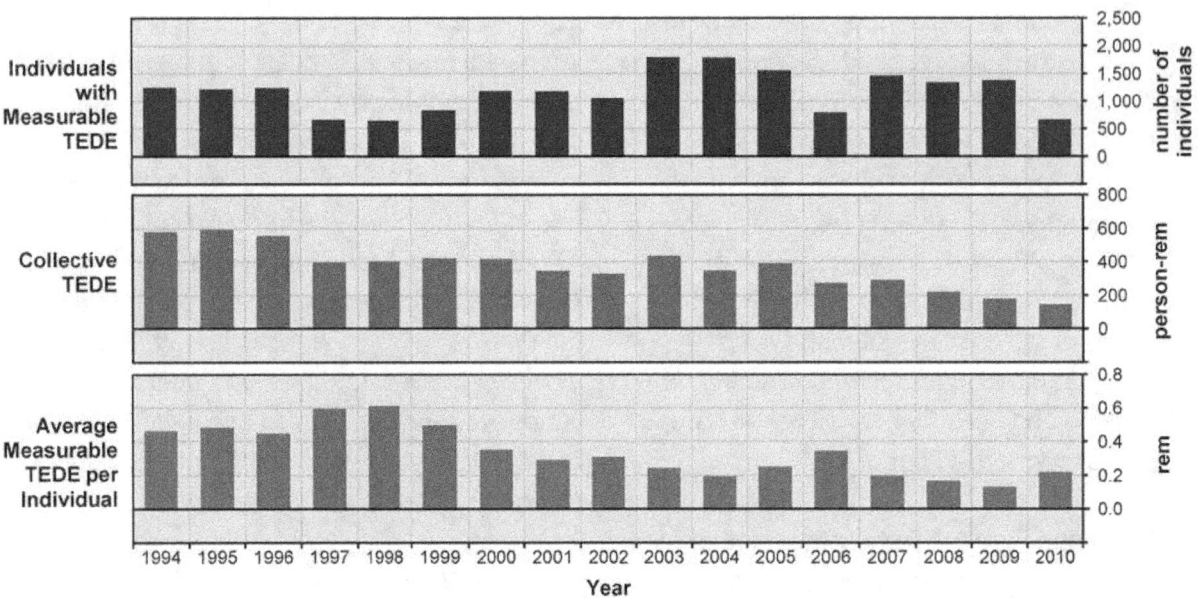

**FIGURE 3.2.** Average Annual Values for Manufacturing and Distribution Licensees 1994–2010

### 3.3.3 Low-Level Waste Disposal Licenses

Low-level waste disposal licenses are issued to allow the receipt, possession, and disposal of low-level radioactive wastes at a land disposal facility. The licensee has the appropriate facilities to receive wastes from such places as hospitals and laboratories, store them for a short time, and dispose of them in a properly prepared burial ground. Since 1999, all licensees that have conducted these activities have been located in Agreement States, which have primary regulatory authority over the licensees' activities; therefore, there are no NRC low-level waste licensees who report radiation exposure data to REIRS.

### 3.3.4 Independent Spent Fuel Storage Installation Licenses

Independent spent fuel storage installation (ISFSI) licenses are issued to allow the possession of commercial nuclear power reactor spent fuel and other associated

radioactive materials for the purpose of storage. The spent fuel, which has undergone at least one year of decay since being used as a source of energy in a commercial nuclear power reactor, is provided interim storage, protection, and safeguarding for a limited time, pending its final disposal.

The majority of ISFSI facilities are located onsite at commercial nuclear power reactors. The dose information from ISFSI facilities is usually included with the dose information reported by the commercial nuclear power reactors and is not reported separately to NRC. In 2010, two ISFSI licensees reported dose information to NRC. One is the GE Morris facility located in Illinois and the second is the Trojan ISFSI located in Oregon. The GE Morris facility is the only spent fuel pool that is not located at an existing reactor site. The GE ISFSI license has been renewed by the NRC until 2022. The Trojan commercial nuclear power reactor is no longer in commercial operation and has been decommissioned. However, the ISFSI facility at Trojan remains in operation and the occupational dose information is reported to NRC under the ISFSI license. Appendix A summarizes the occupational dose information reported by these licensees.

Figure 3.3 shows the number of individuals with measurable dose, the total collective dose, and the average measurable dose per individual for ISFSI facilities. The relatively high values for the collective dose and number of individuals from 1994 to 1996 was mainly because only one licensee reported separately for 1994 through 1998. Table 3.1 shows the number of individuals with measurable dose increased by 15%, while the collective TEDE decreased by 9% from 2009 to 2010.

### 3.3.5 Fuel Cycle Licenses

Fuel cycle licenses are issued to allow the processing, enrichment, and fabrication of reactor fuels. In most uranium facilities where light water reactor (LWR) fuels are fabricated, enriched uranium hexafluoride is converted to solid uranium dioxide pellets and inserted into zirconium alloy tubes. The tubes are fabricated into fuel assemblies that are shipped to commercial nuclear power reactors. Some facilities also perform chemical operations to recover the uranium from scrap and other off-specification materials prior to disposal of these materials. In 1997, the regulatory oversight for the uranium enrichment facilities at Portsmouth, Ohio, and Paducah, Kentucky, was transferred from DOE to NRC and was added to the NRC's fuel cycle license category. In 2005, a third uranium enrichment facility, the Lead Cascade, operated by the United States Enrichment

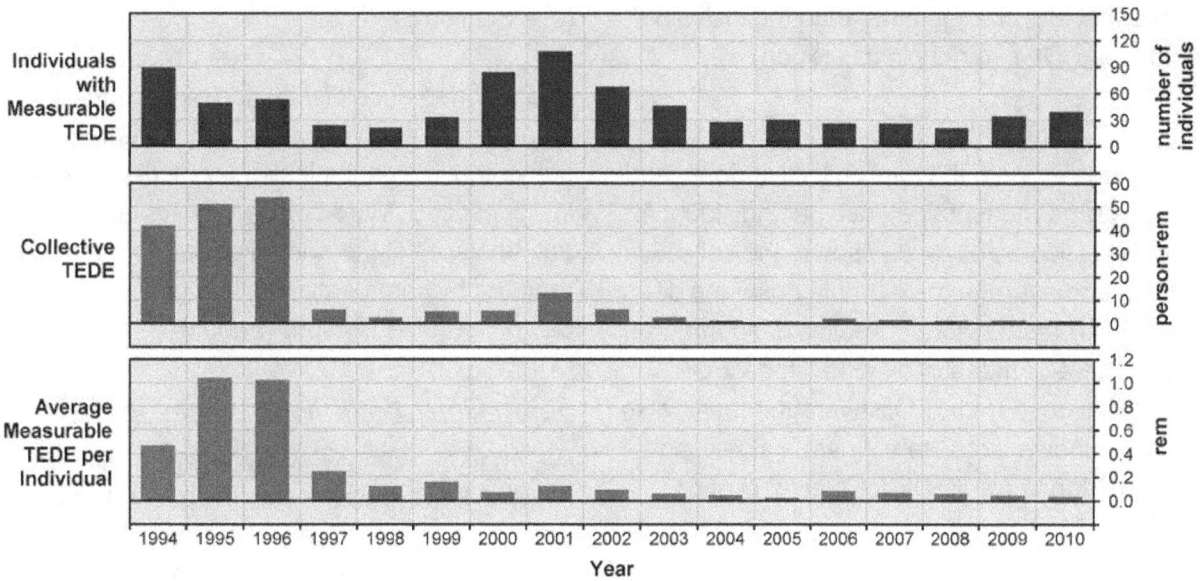

**FIGURE 3.3.** Average Annual Values for Independent Spent Fuel Storage Installations 1994–2010

Corporation, Inc., was added to this category. In 2009, Louisiana Energy Services (LES) joined this category as the fourth uranium enrichment facility. It should be noted that LES was performing construction during 2009 and 2010 and therefore did not significantly contribute to the collective radiation exposure for this licensee category. LES will continue to construct facilities into 2012 and as more operations are brought on-line, it can be expected that exposures at this facility will increase.

For the 2010 report, the decision was made to add Honeywell International, Inc., a uranium hexafluoride (UF6) production plant, to the analysis of fuel cycle licensees. The data for Honeywell from 2000 through 2010 has been added to the tables and figures in this report. Honeywell has reported under their license for UF6 production since 1994, but this activity was not included under the fuel cycle category until 2010, so the addition of this licensee does not represent any change other than the inclusion into fuel cycle category in this report.

Figure 3.4 shows the number of individuals with measurable dose, the total collective dose, and the average measurable dose per individual for fuel cycle licensees. In addition to the collective TEDE and average measurable dose, the deep dose equivalent (DDE) collective dose and DDE average measurable dose and CEDE collective dose and CEDE average measurable dose is shown because they are a significant contribution to the TEDE for fuel fabrication facilities.

As shown in Table 3.5, the collective TEDE and CEDE both increased by 1% and 5%, respectively, while the collective DDE decreased by 3% from 2009.

### 3.3.6 *Light Water Reactor Licenses*

Light water reactor licenses are issued to utilities to allow them to use special nuclear material in a reactor that produces heat to generate electricity to be sold to consumers. There are two major types of commercial LWRs in the U.S., pressurized water reactors and boiling water reactors, each of which uses water as the primary coolant.

Table 3.1 shows the number of licensees, number of monitored individuals, number of individuals with measurable dose, total collective dose, and average dose per individual for reactor facilities that were in commercial operation for at least one full year for each of the years 2000 through 2010. The values do not include reactors that have been permanently shut down or reactors that have not been in commercial operation for one full year. The figures for reactors have not been adjusted for the multiple counting of transient individuals (see Section 5).

The reported dose distribution of individuals monitored at each plant site for the year 2010 is presented in alphabetical order by plant name in Appendix B. More detailed presentations and analyses of the annual dose information reported by commercial nuclear power reactors can be found in Sections 4 and 5.

### 3.3.7 *Other Facilities Reporting to NRC*

Appendix A, Table A2 contains additional facilities that reported occupational radiation dose reports to NRC in 2010. These facilities are not among the seven categories of licensees required to report under 10 CFR 20.2206 and are not included in the analysis presented in this report. However, these facilities may be of interest to researchers and are included in this report for completeness.

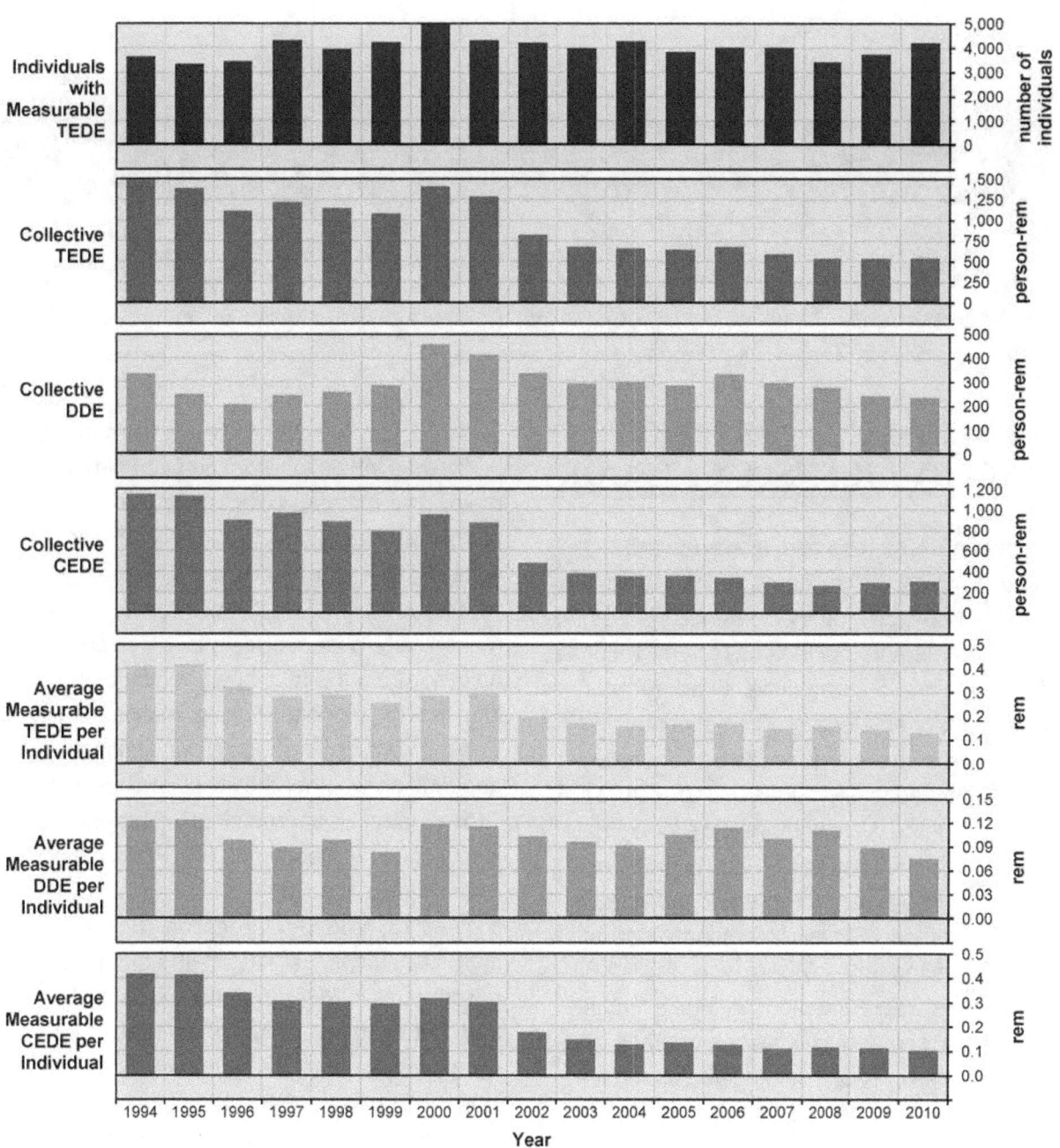

**FIGURE 3.4.** Average Annual Values for Fuel Cycle Licensees
1994–2010

**TABLE 3.5**
Annual Exposure Information for Fuel Cycle Licensees*
2008–2010

| Year | Type of License | Number of Licensees | Number of Monitored Individuals | Individuals with Meas. TEDE | Collective TEDE (person-rem) | Average Meas. TEDE (rem) | Individuals with Meas. DDE | Collective DDE (person-rem) | Average Meas. DDE (rem) | Individuals with Meas. CEDE | Collective CEDE (person-rem) | Average Meas. CEDE (rem) |
|---|---|---|---|---|---|---|---|---|---|---|---|---|
| 2008 | Fuel Cycle | 10 | 7,867 | 3,424 | 538 | 0.16 | 2,493 | 277 | 0.11 | 2,260 | 262 | 0.12 |
| 2009 | Fuel Cycle | 11 | 8,918 | 3,738 | 534 | 0.14 | 2,737 | 243 | 0.09 | 2,598 | 291 | 0.11 |
| 2010 | Fuel Cycle | 11 | 9,362 | 4,212 | 542 | 0.13 | 3,129 | 235 | 0.08 | 2,966 | 307 | 0.10 |

* All data for this table includes program code 11400 for $UF_6$ Production Plants that have not been included in previous years for this table.

## 3.4 SUMMARY OF INTAKE DATA BY LICENSEE CATEGORY

For each intake recorded, licensees are required to list the radionuclide that was taken into the body, pulmonary clearance class, intake mode, and amount of the intake in microcuries. An NRC Form 5, its equivalent paper document or an electronic format containing this information, is required to be completed and submitted to NRC under 10 CFR 20.2206. Tables 3.6 and 3.7 summarize the intake data reported to NRC during 2010. The data are categorized by licensee type and are listed in order of radionuclide and pulmonary clearance class or pulmonary solubility type. Table 3.6 lists the intakes where the mode of intake into the body was recorded as ingestion or other. These other modes of intake can include absorption through the skin and injection through a puncture or wound.

Table 3.7 lists the intakes where the mode of intake was inhalation from ambient airborne radioactive material in the workplace. The pulmonary clearance class or pulmonary solubility type is recorded as D, W, Y (days, weeks, years) or F, M, S (fast, medium, slow), respectively, corresponding to the clearance half-time from the pulmonary region of the lung into the blood and gastrointestinal tract. The pulmonary clearance class designation depends on whether the licensee is using the nomenclature in International Commission on Radiological Protection (ICRP) Publication 30, which is described in 10 CFR Part 20 (D, W, Y) [Ref. 13] or ICRP Publication 68 (F, M, S) [Ref. 14]. Licensees that use the methodology described in ICRP Publication 30 utilize D, W, and Y pulmonary classes to determine dose. Licensees that use the methodology described in ICRP Publication 68 utilize F, M, and S pulmonary solubility types to determine dose.

**TABLE 3.6**
Intake by Licensee Category and Radionuclide Mode of Intake—Ingestion and Other
2010

| Mode | Licensee Category | Program Code | Radionuclide | Number of Intake Records | Collective Intake in Microcuries (sci. notation) |
|---|---|---|---|---|---|
| Ingestion | Fuel Fabrication | 21210 | U-234 | 1 | 3.57E-04 |

NOTE: This intake was a result of an inhalation of large particles that was more properly modeled as an ingestion.

**TABLE 3.7**
Intake by Licensee Category and Radionuclide Mode of Intake—Inhalation
2010

| Licensee Category | Program Code | Radionuclide | Pulmonary Clearance Class or Solubility Type | Number of Intake Records * | Collective Intake in Microcuries (sci. notation) |
|---|---|---|---|---|---|
| Nuclear Pharmacies | 02500 | I-123 | W | 5 | 1.30E+00 |
| | 02500 | I-131 | D | 3 | 2.99E-01 |
| | 02500 | I-131 | W | 44 | 1.82E+01 |
| Manufacturing and Distribution | 03211 | I-131 | D | 3 | 1.70E-01 |
| Uranium Hexafluoride (UF₆) Production Plants | 11400 | AC-227 | D | 36 | 3.70E-05 |
| | 11400 | AC-227 | W | 2 | 2.00E-06 |
| | 11400 | AC-227 | Y | 125 | 1.38E-04 |
| | 11400 | PA-231 | D | 36 | 3.70E-05 |
| | 11400 | PA-231 | W | 2 | 2.00E-06 |
| | 11400 | PA-231 | Y | 125 | 1.38E-04 |
| | 11400 | PB-210 | D | 20 | 2.00E-05 |
| | 11400 | PB-210 | W | 1 | 1.00E-06 |
| | 11400 | PB-210 | Y | 82 | 8.60E-05 |
| | 11400 | PO-210 | D | 13 | 1.30E-05 |
| | 11400 | PO-210 | Y | 65 | 6.60E-05 |
| | 11400 | RA-226 | D | 152 | 1.89E-04 |
| | 11400 | RA-226 | W | 10 | 1.10E-05 |
| | 11400 | RA-226 | Y | 350 | 5.14E-04 |
| | 11400 | RA-228 | D | 9 | 9.00E-06 |
| | 11400 | RA-228 | Y | 54 | 5.50E-05 |
| | 11400 | TH-228 | D | 9 | 9.00E-06 |
| | 11400 | TH-228 | Y | 54 | 5.50E-05 |
| | 11400 | TH-230 | D | 811 | 2.74E-03 |
| | 11400 | TH-230 | W | 21 | 1.16E-04 |
| | 11400 | TH-230 | Y | 897 | 5.61E-03 |
| | 11400 | TH-232 | D | 9 | 9.00E-06 |
| | 11400 | TH-232 | Y | 54 | 5.50E-05 |
| | 11400 | U-234 | D | 979 | 2.55E-01 |
| | 11400 | U-234 | W | 54 | 1.10E-02 |
| | 11400 | U-234 | Y | 978 | 5.18E-01 |
| | 11400 | U-235 | D | 935 | 1.19E-02 |
| | 11400 | U-235 | W | 31 | 5.07E-04 |
| | 11400 | U-235 | Y | 952 | 2.42E-02 |
| | 11400 | U-238 | D | 979 | 2.12E-01 |
| | 11400 | U-238 | W | 54 | 9.15E-03 |
| | 11400 | U-238 | Y | 978 | 4.32E-01 |
| Uranium Enrichment | 21200 | NP-237 | W | 1 | 8.10E-06 |
| | 21200 | TC-99 | W | 5 | 5.85E-01 |
| | 21200 | TH-230 | W | 1 | 2.08E+00 |
| | 21200 | U-234 | D | 48 | 1.05E+00 |
| | 21200 | U-234 | Y | 1 | 6.41E+00 |
| Fuel Fabrication | 21210 | AM-241 | M | 33 | 6.26E-05 |
| | 21210 | CO-60 | Y | 4 | 8.39E-04 |
| | 21210 | PU-239 | M | 59 | 2.01E-04 |
| | 21210 | RA-224 | M | 33 | 7.52E-05 |
| | 21210 | RN-220 | D | 140 | 1.70E+02 |
| | 21210 | SR-90 | S | 194 | 4.53E-01 |
| | 21210 | TH-228 | M | 74 | 1.26E-04 |
| | 21210 | TH-232 | M | 39 | 1.13E-04 |
| | 21210 | TH-232 | S | 7 | 1.02E-05 |

NOTE: The data values shown bolded and in boxes represent the highest value in each category.
* An intake event may involve multiple nuclides, and individuals may incur multiple intakes during he year. The number of intake records given here indicates the number of separate intake reports hat were submitted on NRC Form 5 reports under 10 CFR 20.2206.

**TABLE 3.7**
Intake by Licensee Category and Radionuclide Mode of Intake—Inhalation (continued)
2010

| Licensee Category | Program Code | Radionuclide | Pulmonary Clearance Class or Solubility Type | Number of Intake Records * | Collective Intake in Microcuries (sci. notation) |
|---|---|---|---|---|---|
| Fuel Fabrication | 21210 | U-232 | D | 141 | 0.00E+00 |
| (continued) | 21210 | U-232 | Y | 247 | 5.23E-05 |
| | 21210 | U-234 | D | 199 | 1.95E-01 |
| | 21210 | U-234 | F | 572 | 1.07E-01 |
| | 21210 | U-234 | M | 549 | 2.77E-02 |
| | 21210 | U-234 | S | 1,626 | 2.38E+00 |
| | 21210 | U-234 | W | 74 | 5.20E-02 |
| | 21210 | U-234 | Y | 1,005 | 3.63E+00 |
| | 21210 | U-235 | D | 141 | 6.29E-03 |
| | 21210 | U-235 | M | 3 | 1.81E-08 |
| | 21210 | U-235 | S | 413 | 6.59E-02 |
| | 21210 | U-235 | W | 74 | 1.94E-03 |
| | 21210 | U-235 | Y | 273 | 8.81E-02 |
| | 21210 | U-236 | D | 141 | 2.64E-04 |
| | 21210 | U-236 | F | 483 | 3.69E-03 |
| | 21210 | U-236 | M | 3 | 2.26E-07 |
| | 21210 | U-236 | S | 214 | 6.67E-03 |
| | 21210 | U-236 | W | 74 | 8.13E-05 |
| | 21210 | U-236 | Y | 273 | 3.85E-02 |
| | 21210 | U-238 | D | 199 | 2.72E-02 |
| | 21210 | U-238 | F | 30 | 8.62E-07 |
| | 21210 | U-238 | M | 493 | 1.38E-03 |
| | 21210 | U-238 | S | 419 | 2.32E-01 |
| | 21210 | U-238 | W | 74 | 7.10E-03 |
| | 21210 | U-238 | Y | 1,005 | 5.23E-01 |
| Commercial Light | 41111 | AM-241 | W | 13 | 3.00E-01 |
| Water Reactors | 41111 | AM-241 | Y | 1 | 5.29E-06 |
| | 41111 | CM-242 | W | 7 | 1.00E-06 |
| | 41111 | CM-243 | W | 10 | 4.60E-05 |
| | 41111 | CO-58 | Y | 15 | 8.62E-01 |
| | 41111 | CO-60 | Y | 22 | 1.16E+00 |
| | 41111 | CS-134 | D | 8 | 3.72E-02 |
| | 41111 | CS-137 | D | 11 | 2.51E-01 |
| | 41111 | FE-55 | W | 1 | 4.88E-01 |
| | 41111 | FE-59 | D | 1 | 2.45E-02 |
| | 41111 | H-3 ** | W | 9 | 1.73E+03 |
| | 41111 | I-131 | D | 5 | 8.71E-01 |
| | 41111 | MN-54 | W | 2 | 3.32E-01 |
| | 41111 | MN-54 | Y | 1 | 3.30E-02 |
| | 41111 | NB-95 | Y | 9 | 6.17E-03 |
| | 41111 | PU-238 | Y | 11 | 1.76E-04 |
| | 41111 | PU-239 | Y | 11 | 8.22E-05 |
| | 41111 | PU-241 | Y | 1 | 1.80E-04 |
| | 41111 | ZN-65 | Y | 1 | 2.74E-02 |
| | 41111 | ZR-95 | W | 8 | 3.64E-03 |

NOTE: The data values shown bolded and in boxes represent the highest value in each category.
*  An intake event may involve multiple nuclides, and individuals may incur multiple intakes during the year.  The number of intake records given here indicates the number of separate intake reports that were submitted on NRC Form 5 reports under 10 CFR 20.2206.
** Additional information on tritium can be found on NRC's public website at http://www.nrc.gov/reactors/operating/ops-experience/tritium/faqs.html

The amount of material taken into the body is given in microcuries, a unit of measure of the quantity of radioactive material. For each licensee category, the maximum number of intake records and the maximum intake are highlighted in the table in bold and boxed for ease of reference.

Table 3.8 lists the number of individuals with measurable CEDE, the collective CEDE, and the average measurable CEDE per individual for each licensee category. Fuel fabrication facilities and the UF6 production facility had the majority of internal dose (99%) in 2010 and the highest average CEDE per individual. This is due to the individuals' exposures to uranium during the processing and fabrication of the uranium fuel.

Table 3.9 shows the distribution of internal dose (CEDE) from 1994 to 2010 for licensees required to report under 10 CFR 20.2206. For the purposes of this table, the definition of a "measurable CEDE" is any reported value greater than zero. As noted above, the vast majority of the internal doses are received by individuals working at fuel fabrication facilities. The collective CEDE has decreased nearly every year since 2000 but increased in 2010. While the collective CEDE increased by 5% in 2010, the average measurable CEDE decreased by 6% indicating that while more individuals performed work receiving dose, the average dose received by the workers did not increase.

## TABLE 3.8
### Collective and Average CEDE by Licensee Category
### 2010

| Licensee Category | Licensee Name | License Number | Number with Meas. CEDE | Collective CEDE (person-rem) | Average Meas. CEDE (rem) |
|---|---|---|---|---|---|
| **MANUFACTURING AND DISTRIBUTION** | | | | | |
| 02500 | CARDINAL HEALTH | 04-26507-01MD | 3 | 0.183 | 0.061 |
| 02500 | CARDINAL HEALTH | 11-27664-01MD | 2 | 0.003 | 0.002 |
| 02500 | CARDINAL HEALTH | 34-29200-01MD | 30 | 0.269 | 0.009 |
| 03211 | COVIDIEN | 24-04206-01 | 1 | 0.002 | 0.002 |
| 03211 | INTERNATIONAL ISOTOPES IDAHO INC. | 11-27680-01 | 2 | 0.004 | 0.002 |
| | Totals and Averages | | 38 | 0.461 | 0.012 |
| **UF₆ PRODUCTION** | | | | | |
| 11400 | HONEYWELL INTERNATIONAL, INC. | SUB-526 | 977 | 128.355 | 0.131 |
| | Totals and Averages | | 977 | 128.355 | 0.131 |
| **URANIUM ENRICHMENT** | | | | | |
| 21200 | U. S. ENRICHMENT CORP. - PADUCAH | GDP-1 | 32 | 0.091 | 0.003 |
| 21200 | U. S. ENRICHMENT CORP. - PORTSMOUTH | GDP-2 | 2 | 0.011 | 0.006 |
| | Totals and Averages | | 34 | 0.102 | 0.003 |
| **FUEL FABRICATION** | | | | | |
| 21210 | AREVA NP, INC. - LYNCHBURG | SNM-1168 | 20 | 0.650 | 0.033 |
| 21210 | AREVA NP, INC. - RICHLAND | SNM-1227 | 242 | 73.319 | 0.303 |
| 21210 | B & W NUCLEAR OPERATIONS GROUP | SNM-0042 | 191 | 13.257 | 0.069 |
| 21210 | GLOBAL NUCLEAR FUEL - AMERICAS, LLC | SNM-1097 | 559 | 37.892 | 0.068 |
| 21210 | NUCLEAR FUEL SERVICES, INC. | SNM-0124 | 605 | 4.511 | 0.007 |
| 21210 | WESTINGHOUSE ELECTRIC COMPANY, LLC | SNM-1107 | 338 | 49.065 | 0.145 |
| | Totals and Averages | | 1,955 | 178.694 | 0.091 |
| **COMMERCIAL LIGHT WATER REACTORS** | | | | | |
| 41111 | BRAIDWOOD | NPF-72 | 9 | 0.112 | 0.012 |
| 41111 | BROWNS FERRY | DPR-33 | 27 | 0.034 | 0.001 |
| 41111 | BRUNSWICK | DPR-62 | 1 | 0.010 | 0.010 |
| 41111 | CALLAWAY | NPF-30 | 1 | 0.005 | 0.005 |
| 41111 | DIABLO CANYON | DPR-80 | 1 | 0.014 | 0.014 |
| 41111 | DUANE ARNOLD | DPR-49 | 1 | 0.011 | 0.011 |
| 41111 | GRAND GULF | NPF-29 | 1 | 0.062 | 0.062 |
| 41111 | HARRIS | NPF-63 | 1 | 0.011 | 0.011 |
| 41111 | HUMBOLDT BAY | DPR-07 | 2 | 0.004 | 0.002 |
| 41111 | MILLSTONE | NPF-49 | 1 | 0.009 | 0.009 |
| 41111 | MONTICELLO | DPR-22 | 1 | 0.010 | 0.010 |
| 41111 | OCONEE | DPR-38 | 10 | 0.221 | 0.022 |
| 41111 | PALO VERDE | NPF-41 | 3 | 0.036 | 0.012 |
| 41111 | SAN ONOFRE | DPR-13 | 1 | 0.001 | 0.001 |
| 41111 | SEQUOYAH | DPR-77 | 8 | 0.064 | 0.008 |
| 41111 | ST. LUCIE | DPR-67 | 3 | 0.005 | 0.002 |
| 41111 | SURRY | DPR-32 | 1 | 0.025 | 0.025 |
| 41111 | VERMONT YANKEE | DPR-28 | 1 | 0.018 | 0.018 |
| 41111 | VOGTLE | NPF-68 | 5 | 0.068 | 0.014 |
| | Totals and Averages | | 78 | 0.720 | 0.009 |
| **Grand Totals and Averages** | | | 3,082 | 308.332 | 0.100 |

NOTE: The data values shown bolded and in boxes represent the highest value in each category.

## TABLE 3.9
### Internal Dose (CEDE) Distribution
### 1994–2010

| Year | Number of Individuals with CEDE in the Ranges (rem) * | | | | | | | | | | Total with Meas. CEDE | Collective CEDE (person-rem) | Average Meas. CEDE (rem) |
|------|------|------|------|------|------|------|------|------|------|------|------|------|------|
| | Meas. 0.020 | 0.020-0.100 | 0.100-0.250 | 0.250-0.500 | 0.500-0.750 | 0.750-1.000 | 1-2 | 2-3 | 3-4 | 4-5 | | | |
| 1994 | 3,425 | 577 | 287 | 683 | 237 | 141 | 293 | 69 | 2 | - | 5,714 | 1170.453 | 0.205 |
| 1995 | 2,869 | 691 | 338 | 730 | 254 | 147 | 290 | 49 | 2 | - | 5,370 | 1167.105 | 0.217 |
| 1996 | 3,096 | 598 | 305 | 584 | 324 | 138 | 187 | 22 | 2 | 2 | 5,258 | 931.799 | 0.177 |
| 1997 | 3,835 | 869 | 381 | 827 | 267 | 148 | 169 | 30 | - | - | 6,526 | 998.406 | 0.153 |
| 1998 | 3,310 | 932 | 426 | 746 | 246 | 140 | 153 | 21 | 2 | - | 5,976 | 922.935 | 0.154 |
| 1999 | 3,423 | 752 | 466 | 438 | 206 | 117 | 173 | 29 | - | - | 5,604 | 813.605 | 0.145 |
| 2000 | 3,275 | 1001 | 570 | 383 | 216 | 98 | 224 | 58 | 7 | 1 | 5,833 | 988.640 | 0.169 |
| 2001 | 1,774 | 827 | 716 | 364 | 128 | 53 | 146 | 82 | 15 | 1 | 4,106 | 884.134 | 0.215 |
| 2002 | 1,760 | 746 | 647 | 531 | 144 | 33 | 23 | 3 | - | - | 3,887 | 494.821 | 0.127 |
| 2003 | 2,208 | 778 | 726 | 388 | 116 | 17 | 5 | - | - | - | 4,238 | 395.573 | 0.093 |
| 2004 | 1,989 | 838 | 657 | 381 | 105 | 17 | 3 | - | - | - | 3,990 | 375.021 | 0.094 |
| 2005 | 1,205 | 706 | 685 | 341 | 98 | 33 | 2 | - | - | - | 3,070 | 365.258 | 0.119 |
| 2006 | 1,302 | 726 | 686 | 346 | 96 | 18 | 3 | - | - | - | 3,177 | 346.918 | 0.109 |
| 2007 | 1,480 | 805 | 646 | 310 | 52 | 5 | 3 | - | - | - | 3,301 | 300.863 | 0.091 |
| 2008 | 979 | 758 | 526 | 303 | 41 | 8 | 4 | - | - | - | 2,619 | 267.510 | 0.102 |
| 2009 | 1,115 | 711 | 597 | 229 | 80 | 21 | 7 | - | - | - | 2,760 | 293.251 | 0.106 |
| 2010 | 1,216 | 884 | 669 | 210 | 67 | 30 | 6 | - | - | - | 3,082 | 308.332 | 0.100 |

* Dose values exactly equal to the values separating ranges are reported in the next higher range.

# Section 4
# COMMERCIAL LIGHT WATER REACTORS

## 4.1 INTRODUCTION

General trends in occupational radiation exposures at commercial nuclear power reactors are best evaluated within the context of other pertinent information. In this section, some of the tables and appendices that summarize dose data also show the type, capacity, amount of electricity generated, and age of the reactor. Dose data are then presented as a function of these data.

## 4.2 DEFINITION OF TERMS AND SOURCES OF DATA

### 4.2.1 Number of Reactors

The number of reactors shown in Tables 4.1, 4.2, and 4.3 is the number of BWRs, PWRs, and LWRs that were in commercial operation during the year listed. This is the number of reactors on which the average number of individuals with measurable dose and average collective dose per reactor is based. Excluded are reactors that had not yet completed a first full year of commercial operation and those reactors that have been permanently defueled. The date that each reactor was declared to be in commercial operation was taken from Ref. 15.

Three Mile Island (TMI) Unit 2 was included in the compilation of data for commercially operating reactors from 1975 through 1988 and has not been included in the data analyses since 1988. Three Mile Island Unit 1 and TMI Unit 2 reported data separately beginning in 1986, but since 2001, the dose breakdowns for TMI Unit 2 have been reported with those for TMI Unit 1, as there is very little dose from activities at TMI Unit 2.

There were no changes to the count of operating reactors in 2010. The number of operating BWRs remains the same as in 2009 at 35, and the number of operating PWRs remains the same at 69. The dose information for these reactors and for others that are no longer in commercial operation is listed at the end of Appendix B.

### 4.2.2 Electric Energy Generated

The electric energy generated in megawatt years (MW-yr) each year by each reactor is graphically represented in Appendix D. This number was obtained by dividing the megawatt hours of electricity annually produced by each facility by 8,760, the number of hours in the year, except for leap years, when the number is 8,784 hours. The number of megawatt hours of electricity produced each year was obtained from Ref. 15.

For the years 1973 to 1996, the electricity generated is the gross electricity output of the reactor. For 1997 to 2010, the number reflects the net electricity produced, which is the gross electricity minus the amount the plant uses for operations. This change is the result of a change in NRC power generation reporting requirements. The electricity generated (in MW-yr) that is presented in Tables 4.1, 4.2, and 4.3 is the summation of electricity generated by the number of reactors included in each year. These sums are divided by the number of operating reactors included in each year to yield the average amount of electric energy generated per reactor, which is also shown in Tables 4.1, 4.2, and 4.3.

## TABLE 4.1
## Summary of Information Reported by Commercial Boiling Water Reactors
## 1994–2010

| Year | Number of Reactors Included* | No. of Individuals with Measurable Dose** | Annual Collective Dose (person-rem) | Average Measurable Dose per Individual (rem)** | Average Collective Dose per Reactor (person-rem) | Average No. Individuals with Measurable Doses per Reactor** | Electricity Generated*** (MW-yr) | Average Collective Dose per MW-yr (person-rem/ MW-yr) | Average Electricity Generated per Reactor (MW-yr) | Average Maximum Dependable Capacity Net (MWe) | Maximum Dependable Capacity Achieved |
|------|------|------|------|------|------|------|------|------|------|------|------|
| 1994 | 37 | 39,171 | 12,098 | 0.31 | 327 | 1,059 | 22,139.0 | 0.55 | 598 | 801 | 75% |
| 1995 | 37 | 35,686 | 9,471 | 0.27 | 256 | 964 | 24,737.0 | 0.38 | 669 | 835 | 80% |
| 1996 | 37 | 37,792 | 9,466 | 0.25 | 256 | 1,021 | 24,322.2 | 0.39 | 657 | 838 | 78% |
| 1997 | 37 | 34,021 | 7,603 | 0.22 | 205 | 919 | 22,866.1 | 0.33 | 618 | 845 | 73% |
| 1998 | 36 | 32,899 | 6,829.296 | 0.21 | 190 | 914 | 23,781.2 | 0.29 | 661 | 874 | 76% |
| 1999 | 35 | 31,482 | 6,434.430 | 0.20 | 184 | 899 | 26,962.6 | 0.24 | 770 | 885 | 87% |
| 2000 | 35 | 31,186 | 6,089.676 | 0.20 | 174 | 891 | 28,476.9 | 0.21 | 814 | 893 | 91% |
| 2001 | 35 | 28,797 | 4,835.397 | 0.17 | 138 | 823 | 28,730.4 | 0.17 | 821 | 895 | 92% |
| 2002 | 35 | 30,978 | 6,107.767 | 0.20 | 175 | 885 | 29,460.0 | 0.21 | 842 | 907 | 93% |
| 2003 | 35 | 30,759 | 5,659.434 | 0.18 | 162 | 879 | 29,094.4 | 0.19 | 831 | 912 | 91% |
| 2004 | 35 | 33,948 | 5,450.982 | 0.16 | 156 | 970 | 29,424.8 | 0.19 | 841 | 893 | 94% |
| 2005 | 35 | 33,544 | 5,995.975 | 0.18 | 171 | 958 | 29,386.8 | 0.20 | 840 | 946 | 89% |
| 2006 | 35 | 34,159 | 4,989.761 | 0.15 | 143 | 976 | 30,238.4 | 0.17 | 864 | 954 | 91% |
| 2007 | 35 | 37,515 | 5,388.416 | 0.14 | 154 | 1,072 | 30,189.3 | 0.18 | 863 | 955 | 90% |
| 2008 | 35 | 34,642 | 4,522.413 | 0.13 | 129 | 990 | 31,248.3 | 0.14 | 893 | 957 | 93% |
| 2009 | 35 | 36,207 | 5,282.869 | 0.15 | 151 | 1,034 | 30,762.7 | 0.17 | 879 | 959 | 92% |
| 2010 | 35 | 37,214 | 4,807.656 | 0.13 | 137 | 1,063 | 31,274.6 | 0.15 | 894 | 961 | 93% |

\* Includes only those reactors that had been in commercial operation for at least one full year as of December 31 of each of the indicated years.
\** Figures are not adjusted for he multiple reporting of transient individuals (see section 5).
\*** Beginning in 1997, the electricity reflects the net electricity generated.

## TABLE 4.2
## Summary of Information Reported by Commercial Pressurized Water Reactors 1994–2010

| Year | Number of Reactors Included* | No. of Individuals with Measurable Dose** | Annual Collective Dose (person-rem) | Average Measurable Dose per Individual (rem)** | Average Collective Dose per Reactor (person-rem) | Average No. Individuals with Measurable Doses per Reactor** | Electricity Generated*** (MW-yr) | Average Collective Dose per MW-yr (person-rem/MW-yr) | Average Electricity Generated per Reactor (MW-yr) | Average Maximum Dependable Capacity Net (MWe) | Maximum Dependable Capacity Achieved |
|---|---|---|---|---|---|---|---|---|---|---|---|
| 1994 | 70 | 44,283 | 9,574 | 0.22 | 137 | 633 | 52,397.6 | 0.18 | 749 | 928 | 81% |
| 1995 | 70 | 49,985 | 11,762 | 0.24 | 168 | 714 | 54,138.2 | 0.22 | 773 | 929 | 83% |
| 1996 | 72 | 46,852 | 9,417 | 0.20 | 131 | 651 | 55,337.8 | 0.17 | 769 | 935 | 82% |
| 1997 | 72 | 50,690 | 9,546 | 0.19 | 133 | 704 | 48,985.3 | 0.19 | 680 | 943 | 72% |
| 1998 | 69 | 38,586 | 6,358.096 | 0.16 | 92 | 559 | 53,288.7 | 0.12 | 772 | 942 | 82% |
| 1999 | 69 | 43,938 | 7,231.281 | 0.16 | 105 | 637 | 56,235.0 | 0.13 | 815 | 942 | 86% |
| 2000 | 69 | 42,922 | 6,562.006 | 0.15 | 95 | 622 | 57,529.9 | 0.11 | 834 | 943 | 88% |
| 2001 | 69 | 38,773 | 6,273.155 | 0.16 | 91 | 562 | 58,822.4 | 0.11 | 852 | 946 | 90% |
| 2002 | 69 | 42,264 | 6,018.423 | 0.14 | 87 | 613 | 59,369.7 | 0.10 | 860 | 947 | 91% |
| 2003 | 69 | 44,054 | 6,296.136 | 0.14 | 91 | 638 | 57,920.6 | 0.11 | 839 | 949 | 88% |
| 2004 | 69 | 35,901 | 4,916.915 | 0.14 | 71 | 520 | 60,398.7 | 0.08 | 875 | 943 | 93% |
| 2005 | 69 | 44,583 | 5,459.832 | 0.12 | 79 | 646 | 59,790.9 | 0.09 | 867 | 955 | 91% |
| 2006 | 69 | 46,106 | 6,031.425 | 0.13 | 87 | 668 | 59,751.3 | 0.10 | 866 | 960 | 90% |
| 2007 | 69 | 42,015 | 4,731.597 | 0.11 | 69 | 609 | 61,955.6 | 0.08 | 898 | 961 | 93% |
| 2008 | 69 | 44,808 | 4,673.527 | 0.10 | 68 | 649 | 60,586.0 | 0.08 | 878 | 964 | 91% |
| 2009 | 69 | 45,547 | 4,741.935 | 0.10 | 69 | 660 | 60,467.9 | 0.08 | 876 | 966 | 91% |
| 2010 | 69 | 37,796 | 3,823.728 | 0.10 | 55 | 548 | 60,859.4 | 0.06 | 882 | 967 | 91% |

\* Includes only those reactors that had been in commercial operation for at least one full year as of December 31 of each of the indicated years.
\*\* Figures are not adjusted for he multiple reporting of transient individuals (see section 5).
\*\*\* Beginning in 1997, the electricity reflects the net electricity generated.

## TABLE 4.3
### Summary of Information Reported by Commercial Light Water Reactors 1994–2010

| Year | Number of Reactors Included* | No. of Individuals with Measurable Dose** | Annual Collective Dose (person-rem) | Average Measurable Dose per Individual (rem)*** | Average Collective Dose per Reactor (person-rem) | Average No. Individuals with Measurable Doses per Reactor** | Electricity Generated*** (MW-yr) | Average Collective Dose per MW-yr (person-rem/MW-yr) | Average Electricity Generated per Reactor (MW-yr) | Average Maximum Dependable Capacity Net (MWe) | Maximum Dependable Capacity Achieved |
|---|---|---|---|---|---|---|---|---|---|---|---|
| 1994 | 107 | 83,454 | 21,672 | 0.26 | 203 | 780 | 74,536.6 | 0.29 | 697 | 884 | 79% |
| 1995 | 107 | 85,671 | 21,233 | 0.25 | 198 | 801 | 78,875.2 | 0.27 | 737 | 896 | 82% |
| 1996 | 109 | 84,644 | 18,883 | 0.22 | 173 | 777 | 79,660.0 | 0.24 | 731 | 902 | 81% |
| 1997 | 109 | 84,711 | 17,149 | 0.20 | 157 | 777 | 71,851.4 | 0.24 | 659 | 910 | 72% |
| 1998 | 105 | 71,485 | 13,187.392 | 0.18 | 126 | 681 | 77,069.9 | 0.17 | 734 | 918 | 80% |
| 1999 | 104 | 75,420 | 13,665.711 | 0.18 | 131 | 725 | 83,197.6 | 0.16 | 800 | 923 | 87% |
| 2000 | 104 | 74,108 | 12,651.682 | 0.17 | 122 | 713 | 86,006.8 | 0.15 | 827 | 926 | 89% |
| 2001 | 104 | 67,570 | 11,108.552 | 0.16 | 107 | 650 | 87,552.8 | 0.13 | 842 | 929 | 91% |
| 2002 | 104 | 73,242 | 12,126.190 | 0.17 | 117 | 704 | 88,829.7 | 0.14 | 854 | 934 | 91% |
| 2003 | 104 | 74,813 | 11,955.570 | 0.16 | 115 | 719 | 87,015.0 | 0.14 | 837 | 936 | 89% |
| 2004 | 104 | 69,849 | 10,367.897 | 0.15 | 100 | 672 | 89,823.5 | 0.12 | 864 | 926 | 93% |
| 2005 | 104 | 78,127 | 11,455.807 | 0.15 | 110 | 751 | 89,177.7 | 0.13 | 857 | 952 | 90% |
| 2006 | 104 | 80,265 | 11,021.186 | 0.14 | 106 | 772 | 89,989.7 | 0.12 | 865 | 958 | 90% |
| 2007 | 104 | 79,530 | 10,120.013 | 0.13 | 97 | 765 | 92,144.9 | 0.11 | 866 | 959 | 92% |
| 2008 | 104 | 79,450 | 9,195.940 | 0.12 | 88 | 764 | 91,834.3 | 0.10 | 883 | 961 | 92% |
| 2009 | 104 | 81,754 | 10,024.804 | 0.12 | 96 | 786 | 91,230.6 | 0.11 | 877 | 964 | 91% |
| 2010 | 104 | 75,010 | 8,631.384 | 0.12 | 83 | 721 | 92,134.0 | 0.09 | 886 | 965 | 92% |

\* Includes only those reactors that had been in commercial operation for at least one full year as of December 31 of each of he indicated years.

\*\* Figures are not adjusted for he multiple reporting of transient individuals (see section 5).

\*\*\* Beginning in 1997, the electricity reflects the net electricity generated.

As shown in Table 4.3, in 2010, there was a 1% increase in the net electricity generated at LWRs. Fifty-five of the LWRs (53%) increased power production in 2010. From 2009 to 2010, Cook Unit 1 had the largest increase in power production for PWRs, primarily because this plant was in an extended outage for most of 2009 due to high vibrations in the low pressure turbine. From 2009 to 2010, Perry had the largest increase in power production for BWRs, primarily because this plant had a long outage in 2009 due to refueling, and repairs to cables and the moisture separator reheater but returned to full power production for almost all of 2010. For PWRs, Crystal River 3 had the largest decrease in power production from 2009 to 2010, as this plant had a refueling outage that included a steam generator replacement in 2010. For BWRs, Grand Gulf had the largest decrease in power production from 2009 to 2010, as this plant was online all year in 2009 but had a refueling outage in 2010.

### 4.2.3 Collective Dose per Megawatt-Year

The number of megawatt-years of electricity generated was used in determining the ratio of the average value of the annual collective dose (TEDE) to the number of MW-yr of electricity generated. The ratio was calculated by dividing the total collective dose in person-rem by the electric energy generated in MW-yr and is a measure of the dose incurred by individuals at commercial nuclear power reactors in relation to the electric energy produced.

For the years 1973 to 1996, the electricity generated is the gross electricity output of the reactor. For 1997 to 2010, the number reflects the net electricity produced. The ratio of collective dose to the number of MW-yr

is calculated by year for BWRs, PWRs, and LWRs, and is presented in Tables 4.1, 4.2, and 4.3. This ratio was also calculated for each reactor site (see Appendix C). The average collective dose per MW-yr for LWRs decreased to a value of 0.09 rem/MW-yr in 2010 from a value of 0.11 rem/MW-yr in 2009 due to the combination of a 14% decrease in the collective dose and a 1% increase in power production.

### 4.2.4 Average Maximum Dependable Capacity

Average maximum dependable capacity, as shown in Tables 4.1, 4.2, and 4.3, was calculated by dividing the sum of the net maximum dependable capacities of the reactors in megawatts (net megawatts electric [MWe]) by the number of reactors included each year. The net maximum dependable capacity is defined as the gross electrical output as measured at the output terminals of the turbine generator during the most restrictive seasonal conditions less the normal station service loads. This "capacity" of each plant was found in Ref. 15.

### 4.2.5 Percent of Maximum Dependable Capacity Achieved

The percent of maximum dependable capacity achieved is shown for all LWRs in Table 4.3. This parameter gives an indication of the overall power generation performance of LWRs as compared with the maximum dependable capacity that could be obtained in a given year. It is calculated by dividing the average electricity generated per reactor by the average maximum dependable capacity for each year.

The decrease in maximum dependable capacity from 1996 to 1997 was due to the change from measuring the gross electricity generated to the net electricity generated. The percent of maximum dependable capacity for LWRs increased to 92% in 2010 from 91% in 2009. This increase in capacity was due to an 11% decrease in outage hours from refueling and equipment outages in 2010, reducing the number of hours of power generation.

## 4.3 ANNUAL TEDE DISTRIBUTIONS

Table 4.4 summarizes the distribution of the annual TEDE doses received by individuals at all commercial LWRs during each of the years 1994 through 2010. This distribution is the sum of the annual dose distributions reported by each licensed LWR each year. As previously noted, the distribution reported by each LWR site for 2010 is shown in Appendix B. Table 4.4 includes only those reactors in operation for one full year for each year presented in the table. In 2010, the total collective dose decreased by 14% to a value of 8,631 person-rem.

Each year, this report identifies the reactors with the largest increases and decreases in collective dose from the previous year and identifies the main reasons for these changes. The changes generally are driven by whether the sites had an increase or decrease in outages from one year to the next. During an outage, more work is performed by individuals in radiation areas, thereby resulting in increased collective dose. This is particularly true during a refueling outage, which entails the opening of the reactor vessel and transferring spent fuel to a storage area. In addition, the sites usually schedule maintenance and inspections during a refueling outage, which also tends to increase collective dose. If a site does not have a refueling outage during a year, the collective dose tends to be lower.

From 2009 to 2010, Waterford was the PWR that had the largest decrease in collective dose. This site had a refueling outage in 2009 but no outages in 2010. Davis-Besse was the PWR with the largest increase in collective dose from 2009 to 2010. Davis-Besse had very little outage time during 2009 but had a four-month refueling outage in 2010.

From 2009 to 2010, Perry was the BWR that had the largest decrease in collective dose. This site had a refueling outage in 2009, in addition to significant outages for equipment repair, but only one minor outage in 2010. Browns Ferry 1, 2, and 3 was the BWR site with the largest increase in collective dose from 2009 to 2010. While this site had several forced outages and a refueling outage for Unit 2 in 2009, Units 1 and 3 underwent refueling outages in 2010, resulting in an increase in collective dose.

## 4.4 AVERAGE ANNUAL TEDE DOSES

Some of the data presented in Tables 4.1, 4.2, and 4.3 are graphically displayed in Figure 4.1, where it can be seen that the average collective dose and average number of individuals per BWR have been higher than those for PWRs for the seventeen years depicted on Figure 4.1. BWRs generally have higher collective doses due to the fact that the steam produced directly from the reactor is used to drive turbines to produce electricity. This results in radioactivity being present in both the reactor and power generation components of the systems, while PWR systems are designed to keep the radioactivity within the reactor vessel and primary system and not in the turbine systems. Between 1994 and 2010, the annual collective dose per LWR dropped by 60%. Since 2002,

## TABLE 4.4
## Summary Distribution of Annual Whole-Body Doses at Commercial Light Water Reactors*
## 1994–2010

| Year | No Measurable Exposure | Measurable <0.1 | Number of Individuals with Whole Body Doses in the Ranges (rem)** | | | | | | | | | | | | | | | Total Number Monitored | Number with Measurable Exposure | Collective Dose (person-rem) |
|---|---|---|---|---|---|---|---|---|---|---|---|---|---|---|---|---|---|---|---|---|
| | | | 0.10-0.25 | 0.25-0.50 | 0.50-0.75 | 0.75-1.0 | 1.0-2.0 | 2.0-3.0 | 3.0-4.0 | 4.0-5.0 | 5.0-6.0 | 6.0-7.0 | 7.0-8.0 | 8.0-9.0 | 9.0-10.0 | 10.0-12.0 | >12 | | | |
| 1994 | 85,145 | 36,528 | 18,633 | 14,246 | 6,800 | 3,502 | 3,323 | 215 | 6 | - | - | - | - | - | - | - | - | 168,398 | 83,253 | 21,534.000 |
| 1995 | 81,032 | 38,575 | 20,245 | 15,279 | 6,884 | 3,336 | 3,077 | 125 | 5 | - | - | - | - | - | - | - | - | 168,558 | 87,526 | 21,674.000 |
| 1996 | 78,197 | 39,426 | 19,955 | 14,201 | 5,809 | 2,648 | 2,342 | 68 | - | - | - | - | - | - | - | - | - | 162,646 | 84,449 | 18,874.000 |
| 1997 | 80,163 | 41,759 | 19,951 | 13,396 | 5,394 | 2,240 | 1,671 | 59 | 3 | - | - | - | - | - | - | - | - | 164,636 | 84,473 | 17,136.000 |
| 1998 | 77,080 | 37,039 | 17,189 | 10,467 | 3,930 | 1,562 | 1,129 | 35 | - | - | - | - | - | - | - | - | - | 148,431 | 71,351 | 13,169.366 |
| 1999 | 74,867 | 39,663 | 18,063 | 10,964 | 3,994 | 1,569 | 1,141 | 24 | 2 | - | - | - | - | - | - | - | - | 150,287 | 75,420 | 13,665.711 |
| 2000 | 73,793 | 40,301 | 17,598 | 10,310 | 3,525 | 1,375 | 976 | 23 | - | - | - | - | - | - | - | - | - | 147,901 | 74,108 | 12,651.682 |
| 2001 | 73,206 | 37,461 | 16,078 | 9,231 | 2,930 | 1,060 | 747 | 63 | - | - | - | - | - | - | - | - | - | 140,776 | 67,570 | 11,108.552 |
| 2002 | 76,270 | 41,588 | 16,752 | 9,426 | 3,121 | 1,245 | 1,003 | 105 | 2 | - | - | - | - | - | - | - | - | 149,512 | 73,242 | 12,126.190 |
| 2003 | 77,889 | 42,720 | 17,231 | 9,589 | 3,139 | 1,233 | 864 | 37 | - | - | - | - | - | - | - | - | - | 152,702 | 74,813 | 11,955.570 |
| 2004 | 80,473 | 41,583 | 15,626 | 8,245 | 2,733 | 978 | 668 | 16 | - | - | - | - | - | - | - | - | - | 150,322 | 69,849 | 10,367.897 |
| 2005 | 82,574 | 46,444 | 17,754 | 9,191 | 2,934 | 1,104 | 683 | 17 | - | - | - | - | - | - | - | - | - | 160,701 | 78,127 | 11,455.807 |
| 2006 | 84,558 | 48,571 | 18,269 | 9,312 | 2,675 | 904 | 532 | 2 | - | - | - | - | - | - | - | - | - | 164,823 | 80,265 | 11,021.186 |
| 2007 | 84,551 | 49,998 | 17,672 | 8,294 | 2,329 | 824 | 402 | 11 | - | - | - | - | - | - | - | - | - | 164,081 | 79,530 | 10,120.013 |
| 2008 | 89,874 | 51,831 | 17,337 | 7,578 | 1,847 | 583 | 269 | 5 | - | - | - | - | - | - | - | - | - | 169,324 | 79,450 | 9,195.940 |
| 2009 | 94,627 | 52,670 | 17,417 | 8,352 | 2,161 | 741 | 413 | - | - | - | - | - | - | - | - | - | - | 176,381 | 81,754 | 10,024.804 |
| 2010 | 104,638 | 49,571 | 16,042 | 6,656 | 1,801 | 602 | 333 | 5 | - | - | - | - | - | - | - | - | - | 179,648 | 75,010 | 8,631.384 |

\* Summary of reports submitted in accordance with 10 CFR 20.407 or 20.2206 by BWRs and PWRs that had been in commercial operation for at least 1 full year as of December 31 of each of the indicated years. Figures shown have not been adjusted for the multiple reporting of transient individuals (see Section 5).

\*\* Dose values exactly equal to the values separating ranges are reported in the next higher range.

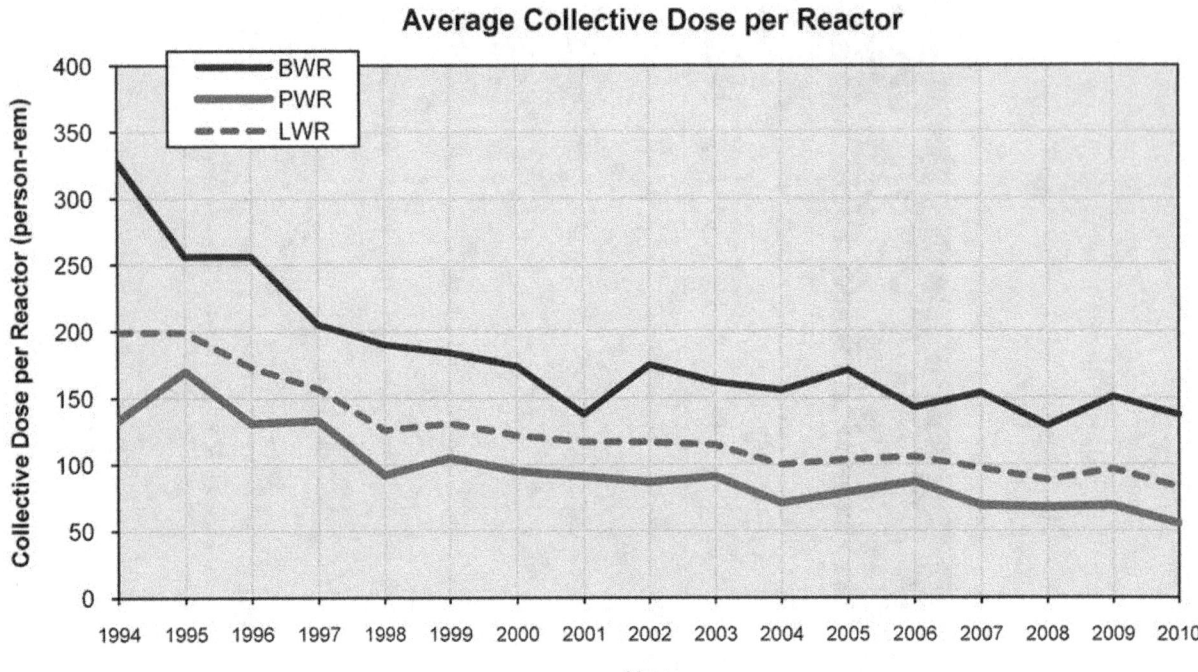

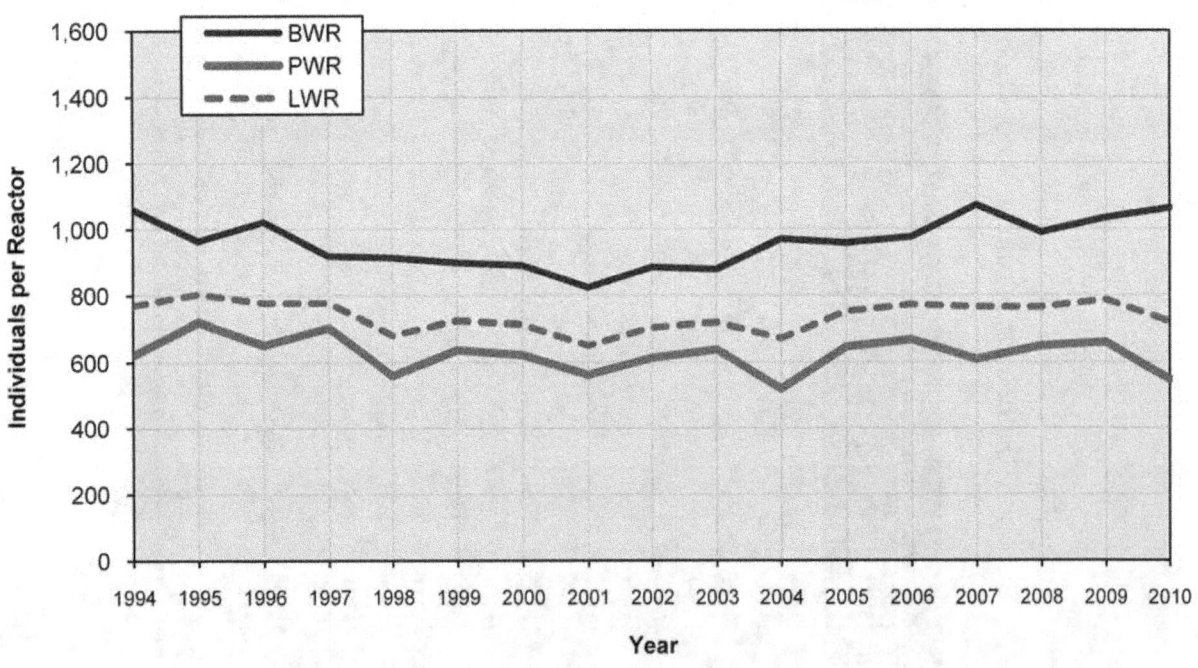

**FIGURE 4.1.** Average Collective Dose per Reactor and
Number of Individuals with Measurable Dose per Reactor
1994–2010

BWR collective doses have decreased by approximately 21% and PWR collective doses have decreased by approximately 36%.

In 2010, the average collective dose per reactor for PWRs decreased by 20% to 55 person-rem and the average collective dose per reactor for BWRs decreased by 9% to 137 person-rem from the 2009 values of 69 person-rem and 151 person-rem respectively. The average collective dose per reactor for LWRs decreased by 14% from 96 person-rem in 2009 to 83 person-rem in 2010. This is the fourth year that the average collective dose per reactor for LWRs has been below 100 person-rem since tracking began in 1973. The overall decreasing trend in average reactor collective doses since 1994 indicates that licensees are continuing to successfully implement as low as is reasonably achievable (ALARA) dose reduction processes at their facilities. In 2010, the number of individuals with measurable dose per reactor decreased to 548 for PWRs and increased to 1,063 for BWRs.

Figures 4.2 and 4.3 are plots of most of the other information that is given in Tables 4.1, 4.2, and 4.3. Table 4.3 shows that in 2010 the net electricity generated decreased slightly to 92,134 MW-yr, while the number of operating reactors has remained constant for the past twelve years. Table 4.3 also shows that the value for the total collective dose for all LWRs decreased by 14% to 8,631 person-rem in 2010 from a value of 10,025 person-rem in 2009. The average measurable dose per individual remained the same at 0.12 rem in 2010 (not adjusted for transient individuals).

The decrease seen in dose trends since 1994 may be attributable to several factors. Utilities have completed the tasks initiated as a result of the lessons learned from the 1979 TMI accident, and they are increasing efforts to avoid and reduce exposure. The concept of keeping exposures to ALARA levels is continually being stressed, and most utilities have established programs to collect and share information relative to exposure control processes, techniques, and procedures.

To further assist in the identification of any trends that might exist, Figure 4.4 displays the average and median[6] values of the collective dose per reactor for BWRs and for PWRs for the years 1994 through 2010. The median values are included here for statistical completeness and are not used in other sections of the report. The ranges of the values reported each year are shown by the vertical lines with a small bar at each end marking the two extreme values. The rectangles indicate the range of values of the collective dose exhibited by those plants ranked in the 25th through the 75th percentiles. The median collective dose for PWRs decreased from 56 person-rem in 2009 to 41 person-rem in 2010. The median collective dose for BWRs decreased from 133 person-rem in 2009 to 123 person-rem in 2010. Figure 4.4 also shows that, in 2010, 50% of the PWRs reported collective doses between 32 and 63 person-rem, while 50% of the BWRs reported collective doses between 88 and 188 person-rem. The middle 50% of BWRs and PWRs in Figure 4.4 are the reactors between the 25% and 75% dose range. These values are based on an annual collective dose

---

[6] The median is the value at which 50% of the reactors reported greater collective doses and the other 50% reported smaller collective doses.

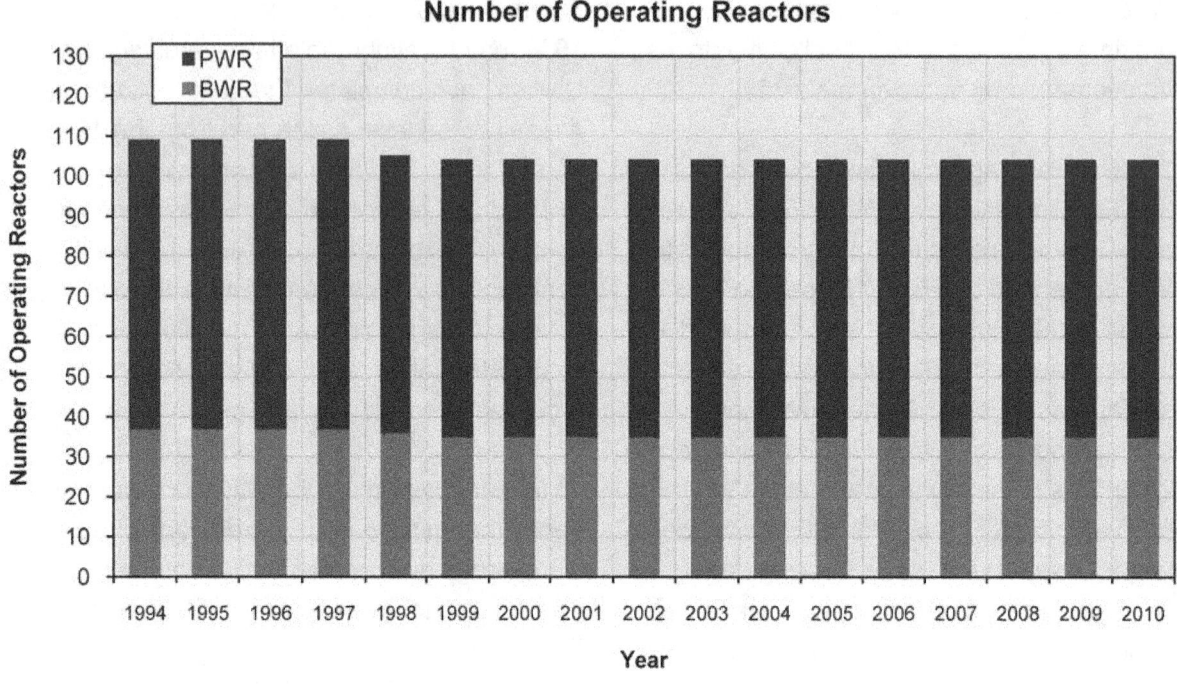

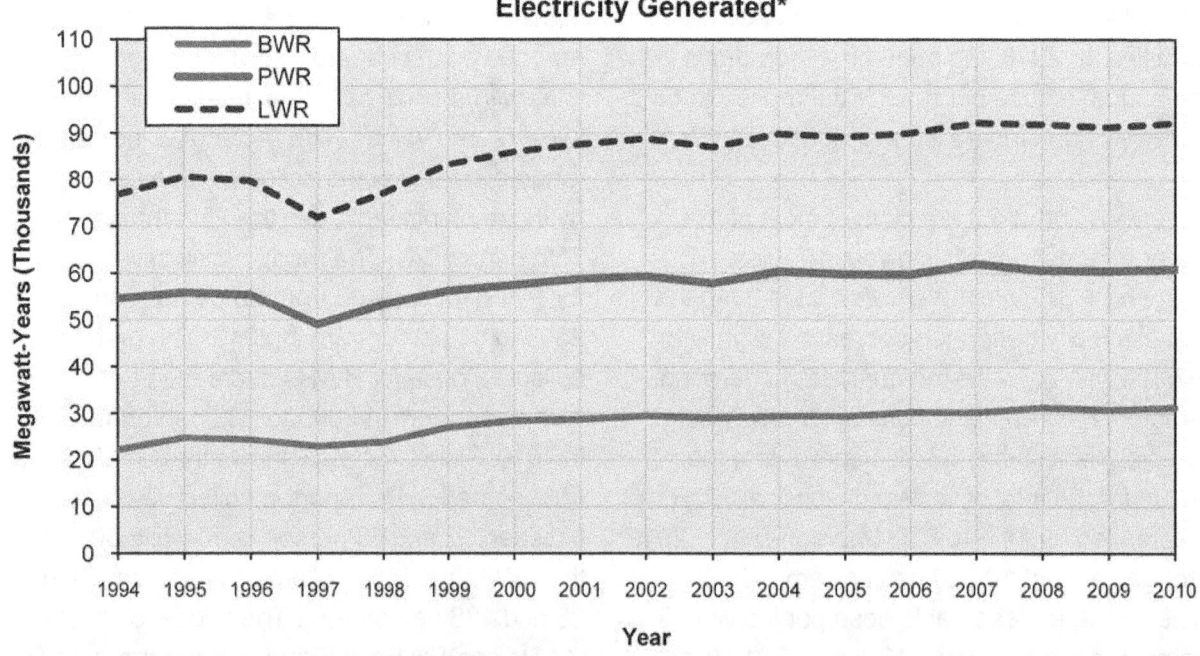

* Gross electricity is shown for 1994–1996, net electricity is shown for 1997–2010.

**FIGURE 4.2.** Number of Operating Reactors and Electricity Generated
1994–2010

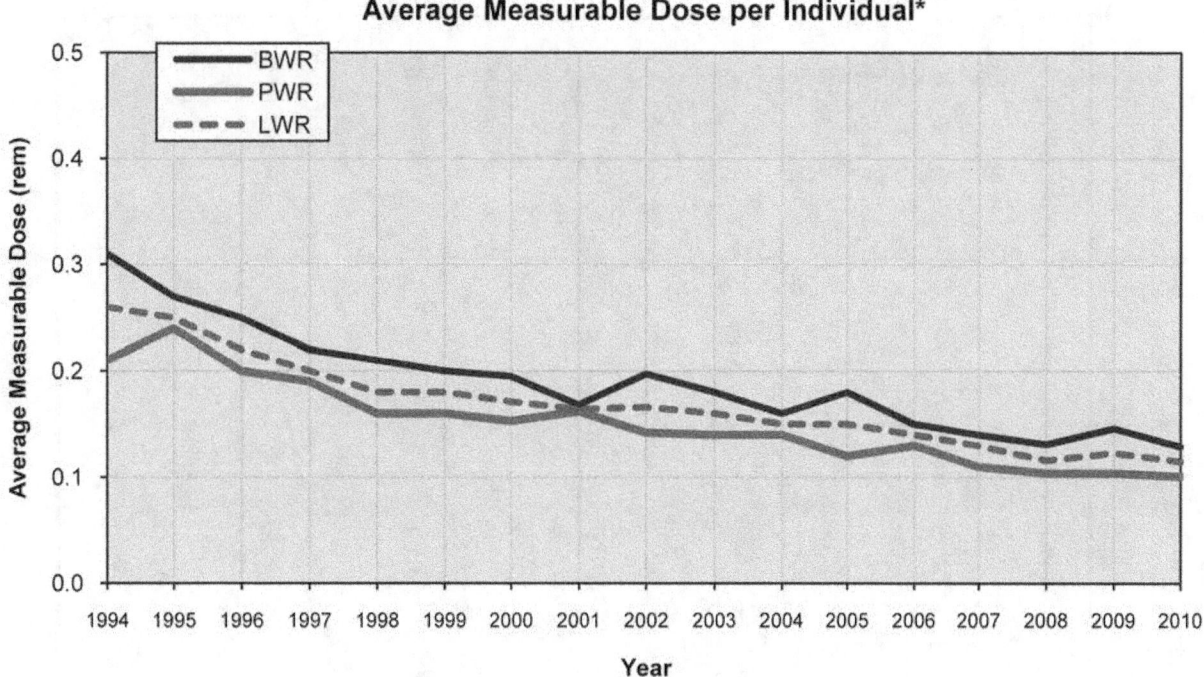

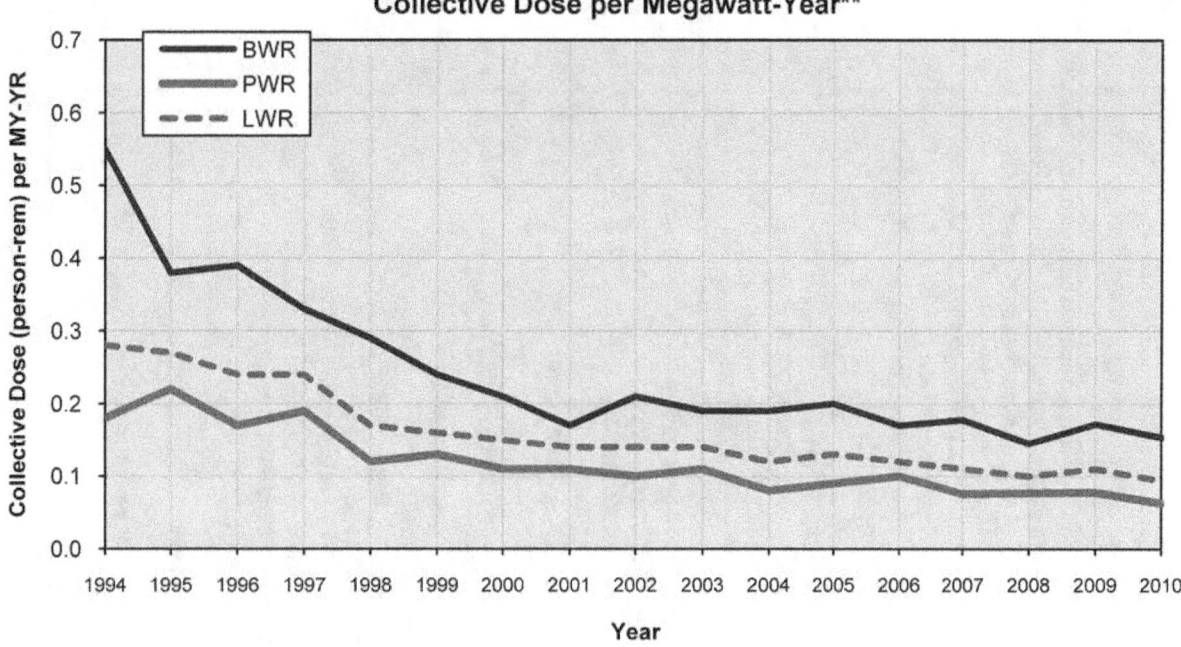

\* Not adjusted for transient workers.  See Section 5.
\*\* Gross electricity is shown for 1994–1996, net electricity is shown for 1997–2010.

**FIGURE 4.3.**  Average Measurable Dose per Individual and Collective Dose per Megawatt-Year
1994–2010

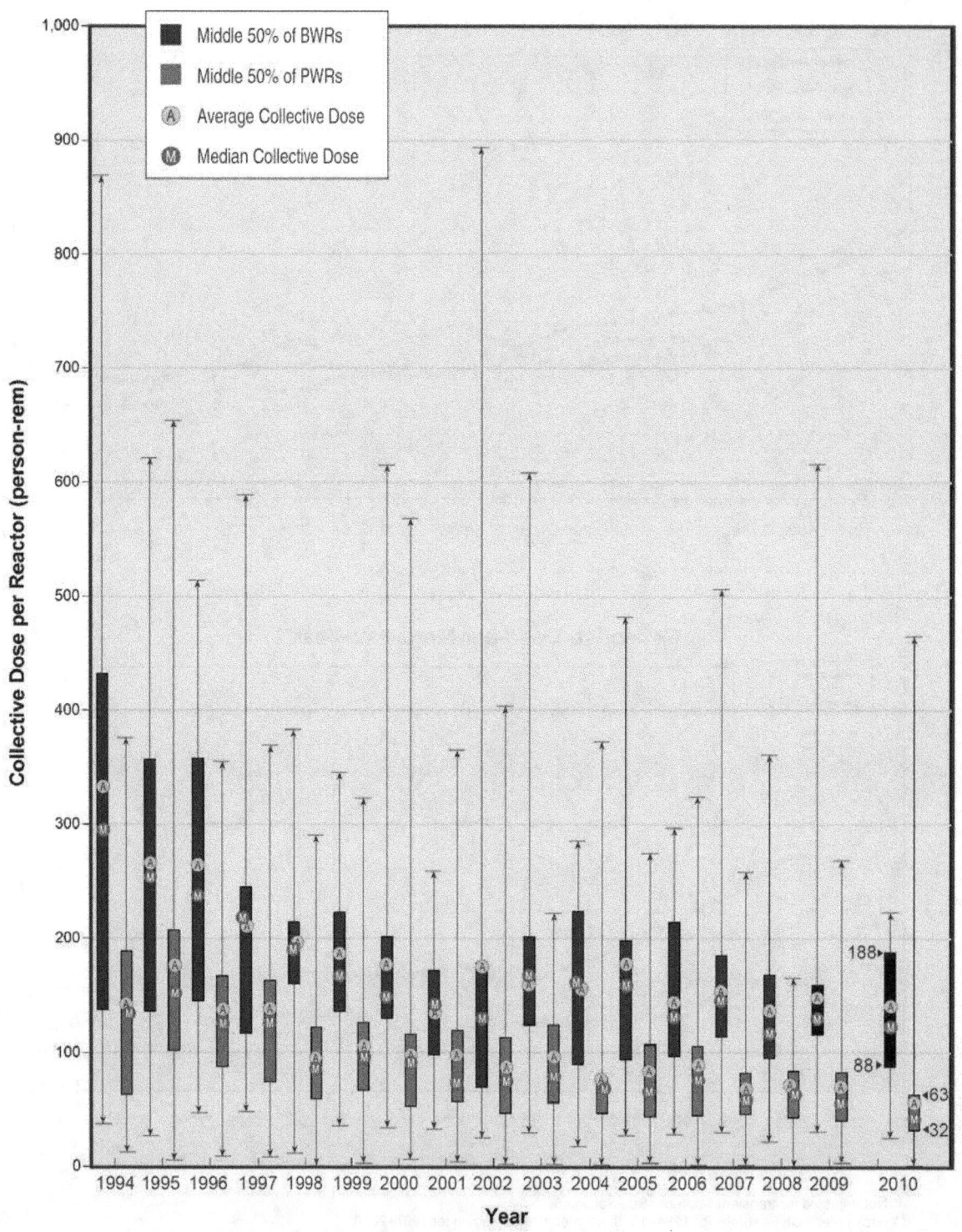

**FIGURE 4.4.** Average, Median, and Extreme Values of the Collective Dose per Reactor 1994–2010

values, not the three-year rolling average that is presented in Section 4.5. Nearly every year the median collective dose is less than the average, which indicates that more of the reactors tend to be at lower collective doses than is reflected by the average. This is a result of the wide difference between the maximum and minimum annual collective doses at power plants and that some plants accrue higher collective dose during refueling outages. These plants that have outages during the year (and thus higher collective doses) increase the value of the average collective dose, while the median (or middle-point of the doses) remains lower.

## 4.5 THREE-YEAR AVERAGE COLLECTIVE TEDE PER REACTOR

The three-year average collective dose per reactor is one of the metrics that the NRC uses in the Reactor Oversight Program to evaluate the effectiveness of the licensee's ALARA program. Tables 4.5 and 4.6 list the sites that had been in commercial operation for at least three years as of December 31, 2010, and show the values of several parameters for each of the sites. These tables also give averages for the two types of reactors.

Based on the 105 reactor-years of operation accumulated over a three-year period by the 35 BWRs listed, the average three-year collective TEDE per reactor was found to be 139 person-rem, the average measurable TEDE per individual was 0.13 rem, and the average collective TEDE per MW-yr was 0.16 person-rem. For BWRs, all values decreased slightly or remained the same from 2009 to 2010.

Based on the 207 reactor-years of operation accumulated over a three-year period at the 69 PWRs listed, the average annual collective TEDE per reactor, average measurable TEDE per individual, and average collective TEDE per MW-yr were found to be 64 person-rem, 0.10 rem, and 0.07 person-rem respectively. For PWRs, all values either decreased slightly or remained the same from 2009 to 2010.

In addition to the listings provided in Tables 4.5 and 4.6, the quartile ranking is used by the NRC as a factor in planning the number of inspection hours assigned per site. For this reason, Tables 4.7 and 4.8 have been included in the 2010 annual report for BWRs and PWRs, respectively. These tables show the plant name, three-year collective TEDE per reactor, the percent change in the three-year average from the previous three-year period, and the quartile ranking from the previous period if the ranking has changed.

## 4.6 INTERNATIONAL OCCUPATIONAL RADIATION EXPOSURE

The NRC must perform certain legislatively mandated international duties. These include licensing the import and export of nuclear materials and equipment and participating in activities supporting U.S. government compliance with international treaties and agreement obligations. In addition, the NRC actively cooperates with multinational organizations, such as the International Atomic Energy Agency (IAEA) and the Nuclear Energy Agency (NEA), a part of the Organisation for Economic Co-operation and Development (OECD).[Ref. 16]

## TABLE 4.5
### Three-Year Totals and Averages Listed in Ascending Order of Collective TEDE per BWR
### 2008–2010

| Plant Name* | Reactor Years | Three-year Collective TEDE per Reactor Year 2008-2010 | Three-year Collective TEDE per Site | Number of Workers with Measurable TEDE | Average TEDE per Worker | Total MW-Yrs | Average TEDE per MW-Yr |
|---|---|---|---|---|---|---|---|
| MONTICELLO | 3 | 91.172 | 273.517 | 2,120 | 0.129 | 1564.4 | 0.17 |
| LIMERICK 1,2 | 6 | 96.561 | 579.364 | 4,666 | 0.124 | 6545.8 | 0.09 |
| HATCH 1,2 | 6 | 103.541 | 621.243 | 4,441 | 0.140 | 4587.4 | 0.14 |
| PILGRIM | 3 | 104.174 | 312.522 | 2,007 | 0.156 | 1959.6 | 0.16 |
| SUSQUEHANNA 1,2 | 6 | 105.942 | 635.650 | 6,130 | 0.104 | 6526.8 | 0.10 |
| DRESDEN 2,3 | 6 | 107.278 | 643.666 | 6,391 | 0.101 | 4932.6 | 0.13 |
| FERMI 2 | 3 | 110.174 | 330.522 | 3,582 | 0.092 | 2827.3 | 0.12 |
| HOPE CREEK 1 | 3 | 121.594 | 364.782 | 5,074 | 0.072 | 3332.5 | 0.11 |
| DUANE ARNOLD | 3 | 121.665 | 364.854 | 2,341 | 0.156 | 1643.6 | 0.22 |
| PEACH BOTTOM 2,3 | 6 | 123.772 | 742.630 | 5,688 | 0.131 | 6391.0 | 0.12 |
| GRAND GULF | 3 | 128.983 | 386.950 | 4,243 | 0.091 | 3429.6 | 0.11 |
| COLUMBIA GENERATING | 3 | 138.277 | 414.832 | 3,406 | 0.122 | 2867.4 | 0.14 |
| QUAD CITIES 1,2 | 6 | 139.051 | 834.306 | 6,698 | 0.125 | 4970.9 | 0.17 |
| FITZPATRICK | 3 | 146.593 | 439.778 | 3,350 | 0.131 | 2332.4 | 0.19 |
| LASALLE 1,2 | 6 | 149.777 | 898.660 | 6,774 | 0.133 | 6470.9 | 0.14 |
| OYSTER CREEK | 3 | 151.829 | 455.488 | 3,682 | 0.124 | 1625.0 | 0.28 |
| NINE MILE POINT 1,2 | 6 | 152.467 | 914.800 | 4,550 | 0.201 | 4975.4 | 0.18 |
| BROWNS FERRY 1,2,3** | 9 | 154.126 | 1,387.133 | 7,646 | 0.181 | 8598.5 | 0.16 |
| CLINTON | 3 | 157.683 | 473.049 | 3,356 | 0.141 | 2970.7 | 0.16 |
| VERMONT YANKEE | 3 | 160.369 | 481.106 | 2,909 | 0.165 | 1717.8 | 0.28 |
| BRUNSWICK 1,2 | 6 | 185.331 | 1,111.983 | 8,456 | 0.132 | 5033.1 | 0.22 |
| RIVER BEND 1 | 3 | 190.500 | 571.499 | 4,684 | 0.122 | 2575.5 | 0.22 |
| COOPER STATION | 3 | 225.087 | 675.261 | 4,198 | 0.161 | 2109.0 | 0.32 |
| PERRY | 3 | 233.068 | 699.203 | 2,624 | 0.266 | 3298.4 | 0.21 |
| **Totals and Averages** | **105** | - | **14,612.798** | **109,016** | **0.134** | **93,285.6** | **0.16** |
| **Average per Reactor-Year** | - | **139.170** | - | **1,038** | - | **888.4** | - |

\* Sites where not all reactors had completed three full years of commercial operations as of December 31, 2010, are not included.

\*\* Although Brown's Ferry 1 was placed on administrative hold in 1985, it remains in the count of operating reactors and has resumed operation as of June, 2007.

## TABLE 4.6
### Three-Year Totals and Averages Listed in Ascending Order of Collective TEDE per PWR 2008–2010

| Plant Name* | Reactor Years | Three-year Collective TEDE per Reactor Year 2008-2010 | Three-year Collective TEDE per Site | Number of Workers with Measurable TEDE | Average TEDE per Worker | Total MW-Yrs | Average TEDE per MW-Yr |
|---|---|---|---|---|---|---|---|
| INDIAN POINT 3 | 3 | 25.049 | 75.147 | 2,243 | 0.034 | 2,948.7 | 0.03 |
| COOK 1,2 | 6 | 33.291 | 199.743 | 2,989 | 0.067 | 4,448.4 | 0.04 |
| FARLEY 1,2 | 6 | 34.000 | 203.997 | 2,647 | 0.077 | 4,701.2 | 0.04 |
| SUMMER 1 | 3 | 35.757 | 107.270 | 1,509 | 0.071 | 2,570.5 | 0.04 |
| CALLAWAY 1 | 3 | 36.431 | 109.294 | 1,748 | 0.063 | 3,267.8 | 0.03 |
| PRAIRIE ISLAND 1,2 | 6 | 39.208 | 235.246 | 2,281 | 0.103 | 2,869.1 | 0.08 |
| PALO VERDE 1,2,3 | 9 | 41.159 | 370.427 | 5,716 | 0.065 | 10,391.8 | 0.04 |
| HARRIS | 3 | 44.778 | 134.335 | 2,003 | 0.067 | 2,543.8 | 0.05 |
| BYRON 1,2 | 6 | 46.780 | 280.677 | 3,390 | 0.083 | 6,721.2 | 0.04 |
| WATTS BAR 1 | 3 | 46.896 | 140.687 | 1,869 | 0.075 | 3,086.3 | 0.05 |
| COMANCHE PEAK 1,2 | 6 | 48.511 | 291.063 | 3,017 | 0.096 | 6,805.8 | 0.04 |
| GINNA | 3 | 48.991 | 146.973 | 1,684 | 0.087 | 1,634.2 | 0.09 |
| CALVERT CLIFFS 1,2 | 6 | 49.748 | 298.486 | 2,470 | 0.121 | 4,928.9 | 0.06 |
| VOGTLE 1,2 | 6 | 51.081 | 306.483 | 3,040 | 0.101 | 6,464.1 | 0.05 |
| SEQUOYAH 1,2 | 6 | 51.244 | 307.462 | 3,203 | 0.096 | 6,235.3 | 0.05 |
| KEWAUNEE | 3 | 51.285 | 153.856 | 1,328 | 0.116 | 1,584.6 | 0.10 |
| BRAIDWOOD 1,2 | 6 | 51.517 | 309.102 | 3,502 | 0.088 | 6,559.4 | 0.05 |
| NORTH ANNA 1,2 | 6 | 53.570 | 321.418 | 2,572 | 0.125 | 4,908.2 | 0.07 |
| ROBINSON 2 | 3 | 53.647 | 160.941 | 1,910 | 0.084 | 1,767.8 | 0.09 |
| MCGUIRE 1,2 | 6 | 54.477 | 326.861 | 4,003 | 0.082 | 6,265.8 | 0.05 |
| POINT BEACH 1,2 | 6 | 55.498 | 332.986 | 2,593 | 0.128 | 2,744.5 | 0.12 |
| SEABROOK | 3 | 55.617 | 166.852 | 2,896 | 0.058 | 3,316.2 | 0.05 |
| SOUTH TEXAS 1,2 | 6 | 57.690 | 346.141 | 3,186 | 0.109 | 7,315.3 | 0.05 |
| TURKEY POINT 3,4 | 6 | 58.387 | 350.323 | 3,451 | 0.102 | 3,805.5 | 0.09 |
| CATAWBA 1,2 | 6 | 58.583 | 351.499 | 3,540 | 0.099 | 6,369.9 | 0.06 |
| BEAVER VALLEY 1,2 | 6 | 59.649 | 357.893 | 3,245 | 0.110 | 4,983.7 | 0.07 |
| WOLF CREEK 1 | 3 | 59.717 | 179.150 | 2,964 | 0.060 | 3,060.1 | 0.06 |
| OCONEE 1,2,3 | 9 | 62.255 | 560.291 | 5,707 | 0.098 | 7,075.3 | 0.08 |
| ARKANSAS 1,2 | 6 | 66.359 | 398.155 | 4,670 | 0.085 | 5,065.1 | 0.08 |
| FORT CALHOUN | 3 | 72.279 | 216.836 | 1,880 | 0.115 | 1,309.6 | 0.17 |
| ST. LUCIE 1,2 | 6 | 73.742 | 442.454 | 3,623 | 0.122 | 4,497.3 | 0.10 |
| SURRY 1,2 | 6 | 75.850 | 455.101 | 3,268 | 0.139 | 4,525.4 | 0.10 |
| SAN ONOFRE 2,3 | 6 | 83.816 | 502.893 | 4,232 | 0.119 | 5,106.4 | 0.10 |
| SALEM 1,2 | 6 | 84.629 | 507.775 | 11,710 | 0.043 | 6,398.8 | 0.08 |
| MILLSTONE 2,3 | 6 | 85.581 | 513.485 | 3,168 | 0.162 | 5,584.0 | 0.09 |
| CRYSTAL RIVER 3 | 3 | 90.125 | 270.376 | 2,653 | 0.102 | 1,411.9 | 0.19 |
| THREE MILE ISLAND 1 | 3 | 94.369 | 283.106 | 2,903 | 0.098 | 2,268.4 | 0.12 |
| INDIAN POINT 2 | 3 | 115.684 | 347.053 | 3,028 | 0.115 | 2,771.9 | 0.13 |
| DIABLO CANYON 1,2 | 6 | 116.387 | 698.322 | 6,291 | 0.111 | 5,940.3 | 0.12 |
| WATERFORD 3 | 3 | 131.407 | 394.222 | 2,964 | 0.133 | 3,227.2 | 0.12 |
| PALISADES | 3 | 170.215 | 510.646 | 2,171 | 0.235 | 2,189.4 | 0.23 |
| DAVIS-BESSE | 3 | 191.440 | 574.319 | 2,749 | 0.209 | 2,244.2 | 0.26 |
| **Totals and Averages** | **207** | **-** | **13,239.346** | **136,016** | **0.097** | **181,913.3** | **0.07** |
| **Average per Reactor-Year** | **-** | **63.958** | **-** | **657** | **-** | **878.8** | **-** |

* Sites where not all reactors had completed three full years of commercial operation as of December 31, 2010, are not included.

**TABLE 4.7**
Three-Year Collective TEDE per Reactor-Year for BWRs
2008-2010

| | Plant Name | Three Year Coll. TEDE per Reactor Year 2008-2010 | Percent Change From 2007-2009 | 2007-2009 Quartile (if changed) |
|---|---|---|---|---|
| 1st Quartile | MONTICELLO | 91.172 | -33% ▼ | 3 |
| | LIMERICK 1,2 | 96.561 | -5% ▼ | - |
| | HATCH 1,2 | 103.541 | 21% ▲ | - |
| | PILGRIM | 104.174 | -41% ▼ | 4 |
| | SUSQUEHANNA 1,2 | 105.942 | -12% ▼ | 2 |
| | DRESDEN 2,3 | 107.278 | -52% ▼ | 4 |
| 2nd Quartile | FERMI 2 | 110.174 | -13% ▼ | - |
| | HOPE CREEK 1 | 121.594 | -8% ▼ | - |
| | DUANE ARNOLD | 121.618 | 3% ▲ | - |
| | PEACH BOTTOM 2,3 | 123.772 | -18% ▼ | 3 |
| | GRAND GULF | 128.983 | 3% ▲ | - |
| | COLUMBIA GENERATING | 138.277 | 46% ▲ | 1 |
| 3rd Quartile | QUAD CITIES 1,2 | 139.051 | -1% ▼ | - |
| | FITZPATRICK | 146.593 | 58% ▲ | 1 |
| | LASALLE 1,2 | 149.777 | 21% ▲ | 2 |
| | OYSTER CREEK | 151.829 | 54% ▲ | 1 |
| | NINE MILE POINT 1,2 | 152.467 | 5% ▲ | - |
| | BROWNS FERRY 1,2,3 | 154.126 | 33% ▲ | 1 |
| 4th Quartile | CLINTON | 157.683 | -5% ▼ | - |
| | VERMONT YANKEE | 160.369 | 8% ▲ | 3 |
| | BRUNSWICK 1,2 | 185.331 | 20% ▲ | 3 |
| | RIVER BEND 1 | 190.500 | -14% ▼ | - |
| | COOPER STATION | 225.087 | 1% ▲ | - |
| | PERRY | 233.068 | -40% ▼ | - |
| | Average per Reactor-Year | 139.170 | -4% ▼ | |

< Average 139.17

## TABLE 4.8
Three-Year Collective TEDE per Reactor-Year for PWRs
2008-2010

| Plant Name | Three-Year Coll. TEDE per Reactor Year 2008-2010 | Percent Change From 2007-2009 | 2007-2009 Quartile (if changed) |
|---|---|---|---|
| **1st Quartile** | | | |
| INDIAN POINT 3 | 25.049 | -57% ▼ | 2 |
| COOK 1,2 | 33.291 | -44% ▼ | 2 |
| FARLEY 1,2 | 34.000 | -8% ▼ | - |
| SUMMER 1 | 35.757 | -1% ▼ | - |
| CALLAWAY 1 | 36.431 | -12% ▼ | - |
| PRAIRIE ISLAND 1,2 | 39.208 | 26% ▲ | - |
| PALO VERDE 1,2,3 | 41.159 | -9% ▼ | - |
| HARRIS | 44.778 | 15% ▲ | - |
| BYRON 1,2 | 46.780 | -20% ▼ | 2 |
| WATTS BAR 1 | 46.896 | 1% ▲ | - |
| COMANCHE PEAK 1,2 | 48.511 | -34% ▼ | 4 |
| **2nd Quartile** | | | |
| GINNA | 48.991 | -1% ▼ | 1 |
| CALVERT CLIFFS 1,2 | 49.748 | -8% ▼ | - |
| VOGTLE 1,2 | 51.081 | -9% ▼ | - |
| SEQUOYAH 1,2 | 51.244 | -18% ▼ | 3 |
| KEWAUNEE | 51.285 | -4% ▼ | - |
| BRAIDWOOD 1,2 | 51.517 | -10% ▼ | - |
| NORTH ANNA 1,2 | 53.570 | -28% ▼ | 4 |
| ROBINSON 2 | 53.647 | 3% ▲ | 1 |
| MCGUIRE 1,2 | 54.477 | -19% ▼ | 3 |
| POINT BEACH 1,2 | 55.498 | 15% ▲ | 1 |
| **3rd Quartile** | | | |
| SEABROOK | 55.617 | 0% | 2 |
| SOUTH TEXAS 1,2 | 57.690 | -3% ▼ | - |
| TURKEY POINT 3,4 | 58.387 | -6% ▼ | - |
| CATAWBA 1,2 | 58.583 | -12% ▼ | - |
| BEAVER VALLEY 1,2 | 59.649 | -9% ▼ | - |
| WOLF CREEK 1 | 59.717 | 4% ▲ | 2 |
| OCONEE 1,2,3 | 62.255 | -10% ▼ | - |
| ARKANSAS 1,2 | 66.359 | -1% ▼ | - |
| FORT CALHOUN | 72.279 | 3% ▲ | - |
| ST. LUCIE 1,2 | 73.742 | -32% ▼ | 4 |
| **4th Quartile** | | | |
| SURRY 1,2 | 75.850 | -17% ▼ | - |
| SAN ONOFRE 2,3 | 83.816 | 27% ▲ | 3 |
| SALEM 1,2 | 84.629 | -7% ▼ | - |
| MILLSTONE 2,3 | 85.581 | -14% ▼ | - |
| CRYSTAL RIVER 3 | 90.125 | -36% ▼ | - |
| THREE MILE ISLAND 1 | 94.369 | -21% ▼ | - |
| INDIAN POINT 2 | 115.684 | 121% ▲ | 2 |
| DIABLO CANYON 1,2 | 116.387 | 2% ▲ | - |
| WATERFORD 3 | 131.407 | -4% ▼ | - |
| PALISADES | 170.215 | -7% ▼ | - |
| DAVIS-BESSE | 191.440 | 390% ▲ | 1 |
| **Average per Reactor-Year** | **63.958** | **-8% ▼** | |

◄ *Average 63.96* (indicated between OCONEE and ARKANSAS rows)

In 1992, the OECD/NEA, with sponsorship from the IAEA, created the Information System on Occupational Exposure (ISOE) Program as an international forum for representatives from nuclear electric utilities and regulatory agencies to share dose reduction information, operational experience, and information to improve the optimization of radiological protection at commercial nuclear power plants. The ISOE database, ISOEDAT, includes occupational exposure information for 401 operating units and 81 units in cold-shutdown or some stage of decommissioning in 29 countries, covering about 91% of the world's operating commercial nuclear power reactors. One of the purposes of ISOEDAT is to allow for comparison of radiation protection effectiveness and trends among the

participating countries and among the various types of commercial nuclear power reactors.

As part of the agency's international cooperative research program initiatives, NRC joined the ISOE Program as a regulatory member in December 1994. NRC's REIRS database is the U.S. system comparable to ISOEDAT on the global scale. Since joining the ISOE Program, NRC has leveraged experience in data management and analysis of the REIRS database, as well as provided input to OECD/NEA and IAEA for streamlining certain elements of how ISOEDAT captures, maintains, and displays data.

Table 4.9 lists the average number of operating PWRs and BWRs included in ISOEDAT during the

### TABLE 4.9
Average Number of Units Reported to ISOE by Country from 1994 – 2010*

| Country | PWR | BWR |
|---|---|---|
| Belgium | 7 | - |
| Brazil | 2 | - |
| China | 4 | - |
| Finland | - | 2 |
| France | 56 | - |
| Germany | 13 | 6 |
| Japan | 22 | 28 |
| Mexico | - | 2 |
| Pakistan | 1 | - |
| Republic of Korea | 13 | - |
| Republic of South Africa | 2 | - |
| Slovenia | 1 | - |
| Spain | 7 | 2 |
| Sweden | 3 | 8 |
| Switzerland | 3 | 2 |
| The Netherlands | 1 | 1 |
| United Kingdom | 1 | - |
| United States | 69 | 36 |

* The average number of units reported to ISOE by country from 1994 – 2010 was determined by counting the number of BWRs and PWRs that had collective dose recorded in ISOEDAT for each country and dividing this total by the number of years reported.

years 1994 to 2010. While there are additional BWRs and PWRs in operation internationally, the reactors included in Table 4.9 had records available in ISOEDAT for comparing the U.S. experience with the international communities. Figures 4.5 and 4.6 show the average collective dose per reactor for PWRs and BWRs for the U.S. and participating reactors from ISOEDAT. For PWRs, the average collective dose per reactor for the ISOE PWRs has been similar to the U.S. experience since 1994 and for BWRs the U.S. and international plants have been similar since 1997. In the last four years, the U.S. PWR average has remained below the average for other countries. The data was compiled from the ISOEDAT online database. The NEA publishes an annual report entitled "Occupational Exposures at Nuclear Power Plants" that is available on the ISOE web site at www.isoe-network.net.

# 4.7 DECONTAMINATION AND DECOMMISSIONING OF COMMERCIAL NUCLEAR POWER REACTORS

The NRC regulates the decontamination and decommissioning (D&D) of commercial nuclear power reactors. The purpose of the NRC's Decommissioning Program is to ensure that NRC-licensed sites are decommissioned in a safe, timely, and effective manner so that they can be returned to beneficial use and to ensure that stakeholders are informed and involved in the process, as appropriate.

The NRC's Office of Federal and State Materials and Environmental Management Programs (FSME) has project management responsibilities for decommissioning commercial nuclear power reactors.

NRC's commercial nuclear power reactor decommissioning activities include project management, technical review of licensee submittals in support of decommissioning, licensing amendments and exemptions in support of the progressive stages of decommissioning, inspections of decommissioning activities, support for the development of rulemaking guidance, public outreach efforts, international activities, and participation in industry conferences and workshops. FSME staff regularly coordinate with other offices on issues affecting all commercial nuclear power reactors, both operating and decommissioning, and specifically with staff in the Office of Nuclear Material Safety and Safeguards (NMSS) regarding the ISFSIs at reactor sites undergoing decommissioning [Ref. 17].

## *4.7.1 Decommissioning Process*

The decommissioning process begins when a licensee decides to permanently cease operations. The major steps that comprise the commercial nuclear power reactor decommissioning process are notification of cessation of operations; submittal and review of the post-shutdown decommissioning activities report (PSDAR); submittal, review and approval of the license termination plan (LTP); implementation of the LTP; and completion of decommissioning. The flowchart in Figure 4.7 illustrates the D&D process.

## *4.7.1.1 Notification*

When a licensee has decided to permanently cease operations, the licensee is required to submit a written notification to NRC. In addition, the licensee is required to notify the NRC in writing once fuel has been permanently removed from the reactor vessel.

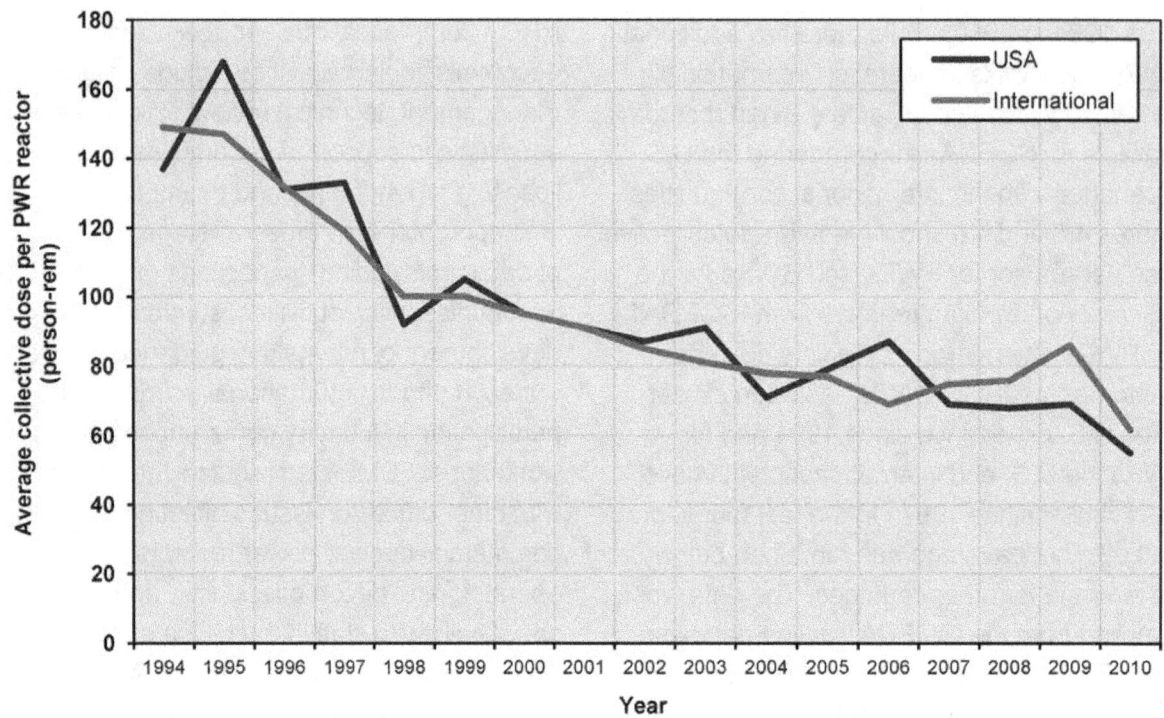

**FIGURE 4.5.** Average Collective Dose per PWR Reactor
1994–2010

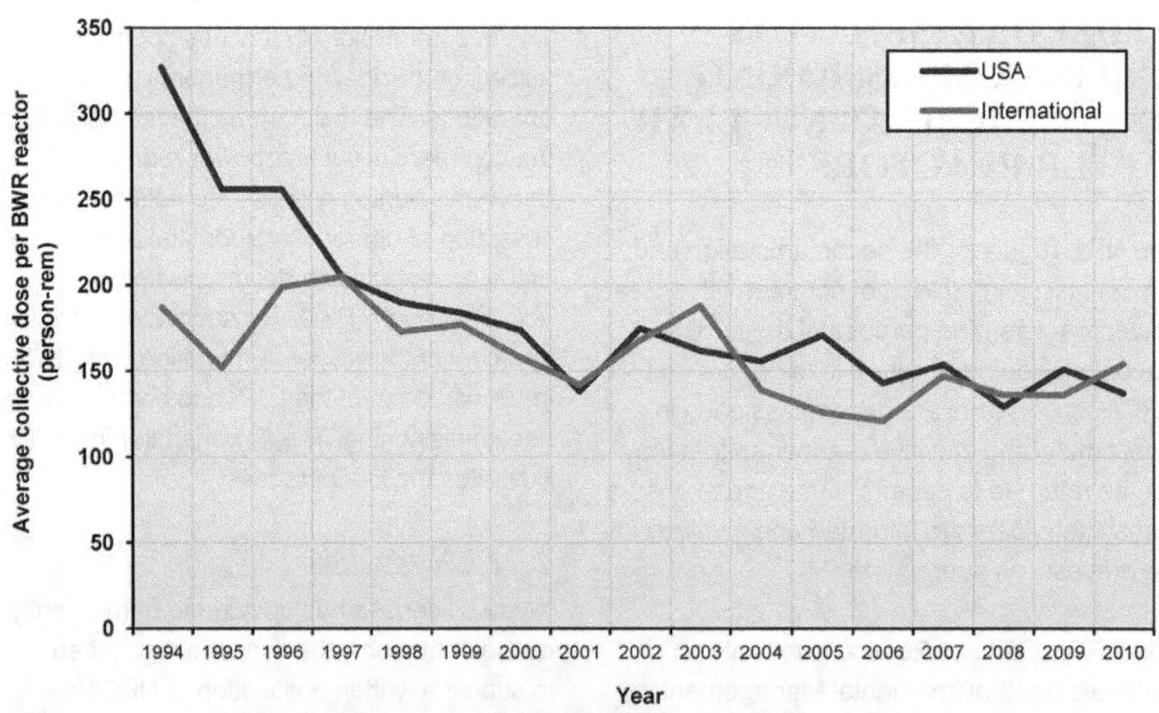

**FIGURE 4.6.** Average Collective Dose per BWR Reactor
1994–2010

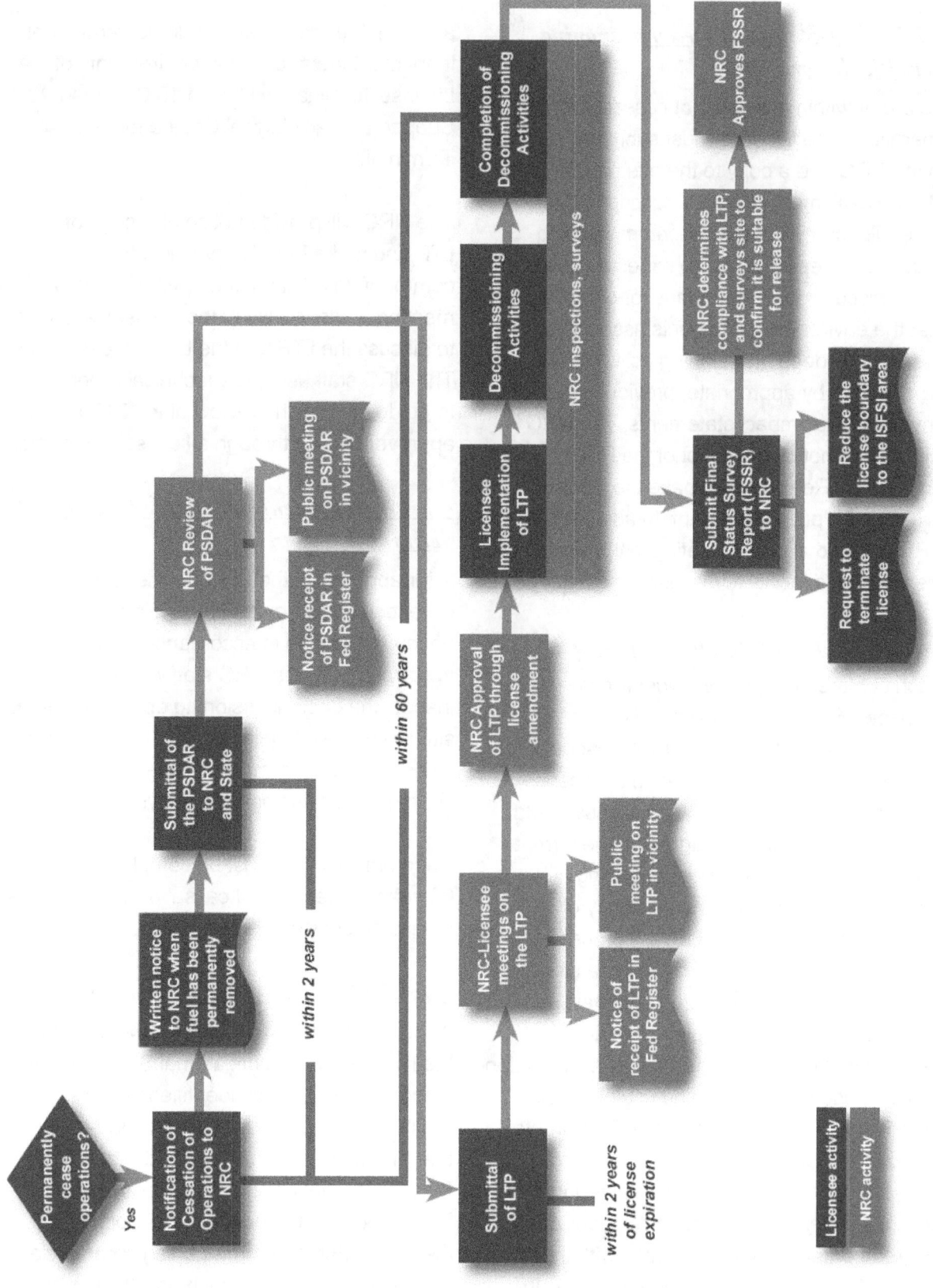

**FIGURE 4.7.** D&D Process Flowchart

## 4.7.1.2 Post-Shutdown Decommissioning Activities Report

Before or within two years of cessation of operations, the licensee must submit a PSDAR to the NRC and a copy to the affected State(s). The PSDAR must include a description and schedule for the planned decommissioning activities; an estimate of the expected costs; and a discussion of the means for concluding that the environmental impacts associated with site-specific decommissioning activities will be bounded by appropriate, previously issued environmental impact statements. The NRC will provide notice of receipt of the PSDAR in the Federal Register and make the PSDAR available for public comment. In addition, the NRC will hold a public meeting in the vicinity of the licensee's facility to discuss the PSDAR.

## 4.7.1.3 License Termination Plan

Each commercial nuclear power reactor licensee must submit an application for termination of its license. An LTP must be submitted at least 2 years before the license termination date. The NRC and licensee hold presubmittal meetings to agree on the format and content of the LTP. These meetings are intended to improve the efficiency of the LTP development and review process. The LTP must include the following: a site characterization; identification of remaining dismantlement activities; plans for site remediation; detailed plans for the final radiation survey; description of the end use of the site, if restricted; an updated site-specific estimate of remaining decommissioning costs; and a supplement to the environmental report describing any new information or significant environmental change associated with the licensee's proposed termination activities. In addition, the licensee must demonstrate that it will meet the applicable requirements of the License Termination Rule in 10 CFR Part 20, Subpart E, "Radiological Criteria for License Termination."

The NRC will provide notice of receipt of the LTP and make the LTP available for public comment. In addition, the NRC will hold a public meeting in the vicinity of the licensee's facility to discuss the LTP and the LTP review process. The NRC staff use three technical reports to guide them in the review of the LTP and approve the LTP through a license amendment.

## 4.7.1.4 Implementation of the License Termination Plan

After approval of the LTP, the licensee or responsible party must complete decommissioning in accordance with the approved LTP. The NRC staff will periodically inspect the decommissioning operations at the site to ensure compliance with the LTP. These inspections will normally include in-process and confirmatory radiological surveys.

Decommissioning must be completed within 60 years of permanent cessation of operations, unless otherwise approved by the Commission.

## 4.7.1.5 Completion of Decommissioning

At the conclusion of decommissioning activities, the licensee will submit a Final Status Survey Report (FSSR), which identifies the final radiological conditions of the site and requests that the NRC either: (1) terminate the 10 CFR Part 50 license; or (2) reduce the 10 CFR Part 50 license boundary to the footprint of the ISFSI. For decommissioning commercial nuclear power reactors with no ISFSI or an

ISFSI holding a specific license under 10 CFR Part 72, completion of reactor decommissioning will result in the termination of the 10 CFR Part 50 license. The NRC will approve the FSSR and the licensee's request if it determines that the licensee has met both of the following conditions: the remaining dismantlement has been performed in accordance with the approved LTP, and the final radiation survey and associated documentation demonstrate that the facility and site are suitable for release in accordance with the License Termination Rule.

## 4.7.2 Status of Decommissioning Activities at Commercial Nuclear Power Reactors

While 104 commercial nuclear power reactors are currently in operation, several shutdown power reactors have undergone the process of D&D. As more commercial nuclear power reactors reach the end of their operating license, there will be a commensurate increase in activities involving radiation exposure related to D&D. For this reason, there is an increased need to provide further information on plants undergoing D&D.

Appendix B contains a list of the plants that are no longer in commercial operation, along with the dose distribution and collective dose for these plants. It should be noted that these plants may be in different stages of D&D, so that a comparison of dose at one plant versus another would not be meaningful. In addition, Appendix B lists the plant units that are no longer in commercial operation but report along with other units at the site. Under the licensing conditions and reporting requirements, it is permitted to report this information together in one report. Table 4.10 lists the plants that

have ceased operation and have changed the operational status as of the date shown [Ref. 16]. In addition, Appendix E provides descriptions of the decommissioning activities currently underway at these commercial nuclear power reactors, as well as the total collective TEDE for each plant, from the year megawatt production stopped through 2010.

## TABLE 4.10*
### Plants No Longer in Operation
### 2010

| Plant Name | Date of First Commercial Operation | Plant Shutdown/ Notification to NRC | License Termination Plan Approved by NRC | PSDAR Submitted | Plant Status | Completion of Decommissioning |
|---|---|---|---|---|---|---|
| BIG ROCK POINT | 3/29/1963 | 8/1997 | 3/2005 | 9/1997 | ISFSI only | 2007 |
| DRESDEN 1 | 8/1/1960 | 10/1978 | 9/1993 | 6/1998 | SAFSTOR** | 2036 |
| FERMI 1 | 5/10/1963 | 9/1972 | 2010 | 4/1998 | DECON | 2012 |
| HADDAM NECK | 12/27/1974 | 12/1996 | 11/2002 | 8/1997 | ISFSI only | 2007 |
| HUMBOLDT BAY 3 | 8/1/1963 | 7/1976 | 4/1987 | 2/1998 | DECON*** | 2015 |
| INDIAN POINT 1 | 3/26/1962 | 10/1974 | 1/1996 | 1/1996 | SAFSTOR | 2026 |
| LACROSSE | 11/1/1969 | 4/1987 | 8/1991 | 5/1991 | SAFSTOR | 2026 |
| MAINE YANKEE | 6/29/1973 | 8/1997 | 2/2003 | 8/1997 | ISFSI only | 2005 |
| MILLSTONE 1 | 12/28/1970 | 7/1998 | TBD | 6/1999 | SAFSTOR | TBD |
| PEACH BOTTOM 1 | 1/24/1966 | 10/1974 | TBD | 6/1998 | SAFSTOR | 2034 |
| RANCHO SECO | 4/17/1975 | 6/1989 | 3/1995 | - | DECON | 2009 |
| SAN ONOFRE 1 | 1/1/1968 | 11/1992 | 11/1994 | 12/1998 | DECON | 2030 |
| THREE MILE ISLAND 2 | 12/30/1978 | 3/1979 | TBD | 2/1979 | Post-Defueling Monitored Storage | 2036 |
| TROJAN | 5/20/1976 | 11/1992 | 2/2001 | - | ISFSI only | 2004 |
| YANKEE ROWE | 12/24/1963 | 10/1991 | 2005 | - | ISFSI only | 2006 |
| ZION 1 | 12/31/1973 | 2/1997 | TBD | 2/2000 | SAFSTOR | 2020 |
| ZION 2 | 9/17/1974 | 9/1996 | TBD | 2/2000 | SAFSTOR | 2020 |

\* Information regarding the latest decommissioning status of plants listed in this table can be found in Status of the Decommissioning Program: 2011 Annual Report from the NRC's public library under ADAMS Accession No. ML112700498.
\*\* SAFSTOR - (often considered 'delayed DECON'): a nuclear facility that is maintained and monitored in a condition that allows the radioactivity to decay; afterwards, it is dismantled.
\*\*\* DECON - (immediate dismantlement): soon after he nuclear facility closes, equipment, structures, and portions of the facility containing radioactive contaminants are removed or decontaminated to a level that permits release of the property and termination of the NRC license.

# Section 5
# TRANSIENT INDIVIDUALS AT NRC-LICENSED FACILITIES

The following analysis examines the individuals who had more than one Form 5 dose record at more than one NRC-licensed facility during the monitoring year. These individuals are defined as "transient" because they worked at more than one facility during the monitoring year.

The term "monitoring year" is used here in accordance with the definition of a year given in 10 CFR 20.1003, which defines a year as "the period of time beginning in January used to determine compliance with the provisions of 10 CFR Part 20. The licensee may change the start date of the monitoring year used to determine compliance provided that the change is made at the beginning of the monitoring/ calendar year and that no day is omitted or duplicated in consecutive years."

Examination of the data reported for individuals who began and terminated two or more periods of employment with two or more different facilities within one monitoring year is useful in many ways. For example, the number of transients and the individual doses received by them can be determined from examining these data.

Additionally, the distribution of the doses received by transient individuals can be useful in determining the impact that the inclusion of these individuals in each of two or more licensees' annual reports has on the annual summary (as reported in Appendix B) for all commercial nuclear power reactors and all NRC licensees combined (one of the issues mentioned in Section 2). Table 5.1 shows the actual distribution of transient individual doses as determined from the NRC Form 5 termination reports and compares it with the reported distribution of the doses of these individuals as they would have appeared in a summation of the annual reports submitted by each of the licensees.

In 2010, over 99% of the transient individuals were reported by commercial nuclear power reactors. For this reason, these data are shown separately in Table 5.1.

Table 5.1 illustrates the impact that the multiple reporting of these transient individuals had on the summation of the dose reports for 2010. Each licensee reports the radiation dose received by individuals monitored at their facility. Many of these individuals are monitored at more than one facility during the year. When these dose records are summed for all licensees, they appear to be separate individuals reported by each facility. If an individual visited five facilities during a year, this individual would appear in the summation to be five different people, with one dose record for each of the five facilities. When these dose records are summed per individual, these records appear as one person, with a total annual dose that accurately represents the dose received for the entire monitoring year. Thus, while the total collective dose would remain the same, the number of individuals, their dose distributions, and average doses would be affected by this multiple reporting.

## TABLE 5.1
### Effects of Transient Individuals on Annual Statistical Compilations 2010

| License Category | Number of Individuals with TEDE in the Ranges (rem) * | | | | | | | | | | | | Total Number Monitored | Number with Measurable TEDE | Collective TEDE (person-rem) | Average Meas. TEDE (rem) |
|---|---|---|---|---|---|---|---|---|---|---|---|---|---|---|---|---|
| | No Measurable Exposure | Measurable <0.10 | 0.10-0.25 | 0.25-0.50 | 0.50-0.75 | 0.75-1.0 | 1.0-2.0 | 2.0-3.0 | 3.0-4.0 | 4.0-5.0 | 5.0-6.0 | >6 | | | | |
| **COMMERCIAL LIGHT WATER REACTORS** | | | | | | | | | | | | | | | | |
| (1) Form 5 Summation | 104,638 | 49,571 | 16,042 | 6,656 | 1,801 | 602 | 333 | 5 | - | - | - | - | 179,648 | 75,010 | 8,631.384 | 0.12 |
| (2) Transients, As Reported | 39,691 | 24,283 | 9,251 | 3,824 | 1,076 | 369 | 181 | 4 | - | - | - | - | 78,679 | 38,988 | 4,926.989 | 0.13 |
| (3) Transients, Actual | 9,271 | 8,586 | 4,879 | 3,524 | 1,506 | 713 | 680 | 41 | 3 | - | - | - | 29,203 | 19,932 | 4,926.989 | 0.25 |
| Corrected Distribution (1-[2-3]) ** | 74,218 | 33,874 | 11,670 | 6,356 | 2,231 | 946 | 832 | 42 | 3 | - | - | - | 130,172 | 55,954 | 8,631.384 | 0.15 |
| **ALL LICENSEES** | | | | | | | | | | | | | | | | |
| (1) Form 5 Summation | 110,463 | 53,284 | 17,286 | 7,439 | 2,233 | 843 | 732 | 109 | 34 | 1 | - | - | 192,424 | 81,961 | 10,617.253 | 0.13 |
| (2) Transients, As Reported | 40,018 | 24,461 | 9,289 | 3,848 | 1,095 | 377 | 193 | 4 | 1 | - | - | - | 79,286 | 39,268 | 4,985.525 | 0.13 |
| (3) Transients, Actual | 9,244 | 8,656 | 4,897 | 3,550 | 1,523 | 720 | 697 | 42 | 4 | - | - | - | 29,333 | 20,089 | 4,985.525 | 0.25 |
| Corrected Distribution (1-[2-3]) ** | 79,689 | 37,479 | 12,894 | 7,141 | 2,661 | 1,186 | 1,236 | 147 | 37 | 1 | - | - | 142,471 | 62,782 | 10,617.253 | 0.17 |

* Dose values exactly equal to the values separating ranges are reported in the next higher range.
** The corrected distribution only applies to the number of individuals.

For example, in 2010, Table 5.1 shows that the initial summation of the Form 5 reports for reactor licensees indicated that five individuals received a dose greater than 2 rem. After accounting for those individuals who were reported more than once, the corrected distribution indicated that there were 42 transient individuals who received doses between 2.0 rem and 3.0 rem, and three transient individuals who received doses between 3.0 rem and 4.0 rem. Correcting for the multiple counting of individuals also has a significant effect on the average measurable dose for these individuals. The corrected average measurable dose for transient individuals is twice as high as the value calculated by the summation of the Form 5 records. The transient individuals represent 32% of the workforce that receives measurable dose. The correction for the transient individuals increases the average measurable dose by about a factor of two from 0.13 rem to 0.25 rem for the transient workforce for all licensees. It should be noted that the analysis of transient individuals does not include individuals who

may have been exposed at facilities that are not required to report to the NRC (see Section 1), such as Agreement State licensees and DOE facilities.

One purpose of the REIRS database, which tracks occupational radiation exposures at NRC-licensed facilities, is to identify individuals who may have exceeded the occupational radiation dose limits because of multiple exposures at different facilities throughout the year. The REIRS database stores the radiation dose information for an individual by his/her unique identification number and identification type [Ref. 12, Section 1.5] and sums the dose for all facilities during the monitoring year. An individual exceeding the 5 rem per year regulatory limit (TEDE) would be identified in Table 5.1 in one of the dose ranges >5 rem. In 2010, there were no individuals reported by NRC licensees that exceeded this limit.

# Section 6
# EXPOSURES TO PERSONNEL IN EXCESS OF REGULATORY LIMITS

## 6.1 REPORTING CATEGORIES

Doses in excess of regulatory limits are sometimes referred to as "overexposures." The phrase "doses in excess of regulatory limits" is preferred to "overexposures" because the latter suggests that an individual has been subjected to an unacceptable biological risk, which may or may not be the case.

10 CFR 20.2202 and 10 CFR 20.2203 require that all licensees submit reports of all incidents involving personnel radiation doses that exceed certain levels, thus providing for investigations and corrective actions as necessary. Based on the magnitude of the dose, the occurrence may be placed into one of three categories as follows:

1. Category A
   10 CFR 20.2202(a)(1) — a TEDE to any individual of 25 rem or more, a lens dose equivalent of 75 rem or more, or a shallow-dose equivalent to the skin or extremities of 250 rads or more. The Commission must be notified immediately of these events.

2. Category B
   10 CFR 20.2202(b)(1) — In a 24-hour period, the Commission must be notified of the following events: a TEDE to any individual exceeding 5 rem, a lens dose equivalent exceeding 15 rem, or a shallow-dose equivalent to the skin or extremities exceeding 50 rem.

3. Category C
   10 CFR 20.2203 — In addition to the notification required by 10 CFR 20.2202

(Category A or B events), each licensee must submit a written report within 30 days after learning of any of the following occurrences:

a. Any incident for which notification is required by 10 CFR 20.2202

b. Doses that exceed the limits in §20.1201, §20.1207, §20.1208, or §20.1301 (for adults, minors, the embryo/fetus of a declared pregnant woman, and the public, respectively) or any applicable limit in the license

c. Levels of radiation or concentrations of radioactive material that exceed any applicable license limit for restricted areas or that, for unrestricted areas, are in excess of 10 times any applicable limit set forth in 10 CFR Part 20 or in the license (whether or not involving dose of any individual in excess of the limits in §20.1301)

d. For licensees subject to the provisions of the Environmental Protection Agency's generally applicable environmental radiation standards in 40 CFR 190, levels of radiation or releases of radioactive material in excess of those standards or license conditions related to those standards

Exposure events reported as either Category A, B, or C typically undergo a long review and evaluation process by the licensee, NRC inspectors, and NRC Headquarters staff. Preliminary dose estimates submitted by licensees are often conservatively high and do not represent the final (record) dose assigned for the event. It is, therefore, not uncommon for a dose in excess of a regulatory limit event to be reassessed and the final assigned dose to be categorized as not having been in excess of a regulatory limit. In other cases, the exposure event may not be identified until a later date, such as during the next scheduled audit or inspection of the licensee's event records.

## 6.2 SUMMARY OF OCCUPATIONAL RADIATION DOSES IN EXCESS OF NRC REGULATORY LIMITS

The exposure events summary presented here are for events that occurred between 2000 through 2010. An event that has been reassessed and determined not to be a dose in excess of a regulatory limit is not included in this report. In addition, events that occurred in prior years are added to the summary in the appropriate year of occurrence. The reader should note that the summary presented here represents a snapshot of the status of events as of the publication date of this report. Previous or future reports may not correlate in the exact number of events because of the review cycle and reassessment of the events.

It is important to note that this summary of events includes only
• Occupational radiation doses in excess of the annual 5 rem regulatory limit
• Events at NRC-licensed facilities
• Final dose of record assigned to an individual

It **does not** include
• Medical events as defined in 10 CFR Part 35
• Doses in excess of the regulatory limits to the general public
• Agreement State-licensed activities or DOE facilities

• Exposures to dosimeters that, upon evaluation, have been determined to be high dosimeter readings only and are not assigned to an individual as the dose of record by the licensee

In 2010, there were no category A, B, or C occurrences reported under the licensed activities included in this report.

## 6.3 SUMMARY OF ANNUAL DOSE DISTRIBUTIONS FOR CERTAIN NRC LICENSEES

Table 6.1 gives a summary of the annual occupational dose records reported to NRC, as required by 10 CFR 20.2206, by certain categories of NRC licensees. Table 6.1 shows that for the past eleven years, the percentage of individuals with <2 rem has been greater than 99%. The number of individuals receiving an annual dose greater than 5 rem has been <0.01% since 2000. No individual monitored at any of the five NRC licensee categories included in this report received a dose above the 5 rem annual regulatory limit (TEDE) during the past seven years.

## 6.4 MAXIMUM OCCUPATIONAL RADIATION DOSES BELOW NRC REGULATORY LIMITS

Certain researchers have expressed an interest in a listing of the maximum doses received at NRC licensees that do not exceed the regulatory limits. This information allows for an examination of these doses and could possibly provide insights for where certain improvements could be made in the licensee's radiation protection program. Table 6.2 shows the maximum doses for each dose category required to be reported to the NRC. In addition, the number of doses in certain dose ranges is shown to reflect the number of doses that approach NRC regulatory limits. As shown in Table 6.2, few doses exceed half of the NRC occupational annual limits. In 2010, four individuals exceeded 75% of the TEDE dose limit, but no individual exceeded any of the annual occupational dose limits.

## TABLE 6.1
### Summary of Annual Dose Distributions for Certain* NRC Licensees
### 2000–2010

| Year | Total Number of Monitored Individuals | | Individuals with Dose (TEDE) *** | | | | Individuals with Dose >12 rem TEDE *** |
|---|---|---|---|---|---|---|---|
| | Reported Number | Corrected Number ** | < 2 rem % | > 2 rem Number | < 5 rem % | > 5 rem Number | |
| 2000 | 163,345 | 125,368 | 99.5% | 573 | >99.99% | 3 | - |
| 2001 | 154,693 | 118,502 | 99.4% | 734 | >99.99% | 1 | - |
| 2002 | 162,714 | 120,026 | 99.5% | 582 | >99.99% | 1 | - |
| 2003 | 166,347 | 122,575 | 99.7% | 419 | >99.99% | 1 | 1 |
| 2004 | 164,526 | 123,470 | 99.7% | 368 | 100% | - | - |
| 2005 | 174,550 | 127,138 | 99.7% | 370 | 100% | - | - |
| 2006 | 176,623 | 127,391 | 99.8% | 258 | 100% | - | - |
| 2007 | 177,253 | 126,709 | 99.8% | 243 | 100% | - | - |
| 2008 | 182,085 | 130,462 | 99.9% | 167 | 100% | - | - |
| 2009 | 189,955 | 139,448 | 99.9% | 173 | 100% | - | - |
| 2010 | 192,424 | 142,471 | 99.9% | 185 | 100% | - | - |

\* Licensees required to submit radiation exposure reports to the NRC under 10 CFR 20.2206.
\*\* This column lists the actual number of persons who may have been counted more than once because they worked at more than one facility during the calendar year (see Section 5).
\*\*\* Data for 2000–2010 are based on the distribution of individual doses after adjusting for the multiple counting of transient individuals (see Section 5).

## TABLE 6.2
### Maximum Occupational Doses for Each Exposure Category*
### 2010

| Dose Category** | Annual Dose Limit 10CFR20*** | Maximum Dose Reported (rem) | Max Dose Percent of the Limit | Number of Individuals with Measurable Dose | Number of Individuals >25% of the Limit | Number of Individuals >50% of the Limit | Number of Individuals >75% of the Limit | Number of Individuals >95% of the Limit | Number of Individuals > Limit |
|---|---|---|---|---|---|---|---|---|---|
| SDE-ME | 50 rem | 29.820 | 60% | 57,238 | 5 | 1 | - | - | - |
| SDE-WB | 50 rem | 6.271 | 13% | 62,628 | - | - | - | - | - |
| LDE | 15 rem | 4.624 | 31% | 60,723 | 7 | - | - | - | - |
| CEDE | | 1.557 | | 3,045 | | | | | |
| CDE | | 12.538 | | 2,243 | | | | | |
| DDE | | 4.102 | | 61,397 | | | | | |
| TEDE | 5 rem | 4.102 | 82% | 62,782 | 826 | 83 | 4 | - | - |
| TODE | 50 rem | 12.623 | 25% | 61,374 | 1 | - | - | - | - |

\* Only records reported by licensees required to report under 10 CFR 20.2206 are included. Numbers have been adjusted for the multiple reporting of transient individuals.
\*\* SDE-ME = shallow dose equivalent to the maximally exposed extremity
SDE-WB = shallow dose equivalent to the whole body
LDE = lens dose equivalent to the lens of the eye
CEDE = committed effective dose equivalent
CDE = committed dose equivalent
DDE = deep dose equivalent
TEDE = total effective dose equivalent
TODE = total organ dose equivalent
\*\*\* Shaded boxes represent dose categories that do not have specific dose limits defined in 10 CFR 20.

# Section 7
# REFERENCES

1. National Council on Radiation Protection and Measurements, *Ionizing Radiation Exposure of the Population of the United States*, Report No. 160, 2009.

2. United Nations, *Sources and Effects of Ionizing Radiation, United Nations Scientific Committee on the Effects of Atomic Radiation UNSCEAR 2008 Report to the General Assembly, Volume I*, General Assembly of Official Records, United Nations, New York, 2010.

3. U.S. Atomic Energy Commission, *Nuclear Power Plant Operating Experience During 1973*, USAEC Report 00E-ES-004, December 1974.*

4. U.S. Nuclear Regulatory Commission, *Nuclear Power Plant Operating Experience 1974–1975*, USNRC Report NUREG-0227, April 1977.*

5. U.S. Nuclear Regulatory Commission, *Nuclear Power Plant Operating Experience 1976*, USNRC Report NUREG-0366, December 1977.*

6. M. R. Beebe, *Nuclear Power Plant Operating Experience – 1977*, USNRC Report NUREG-0483, February 1979.*

7. *Nuclear Power Plant Operating Experience – 1978*, USNRC Report NUREG-0618, December 1979.*

8. *Nuclear Power Plant Operating Experience – 1979*, USNRC Report NUREG/CR-1496, May 1981.*

9. *Nuclear Power Plant Operating Experience – 1980*, USNRC Report NUREG/CR-2378, ORNL/NSIC-191, October 1982.*

10. *Nuclear Power Plant Operating Experience – 1981*, USNRC Report NUREG/CR-3430, ORNL/NSIC-215, Vol. 1, December 1983.*

11. *Nuclear Power Plant Operating Experience – 1982*, USNRC Report NUREG/CR-3430, ORNL/NSIC-215, Vol. 2, January 1985.*

12. *Instructions for Recording and Reporting Occupational Radiation Exposure Data*, USNRC Regulatory Guide 8.7, Rev. 2, November 2005.

---

*Report is available for purchase from the National Technical Information Service, Springfield, VA, 22161, and/or the Superintendent of Documents, U.S. Government Printing Office, P.O. Box 37082, Washington, DC 20402-9328.

13.  International Commission on Radiological Protection Publication 30, *Limits for Intakes of Radionuclides by Workers*, Annals of the ICRP Volume 2 No 3/4, 1972.

14.  International Commission on Radiological Protection Publication 68, *Dose Coefficients for Intakes of Radionuclides by Workers*, Annals of the ICRP Volume 24/4, December 1994.

15.  *Licensed Operating Reactors, Status Summary Report,* compiled from reactor monthly operating reports submitted to the NRC. Data provided electronically from the Idaho National Engineering and Environmental Laboratory Risk, Reliability and Regulatory Support Department under contract to the NRC in support of NRC's Performance Indicator Project.

16.  U. S. Nuclear Regulatory Commission, *2010-2011 Information Digest*, USNRC Report NUREG-1350, Volume 22, August 2010.

17.  U. S. Nuclear Regulatory Commission, Division of Waste Management and Environmental Protection, Office of Federal and State Materials and Environmental Management Programs, *Status of the Decommissioning Program, 2009 Annual Report.*

18.  http://www.nrc.gov/info-finder/decommissioning/power-reactor/

19.  http://www.nrc.gov/reading-rm/basic-ref/glossary.html

20.  Shleien, Slaback, et al. *Handbook of Health Physics and Radiological Health.* 3rd ed. Lippincott Williams & Wilkins, 1998.

Appendix A

# ANNUAL TEDE FOR NONREACTOR NRC LICENSEES AND OTHER FACILITIES REPORTING TO THE NRC

## 2010

## APPENDIX A
## Table A1 - Annual TEDE for Nonreactor NRC Licensees
## 2010

| PROGRAM CODE - LICENSEE NAME | LICENSE # | No Meas. Exposure | Number of Individuals with Whole Body Doses in the Ranges (rems)* | | | | | | | | | | | | Total Number Monitored | Number with Meas. Dose | Total Collective TEDE (person-rem) | Average Meas. TEDE (rem) |
|---|---|---|---|---|---|---|---|---|---|---|---|---|---|---|---|---|---|---|---|
| | | | Meas. <0.10 | 0.10-0.25 | 0.25-0.50 | 0.50-0.75 | 0.75-1.00 | 1.00-2.00 | 2.00-3.00 | 3.00-4.00 | 4.00-5.00 | 5.00-6.00 | 6.00-12.00 | >12.0 | | | | |
| *INDUSTRIAL RADIOGRAPHY – FIXED LOCATION – 03310* | | | | | | | | | | | | | | | | | | |
| DEPARTMENT OF THE ARMY | 13-18235-01 | 66 | 10 | 1 | - | - | - | - | - | - | - | - | - | - | - | 77 | 11 | 0.417 | 0.038 |
| HARRISON STEEL CASTINGS CO. | 13-02141-01 | 5 | 2 | - | - | - | - | - | - | - | - | - | - | - | - | 7 | 2 | 0.079 | 0.040 |
| Total | 2 | 71 | 12 | 1 | - | - | - | - | - | - | - | - | - | - | - | 84 | 13 | 0.496 | 0.038 |
| *INDUSTRIAL RADIOGRAPHY – TEMPORARY JOB SITE – 03320* | | | | | | | | | | | | | | | | | | |
| ACUREN USA, INC. | 42-32443-01 | 17 | 72 | 40 | 36 | 30 | 14 | 18 | - | - | - | - | - | - | - | 227 | 210 | 75.429 | 0.359 |
| ALASKA INDUSTRIAL X-RAY | 50-16084-01 | - | 3 | - | 1 | - | - | 4 | - | - | - | - | - | - | - | 8 | 8 | 6.143 | 0.768 |
| ALLIED INSPECTION SERVICES, INC. | 21-18428-01 | - | 1 | - | 2 | - | 1 | 1 | - | - | - | - | - | - | - | 5 | 5 | 3.113 | 0.623 |
| ALONSO & CARUS IRON WORKS, INC. | 52-21350-01 | 1 | 1 | 1 | - | - | - | - | - | - | - | - | - | - | - | 3 | 2 | 0.151 | 0.076 |
| AMERICAN ENGINEERING TESTING, INC. | 22-20271-02 | - | 2 | 1 | 1 | 1 | 2 | 1 | 1 | - | - | - | - | - | - | 9 | 9 | 6.324 | 0.703 |
| ANVIL CORPORATION | 46-23236-03 | 11 | 20 | 19 | 5 | 10 | 3 | 2 | - | - | - | - | - | - | - | 70 | 59 | 16.831 | 0.285 |
| BRANCH RADIOGRAPHIC LABS., INC. | 29-03405-02 | 9 | 3 | 6 | 3 | 4 | 1 | 1 | - | - | - | - | - | - | - | 27 | 18 | 6.761 | 0.376 |
| CALUMET TESTING SERVICES, INC. | 13-16347-01 | 5 | 7 | 1 | 2 | 1 | 1 | 4 | 2 | - | - | - | - | - | - | 23 | 18 | 15.842 | 0.880 |
| CANYON STATE INSPECTION | 02-29359-01 | 2 | 5 | 1 | 3 | 1 | - | - | 1 | - | - | - | - | - | - | 12 | 10 | 2.289 | 0.229 |
| CAPITAL X-RAY SERVICES, INC. | 35-11114-01 | 5 | 7 | - | - | 2 | 1 | 4 | 2 | - | - | - | - | - | - | 21 | 16 | 13.361 | 0.835 |
| CENTURY INSPECTION, INC. | 42-08456-02 | 10 | 14 | 6 | 6 | 1 | 2 | - | - | - | - | - | - | - | - | 39 | 29 | 5.959 | 0.205 |
| COLBY & THIELMEIER TESTING CO. | 24-13737-01 | - | - | - | 1 | 1 | 2 | 2 | - | - | - | - | - | - | - | 6 | 6 | 5.701 | 0.950 |
| COMO TECH INSPECTION | 15-26978-01 | - | - | 2 | 1 | - | - | - | - | - | - | - | - | - | - | 3 | 3 | 0.532 | 0.177 |
| CONCRETE IMAGING, INC. | 47-31316-01 | - | 3 | - | 1 | 1 | 1 | 3 | - | - | - | - | - | - | - | 9 | 9 | 7.018 | 0.780 |
| DBI, INC. | 49-29301-01 | - | 4 | 2 | 3 | 6 | 2 | 9 | - | - | - | - | - | - | - | 26 | 26 | 20.174 | 0.776 |
| ENGINEERING & INSPECTIONS - HAWAII | 53-27731-01 | - | 2 | 2 | 1 | - | 1 | 2 | - | - | - | - | - | - | - | 7 | 7 | 4.474 | 0.639 |
| GENERAL DYNAMICS CORP - ELEC BOAT | 06-01781-08 | 3 | 13 | - | - | - | - | - | - | - | - | - | - | - | - | 16 | 13 | 0.311 | 0.024 |
| GENERAL TESTING & INSPECTION CO. | 47-32191-01 | 2 | - | - | - | - | - | - | - | - | - | - | - | - | - | 2 | - | - | - |

NOTE: The data values shown bolded and in boxes represent the highest value in each category. These values have not been adjusted for the multiple counting of transient workers (see section 5).
* Dose values exactly equal to the values separating ranges are reported in the next higher range.

## APPENDIX A
### Table A1 - Annual TEDE for Nonreactor NRC Licensees 2010 (continued)

*INDUSTRIAL RADIOGRAPHY – TEMPORARY JOB SITE – 03320 (Continued)*

| PROGRAM CODE - LICENSEE NAME | LICENSE # | No Meas. Exposure | Meas. <0.10 | 0.10-0.25 | 0.25-0.50 | 0.50-0.75 | 0.75-1.00 | 1.00-2.00 | 2.00-3.00 | 3.00-4.00 | 4.00-5.00 | 5.00-6.00 | 6.00-12.00 | >12.0 | Total Number Monitored | Number with Meas. Dose | Total Collective TEDE (person-rem) | Average Meas. TEDE (rem) |
|---|---|---|---|---|---|---|---|---|---|---|---|---|---|---|---|---|---|---|
| GLOBAL X-RAY & TESTING CORP. | 17-29308-01 | - | 16 | 10 | 11 | 11 | 5 | 14 | 3 | 3 | - | - | - | - | 73 | 73 | 54.997 | 0.753 |
| H & H X-RAY SERVICES, INC. | 17-19236-01 | 2 | 35 | 17 | 32 | 41 | 27 | 50 | 27 | 2 | - | - | - | - | 233 | 231 | 207.431 | 0.898 |
| HIGH COUNTRY FABRICATION | 49-29300-01 | 2 | 2 | 1 | 1 | - | - | - | - | - | - | - | - | - | 6 | 4 | 0.561 | 0.140 |
| HIGH MOUNTAIN INSPECTION SERVICES | 49-26808-02 | - | 10 | 5 | 5 | 7 | 5 | 20 | 6 | 2 | - | - | - | - | 60 | 60 | 63.777 | 1.063 |
| HUNTINGTON TESTING & TECHNOLOGY | 47-23076-01 | - | 10 | 4 | 8 | 4 | 4 | 4 | 3 | 2 | - | - | - | - | 39 | 39 | 30.206 | 0.775 |
| INTEGRITY TESTLAB | 07-30791-01 | 7 | 5 | 4 | 4 | 4 | 4 | 3 | 2 | 1 | - | - | - | - | 34 | 27 | 19.793 | 0.733 |
| JANX INTEGRITY GROUP | 21-16560-01 | 98 | 58 | 40 | 60 | 42 | 25 | 76 | 20 | 5 | - | - | - | - | **424** | **326** | **245.629** | 0.753 |
| KAKIVIK ASSET MANAGEMENT | 50-27667-01 | 2 | 54 | 20 | 24 | 22 | 9 | 13 | - | - | - | - | - | - | 144 | 142 | 51.043 | 0.359 |
| LEHIGH TESTING LABORATORIES, INC. | 07-01173-03 | 2 | 2 | - | - | - | - | - | - | - | - | - | - | - | 4 | 2 | 0.015 | 0.008 |
| LKS INSPECTION SERVICES, LLC | 53-27795-01 | 2 | - | 1 | - | - | 1 | - | 1 | - | - | - | - | - | 5 | 3 | 3.800 | 1.267 |
| MARTIN INDUSTRIAL TESTING, INC. | 45-25452-01 | - | - | - | - | - | 2 | - | - | - | - | - | - | - | 2 | 2 | 1.627 | 0.814 |
| MARYLAND Q.C. LABORATORIES, INC. | 19-28583-01 | 5 | 7 | 2 | 3 | 1 | - | - | - | - | - | - | - | - | 18 | 13 | 2.401 | 0.185 |
| MATERIALS INTEGRITY, INC. | 50-27722-01 | - | 3 | - | - | - | - | - | - | - | - | - | - | - | 3 | 3 | 0.126 | 0.042 |
| MECHNICAL INTEGRITY SOLUTIONS | 52-25615-01 | - | 5 | 2 | 2 | 2 | 1 | 1 | - | - | - | - | - | - | 13 | 13 | 4.558 | 0.351 |
| METALS TESTING SERVICES, INC. | 25-29406-01 | - | 4 | 3 | - | - | - | - | - | - | - | - | - | - | 7 | 7 | 0.612 | 0.087 |
| MID AMERICAN INSPECTION SERVICES, INC. | 21-26060-01 | - | - | 3 | 3 | 4 | 4 | 2 | - | - | - | - | - | - | 16 | 16 | 9.930 | 0.621 |
| MISTRAS GROUP, INC. | 12-16559-02 | 6 | 28 | 20 | 20 | 2 | 3 | 2 | 2 | - | - | - | - | - | 83 | 77 | 23.002 | 0.299 |
| NORTHROP GRUMMAN SHIPBUILDING, INC. | 45-09428-02 | 6 | 27 | 4 | - | - | - | - | - | - | - | - | - | - | 37 | 31 | 1.187 | 0.038 |
| PACIFIC TESTING SERVICES, INC. | 53-29118-01 | 7 | - | - | - | - | - | - | - | - | - | - | - | - | 7 | - | - | - |
| POLE BROTHERS IMAGING COMPANY | 45-25383-01 | 1 | - | 2 | 1 | - | - | - | - | - | - | - | - | - | 4 | 3 | 0.537 | 0.179 |
| PRIME NDT SERVICES, INC. | 37-23370-01 | - | 4 | 2 | 1 | 8 | 5 | 16 | 4 | 1 | - | - | - | - | 41 | 41 | 45.760 | 1.116 |
| QUALITY INSPECTION SERVICES, INC. | 31-30187-01 | - | 10 | 2 | 3 | 3 | 1 | - | - | - | - | - | - | - | 19 | 19 | 4.664 | 0.245 |
| QUALITY INSPECTION & TESTING, INC. | 50-29038-01 | - | 1 | - | 2 | 1 | - | - | - | - | - | - | - | - | 4 | 4 | 1.531 | 0.383 |
| QUALITY TESTING SERVICES, INC. | 24-32292-01 | 3 | 9 | 1 | 1 | - | - | 1 | - | - | - | - | - | - | 15 | 12 | 2.507 | 0.209 |

NOTE: The data values shown bolded and in boxes represent the highest value in each category. These values have not been adjusted for the multiple counting of transient workers (see section 5).
* Dose values exactly equal to the values separating ranges are reported in the next higher range.

## APPENDIX A
### Table A1 - Annual TEDE for Nonreactor NRC Licensees
### 2010 (continued)

| PROGRAM CODE - LICENSEE NAME | LICENSE # | No Meas. Exposure | Meas. <0.10 | 0.10-0.25 | 0.25-0.50 | 0.50-0.75 | 0.75-1.00 | 1.00-2.00 | 2.00-3.00 | 3.00-4.00 | 4.00-5.00 | 5.00-6.00 | 6.00-12.00 | >12.0 | Total Number Monitored | Number with Meas. Dose | Total Collective TEDE (person-rem) | Average Meas. TEDE (rem) |
|---|---|---|---|---|---|---|---|---|---|---|---|---|---|---|---|---|---|---|
| *INDUSTRIAL RADIOGRAPHY – TEMPORARY JOB SITE – 03320 (Continued)* | | | | | | | | | | | | | | | | | | |
| SCIENTIFIC TECHNICAL, INC. | 45-24882-01 | 6 | 1 | 1 | 1 | - | - | - | - | - | - | - | - | - | 9 | 3 | 0.587 | 0.196 |
| SOUTHWEST X-RAY CORP | 49-29277-01 | - | - | - | - | 3 | 1 | - | - | - | - | - | - | - | 4 | 4 | 3.173 | 0.793 |
| SYSTEM ONE SERVICES, INC | 37-27891-02 | - | 6 | 5 | 2 | 2 | 2 | 2 | - | - | - | - | - | - | 19 | 19 | 7.906 | 0.416 |
| T & K INSPECTION, INC. | 33-27678-01 | - | - | - | 1 | 1 | 1 | 8 | 2 | 5 | - | - | - | - | 18 | 18 | 36.365 | **2.020** |
| TEAM INDUSTRIAL SERVICES, INC. | 42-32219-01 | 32 | 41 | 21 | 15 | 6 | 8 | 22 | 7 | 1 | - | - | - | - | 153 | 121 | 70.627 | 0.584 |
| TEI ANALYTICAL SERVICES, INC. | 37-28004-01 | - | 10 | 4 | 3 | 5 | 4 | 15 | 6 | 4 | - | - | - | - | 51 | 51 | 56.700 | 1.112 |
| TESTING TECHNOLOGIES, INC. | 45-25007-01 | 1 | 6 | 7 | 3 | 5 | 2 | - | - | - | - | - | - | - | 24 | 23 | 7.167 | 0.312 |
| TULSA GAMMA RAY, INC. | 35-17178-01 | - | 1 | 6 | 5 | 9 | 4 | 21 | 15 | 6 | - | - | - | - | 67 | 67 | 101.414 | 1.514 |
| TVA - INSPECTION SERVICES ORG | 41-06832-06 | 10 | 5 | - | - | - | - | - | - | - | - | - | - | - | 15 | 5 | 0.143 | 0.029 |
| URS ENERGY AND CONSTRUCTION | 29-27761-01 | 11 | 26 | 8 | 1 | - | - | - | - | - | - | - | - | - | 46 | 35 | 1.992 | 0.057 |
| WELDSONIX, INC. | 42-29354-01 | - | 6 | 7 | 22 | 13 | 8 | 12 | - | 1 | - | - | - | - | 69 | 69 | 43.239 | 0.627 |
| WR NON DESTRUCTIVE TESTING, INC. | 52-25538-01 | 2 | 3 | 3 | - | - | - | - | - | - | - | - | - | - | 8 | 6 | 0.375 | 0.063 |
| **Total** | | 54 | 552 | 285 | 299 | 255 | 153 | 336 | 103 | 34 | - | - | - | - | 2,287 | 2,017 | 1,295.795 | 0.642 |

NOTE: The data values shown bolded and in boxes represent the highest value in each category. These values have not been adjusted for the multiple coun ing of transient workers (see section 5).
* Dose values exactly equal to the values separating ranges are reported in the next higher range.

## APPENDIX A
### Table A1 - Annual TEDE for Nonreactor NRC Licensees 2010 (continued)

| PROGRAM CODE - LICENSEE NAME | LICENSE # | No Meas. Exposure | Meas. <0.10 | 0.10-0.25 | 0.25-0.50 | 0.50-0.75 | 0.75-1.00 | 1.00-2.00 | 2.00-3.00 | 3.00-4.00 | 4.00-5.00 | 5.00-6.00 | 6.00-12.00 | >12.0 | Total Number Monitored | Number with Meas. Dose | Total Collective TEDE (person-rem) | Average Meas. TEDE (rem) |
|---|---|---|---|---|---|---|---|---|---|---|---|---|---|---|---|---|---|---|
| *MANUFACTURING AND DISTRIBUTION – NUCLEAR PHARMACIES – 02500* | | | | | | | | | | | | | | | | | | |
| CARDINAL HEALTH | 04-26507-01MD | 10 | 17 | 1 | - | - | - | - | - | - | - | - | - | - | 28 | 18 | 0.572 | 0.032 |
| CARDINAL HEALTH | 11-27664-01MD | 2 | 10 | 2 | - | - | - | - | - | - | - | - | - | - | 14 | 12 | 0.541 | 0.045 |
| CARDINAL HEALTH | 34-29200-01MD | 89 | 178 | 21 | 6 | 1 | 3 | - | - | - | - | - | - | - | 298 | 209 | 12.203 | 0.058 |
| CARDINAL HEALTH | 47-25322-01MD | 9 | 6 | 1 | - | - | - | - | - | - | - | - | - | - | 16 | 7 | 0.307 | 0.044 |
| GE HEALTHCARE - KENTWOOD | 21-26707-01MD | 18 | 5 | 1 | - | - | - | - | - | - | - | - | - | - | 24 | 6 | 0.339 | 0.057 |
| GE HEALTHCARE - LIVONIA | 21-24828-01MD | 16 | 9 | 1 | - | - | - | - | - | - | - | - | - | - | 26 | 10 | 0.510 | 0.051 |
| GE HEALTHCARE - ST. LOUIS/OVERLAND | 24-32462-01MD | 11 | 5 | 3 | - | - | - | - | - | - | - | - | - | - | 19 | 8 | 0.572 | 0.072 |
| MID-AMERICA ISOTOPES, INC. | 24-26241-01MD | 21 | 3 | 4 | 2 | 1 | - | - | - | - | - | - | - | - | 31 | 10 | 2.276 | 0.228 |
| SPECTRON MRC, LLC | 13-32726-01MD | 10 | 3 | 1 | 1 | 1 | 1 | - | - | - | 1 | - | - | - | 19 | 9 | 7.623 | 0.847 |
| TRIAD ISOTOPES | 24-04206-08MD | 6 | 4 | 1 | - | - | - | - | - | - | - | - | - | - | 11 | 5 | 0.232 | 0.046 |
| TRIAD ISOTOPES - MI | 09-32781-02MD | 8 | 6 | - | - | - | - | - | - | - | - | - | - | - | 14 | 6 | 0.188 | 0.031 |
| TRIAD ISOTOPES - MO | 09-32781-04MD | 12 | 11 | 2 | 1 | - | - | - | - | - | - | - | - | - | 26 | 14 | 0.859 | 0.061 |
| Total | 12 | 212 | 257 | 38 | 10 | 2 | 5 | 1 | - | - | 1 | - | - | - | 526 | 314 | 26.222 | 0.084 |
| *MANUFACTURING AND DISTRIBUTION – TYPE "A" BROAD – 03211* | | | | | | | | | | | | | | | | | | |
| COVIDIEN - MALLINCKRODT, INC. | 24-04206-01 | 47 | 141 | 79 | 37 | 14 | 18 | 37 | 1 | - | - | - | - | - | 373 | 326 | 104.214 | 0.320 |
| INTERNATIONAL ISOTOPES IDAHO, INC. | 11-27680-01 | - | - | 2 | 1 | 3 | 3 | 4 | - | - | - | - | - | - | 13 | 13 | 11.519 | 0.886 |
| Total | 2 | 47 | 141 | 81 | 38 | 17 | 21 | 41 | 1 | - | - | - | - | - | 386 | 339 | 115.733 | 0.341 |
| *MANUFACTURING AND DISTRIBUTION – TYPE "B" BROAD – 03212* | | | | | | | | | | | | | | | | | | |
| BEST MEDICAL INTERNATIONAL, INC. | 45-19757-01 | 29 | 8 | 1 | - | - | - | 1 | - | - | - | - | - | - | 39 | 10 | 3.064 | 0.306 |
| Total | 1 | 29 | 8 | 1 | - | - | - | 1 | - | - | - | - | - | - | 39 | 10 | 3.064 | 0.306 |

NOTE: The data values shown bolded and in boxes represent the highest value in each category. These values have not been adjusted for the multiple counting of transient workers (see section 5).
* Dose values exactly equal to the values separating ranges are reported in the next higher range.

## APPENDIX A
## Table A1 - Annual TEDE for Nonreactor NRC Licensees 2010 (continued)

| PROGRAM CODE - LICENSEE NAME | LICENSE # | Number of Individuals with Whole Body Doses in the Ranges (rems)* | | | | | | | | | | | | | Total Number Monitored | Number with Meas. Dose | Total Collective TEDE (person-rem) | Average Meas. TEDE (rem) |
|---|---|---|---|---|---|---|---|---|---|---|---|---|---|---|---|---|---|---|
| | | No Meas. Exposure | Meas. <0.10 | 0.10-0.25 | 0.25-0.50 | 0.50-0.75 | 0.75-1.00 | 1.00-2.00 | 2.00-3.00 | 3.00-4.00 | 4.00-5.00 | 5.00-6.00 | 6.00-12.00 | >12.0 | | | | |
| *MANUFACTURING AND DISTRIBUTION – OTHER – 03214* | | | | | | | | | | | | | | | | | | |
| BEST THERATRONICS | 45-31299-01 | - | 2 | - | - | - | - | - | - | - | - | - | - | - | 2 | 2 | 0.090 | 0.045 |
| I2S, LLC | 06-21253-01 | 12 | 2 | 2 | - | 1 | - | - | - | - | - | - | - | - | 17 | 5 | 1.266 | 0.251 |
| Total | 2 | 12 | 4 | 2 | - | 1 | - | - | - | - | - | - | - | - | 19 | 7 | 1.346 | 0.192 |
| *INDEPENDENT SPENT FUEL STORAGE INSTALLATION – 23200* | | | | | | | | | | | | | | | | | | |
| GENERAL ELECTRIC - MORRIS ISFSI | SNM-2500 | 8 | 35 | 4 | - | - | - | - | - | - | - | - | - | - | 47 | 39 | 1.337 | 0.034 |
| TROJAN ISFSI | SNM-2509 | 26 | - | - | - | - | - | - | - | - | - | - | - | - | 26 | - | - | - |
| Total | 2 | 34 | 35 | 4 | - | - | - | - | - | - | - | - | - | - | 73 | 39 | 1.337 | 0.034 |
| *URANIUM HEXAFLUORIDE (UF₆) PRODUCTION PLANTS – 11400* | | | | | | | | | | | | | | | | | | |
| HONEYWELL INTERNATIONAL, INC. | SUB-0526 | 112 | 566 | 346 | 168 | 49 | 21 | 8 | - | - | - | - | - | - | 1,270 | 1,158 | 196.893 | 0.170 |
| Total | 1 | 112 | 566 | 346 | 168 | 49 | 21 | 8 | - | - | - | - | - | - | 1,270 | 1,158 | 196.893 | 0.170 |
| *FUEL CYCLE URANIUM ENRICHMENT PLANTS – 21200* | | | | | | | | | | | | | | | | | | |
| LOUISIANA ENERGY SERVICES, LLC | SNM-2010 | 365 | 94 | - | - | - | - | - | - | - | - | - | - | - | 459 | 94 | 0.745 | 0.008 |
| USEC, INC. | SNM-7003 | 317 | 22 | - | - | - | - | - | - | - | - | - | - | - | 339 | 22 | 0.333 | 0.015 |
| USEC - PADUCAH GDP | GDP-1 | 1,611 | 207 | 20 | - | - | - | - | - | - | - | - | - | - | 1,838 | 227 | 8.205 | 0.036 |
| USEC - PORTSMOUTH GDP | GDP-2 | 1,581 | 111 | - | - | - | - | - | - | - | - | - | - | - | 1,692 | 111 | 2.105 | 0.019 |
| Total | 4 | 3,874 | 434 | 20 | - | - | - | - | - | - | - | - | - | - | 4,328 | 454 | 11.388 | 0.025 |
| *FUEL CYCLE FUEL FABRICATION FACILITIES – 21210* | | | | | | | | | | | | | | | | | | |
| AREVA NP, INC. - LYNCHBURG | SNM-1168 | 126 | 47 | 20 | 10 | 2 | - | - | - | - | - | - | - | - | 205 | 79 | 9.437 | 0.119 |
| AREVA NP, INC. - RICHLAND | SNM-1227 | 88 | 134 | 58 | 38 | 38 | 39 | 13 | - | - | - | - | - | - | 408 | 320 | 99.976 | 0.312 |
| B & W NUCLEAR OPERATIONS GROUP | SNM-0042 | 16 | 154 | 33 | 5 | - | - | - | - | - | - | - | - | - | 208 | 192 | 13.594 | 0.071 |
| GLOBAL NUCLEAR FUEL - AMERICAS, LLC | SNM-1097 | 380 | 427 | 175 | 31 | - | - | - | - | - | - | - | - | - | 1,013 | 633 | 49.168 | 0.078 |
| NUCLEAR FUEL SERVICES, INC. | SNM-0124 | 486 | 684 | 42 | - | - | - | - | - | - | - | - | - | - | 1,212 | 726 | 19.520 | 0.027 |
| WESTINGHOUSE ELECTRIC COMPANY | SNM-1107 | 68 | 258 | 138 | 184 | 69 | 1 | - | - | - | - | - | - | - | 718 | 650 | 141.900 | 0.218 |
| Total | 6 | 1,164 | 1,704 | 466 | 268 | 109 | 40 | 13 | - | - | - | - | - | - | 3,764 | 2,600 | 333.595 | 0.128 |

NOTE: The data values shown bolded and in boxes represent the highest value in each category. These values have not been adjusted for the multiple coun ing of transient workers (see section 5).
* Dose values exactly equal to the values separating ranges are reported in the next higher range.

## APPENDIX A
## Table A2 - Other Facilities Reporting to the NRC
### 2010

| PROGRAM CODE - LICENSEE NAME | LICENSE # | No Meas. Exposure | Meas. <0.10 | 0.10-0.25 | 0.25-0.50 | 0.50-0.75 | 0.75-1.00 | 1.00-2.00 | 2.00-3.00 | 3.00-4.00 | 4.00-5.00 | 5.00-6.00 | 6.00-12.00 | >12.0 | Total Number Monitored | Number with Meas. Dose | Total Collective TEDE (person-rem) | Average Meas. TEDE (rem) |
|---|---|---|---|---|---|---|---|---|---|---|---|---|---|---|---|---|---|---|
| *MEASURING SYSTEMS FIXED GAUGES – 03120* | | | | | | | | | | | | | | | | | | |
| TRANSCANADA | 21-29258-01 | - | 1 | - | - | - | - | - | - | - | - | - | - | - | 1 | 1 | 0.010 | 0.010 |
| Total | 1 | - | 1 | - | - | - | - | - | - | - | - | - | - | - | 1 | 1 | 0.010 | 0.010 |
| *INSTRUMENT CALIBRATION SERVICE ONLY – SOURCE <100 CURIES – 03221* | | | | | | | | | | | | | | | | | | |
| NORTHROP GRUMMAN SHIPBUILDING, INC. | 45-09428-03 | 10 | 8 | - | - | - | - | - | - | - | - | - | - | - | 18 | 8 | 0.258 | 0.032 |
| Total | 1 | 10 | 8 | - | - | - | - | - | - | - | - | - | - | - | 18 | 8 | 0.258 | 0.032 |
| *INSTRUMENT CALIBRATION SERVICE ONLY – SOURCE >100 CURIES – 03222* | | | | | | | | | | | | | | | | | | |
| GENERAL DYNAMICS CORP - ELEC BOAT | 06-01781-03 | 2 | 2 | - | - | - | - | - | - | - | - | - | - | - | 4 | 2 | 0.003 | 0.002 |
| Total | 1 | 2 | 2 | - | - | - | - | - | - | - | - | - | - | - | 4 | 2 | 0.003 | 0.002 |
| *OTHER SERVICES - 03225* | | | | | | | | | | | | | | | | | | |
| OHMART/VEGA CORPORATION | 34-00639-04 | 7 | 47 | 10 | 3 | 1 | 1 | - | - | - | - | - | - | - | 69 | 62 | 4.696 | 0.076 |
| Total | 1 | 7 | 47 | 10 | 3 | 1 | 1 | - | - | - | - | - | - | - | 69 | 62 | 4.696 | 0.076 |
| *MULTI-SITE, MULT-REGIONAL MATERIALS LICENSE – 03613* | | | | | | | | | | | | | | | | | | |
| NAVY, DEPARTMENT OF THE | 45-23645-01NA | 113 | 66 | - | - | - | - | - | - | - | - | - | - | - | 179 | 66 | 1.263 | 0.019 |
| Total | 1 | 113 | 66 | - | - | - | - | - | - | - | - | - | - | - | 179 | 66 | 1.263 | 0.019 |
| *CRITICAL MASS MATERIAL - OTHER THAN UNIVERSITIES – 21320* | | | | | | | | | | | | | | | | | | |
| G.E. - HITACHI (VALLECITOS NUCLEAR CENTER) | SNM-0960 | 149 | 171 | 7 | 12 | 6 | 2 | 5 | 1 | - | - | - | - | - | 353 | 204 | 22.377 | 0.110 |
| Total | 1 | 149 | 171 | 7 | 12 | 6 | 2 | 5 | 1 | - | - | - | - | - | 353 | 204 | 22.377 | 0.110 |

NOTE: The data values shown bolded and in boxes represent the highest value in each category. These values have not been adjusted for the multiple counting of transient workers (see section 5).
* Dose values exactly equal to the values separating ranges are reported in the next higher range.

## APPENDIX A
## Table A2 - Other Facilities Reporting to the NRC
### 2010 (continued)

| PROGRAM CODE - LICENSEE NAME | LICENSE # | Number of Individuals with Whole Body Doses in the Ranges (rems)* | | | | | | | | | | | | | Total Number Monitored | Number with Meas. Dose | Total Collective TEDE (person-rem) | Average Meas. TEDE (rem) |
|---|---|---|---|---|---|---|---|---|---|---|---|---|---|---|---|---|---|---|
| | | No Meas. Exposure | Meas. <0.10 | 0.10-0.25 | 0.25-0.50 | 0.50-0.75 | 0.75-1.00 | 1.00-2.00 | 2.00-3.00 | 3.00-4.00 | 4.00-5.00 | 5.00-6.00 | 6.00-12.00 | >12.0 | | | | |
| *TEST REACTOR FACILITIES – 42140**  * | | | | | | | | | | | | | | | | | | |
| NAT'L INSTITUTE OF STANDARDS & TECH | TR-5 | 4 | 134 | 26 | 4 | - | - | - | - | - | - | - | - | - | 168 | 164 | 9.605 | 0.059 |
| Total | 1 | 4 | 134 | 26 | 4 | - | - | - | - | - | - | - | - | - | 168 | 164 | 9.605 | 0.059 |
| *PROGRAM CODE – 42150* | | | | | | | | | | | | | | | | | | |
| AEROTEST OPERATIONS, INC. | R-98 | - | - | 1 | 1 | 1 | 4 | 7 | 2 | 1 | - | - | - | - | 17 | 17 | 23.479 | 1.381 |
| UNIVERSITY OF ARIZONA | R-52 | - | 5 | - | - | - | - | - | - | - | - | - | - | - | 5 | 5 | 0.023 | 0.005 |
| Total | 2 | - | 5 | 1 | 1 | 1 | 4 | 7 | 2 | 1 | - | - | - | - | 22 | 22 | 23.502 | 1.068 |

NOTE: The data values shown bolded and in boxes represent the highest value in each category. These values have not been adjusted for the mul iple coun ing of transient workers (see section 5).
* Dose values exactly equal to the values separating ranges are reported in the next higher range.

## Appendix B

# ANNUAL WHOLE-BODY DOSES AT LICENSED NUCLEAR POWER FACILITIES

# 2010

## APPENDIX B
## Annual Whole-Body Doses at Licensed Nuclear Power Facilities
## 2010

| PLANT NAME | TYPE | No Meas. Exposure | Meas. <0.10 | Number of Individuals with Whole Body Doses in the Ranges (rem)* | | | | | | | | | | | | Total Number Monitored | Number with Meas. Dose | Total Collective TEDE per Site (person-rem) |
|---|---|---|---|---|---|---|---|---|---|---|---|---|---|---|---|---|---|---|
| | | | | 0.10-0.25 | 0.25-0.50 | 0.50-0.75 | 0.75-1.00 | 1.00-2.00 | 2.00-3.00 | 3.00-4.00 | 4.00-5.00 | 5.00-6.00 | 6.00-7.00 | 7.00-12.00 | >12.0 | | | |
| ARKANSAS 1, 2 | PWR | 1,164 | 1,019 | 328 | 41 | - | - | - | - | - | - | - | - | - | - | 2,552 | 1,388 | 99.376 |
| BEAVER VALLEY 1, 2 | PWR | 1,756 | 587 | 141 | 22 | - | - | - | - | - | - | - | - | - | - | 2,506 | 750 | 49.983 |
| BRAIDWOOD 1, 2 | PWR | 2,018 | 665 | 172 | 33 | - | - | - | - | - | - | - | - | - | - | 2,888 | 870 | 63.856 |
| BROWNS FERRY 1, 2, 3 | BWR | 2,114 | 1,340 | 706 | 487 | 197 | 62 | 33 | - | - | - | - | - | - | - | 4,939 | 2,825 | 556.749 |
| BRUNSWICK 1, 2 | BWR | 1,203 | 2,071 | 667 | 330 | 114 | 35 | 10 | - | - | - | - | - | - | - | 4,430 | 3,227 | 407.424 |
| BYRON 1, 2 | PWR | 1,854 | 748 | 160 | 13 | 1 | - | - | - | - | - | - | - | - | - | 2,776 | 922 | 56.425 |
| CALLAWAY 1 | PWR | 1,075 | 627 | 137 | 30 | 5 | 1 | - | - | - | - | - | - | - | - | 1,875 | 800 | 58.735 |
| CALVERT CLIFFS 1, 2 | PWR | 1,366 | 450 | 232 | 97 | 38 | 16 | 1 | - | - | - | - | - | - | - | 2,200 | 834 | 128.581 |
| CATAWBA 1, 2 | PWR | 2,196 | 702 | 268 | 69 | 6 | - | - | - | - | - | - | - | - | - | 3,241 | 1,045 | 97.010 |
| CLINTON | BWR | 1,653 | 804 | 490 | 172 | 59 | 10 | 5 | - | - | - | - | - | - | - | 3,193 | 1,540 | 219.954 |
| COLUMBIA GENERATING | BWR | 917 | 552 | 138 | 35 | 5 | - | 3 | - | - | - | - | - | - | - | 1,650 | 733 | 54.712 |
| COMANCHE PEAK 1, 2 | PWR | 1,073 | 822 | 151 | 52 | 11 | 1 | 1 | - | - | - | - | - | - | - | 2,110 | 1,037 | 70.807 |
| COOK 1, 2 | PWR | 2,336 | 859 | 209 | 46 | - | 1 | 1 | - | - | - | - | - | - | - | 3,452 | 1,116 | 83.276 |
| COOPER STATION | BWR | 548 | 593 | 108 | 61 | 9 | 2 | - | - | - | - | - | - | - | - | 1,321 | 773 | 61.303 |
| CRYSTAL RIVER 3 | PWR | 1,740 | 580 | 85 | 1 | - | - | - | - | - | - | - | - | - | - | 2,406 | 666 | 31.922 |
| DAVIS-BESSE | PWR | 1,298 | 634 | 396 | 301 | 176 | 71 | 67 | 4 | - | - | - | - | - | - | 2,947 | 1,649 | 464.095 |
| DIABLO CANYON 1, 2 | PWR | 1,448 | 949 | 311 | 85 | 15 | 7 | - | - | - | - | - | - | - | - | 2,815 | 1,367 | 125.457 |
| DRESDEN 2, 3 | BWR | 1,465 | 1,492 | 462 | 158 | 25 | 11 | 4 | - | - | - | - | - | - | - | 3,617 | 2,152 | 213.825 |
| DUANE ARNOLD | BWR | 1,123 | 602 | 257 | 129 | 49 | 34 | 21 | 1 | - | - | - | - | - | - | 2,216 | 1,093 | 200.601 |
| FARLEY 1, 2 | PWR | 1,508 | 907 | 316 | 88 | 8 | 1 | 1 | - | - | - | - | - | - | - | 2,829 | 1,321 | 121.313 |
| FERMI 2 | BWR | 1,521 | 1,122 | 376 | 114 | 13 | - | - | - | - | - | - | - | - | - | 3,146 | 1,625 | 146.490 |
| FITZPATRICK | BWR | 850 | 828 | 313 | 192 | 59 | 18 | 19 | - | - | - | - | - | - | - | 2,279 | 1,429 | 219.887 |
| FT CALHOUN | PWR | 1,057 | 144 | 22 | 5 | - | - | - | - | - | - | - | - | - | - | 1,228 | 171 | 9.763 |
| GINNA | PWR | 1,264 | 67 | 8 | - | - | - | - | - | - | - | - | - | - | - | 1,339 | 75 | 3.168 |
| GRAND GULF | BWR | 1,191 | 1,270 | 318 | 181 | 38 | 12 | 3 | - | - | - | - | - | - | - | 3,013 | 1,822 | 188.370 |
| HARRIS | PWR | 1,574 | 794 | 215 | 57 | 3 | - | - | - | - | - | - | - | - | - | 2,643 | 1,069 | 82.578 |
| HATCH 1, 2 | BWR | 1,415 | 999 | 402 | 242 | 75 | 13 | 3 | - | - | - | - | - | - | - | 3,149 | 1,734 | 245.797 |
| HOPE CREEK 1 | BWR | 1,076 | 1,525 | 256 | 150 | 40 | 9 | 5 | - | - | - | - | - | - | - | 3,061 | 1,985 | 160.910 |
| INDIAN POINT 2 | PWR | 12 | 872 | 337 | 150 | 49 | 25 | 13 | - | - | - | - | - | - | - | 1,458 | 1,446 | 197.279 |
| INDIAN POINT 3 | PWR | 140 | 516 | - | - | - | - | - | - | - | - | - | - | - | - | 656 | 516 | 3.103 |
| KEWAUNEE | PWR | 792 | 125 | 7 | 3 | - | - | - | - | - | - | - | - | - | - | 927 | 135 | 4.690 |
| LASALLE 1, 2 | BWR | 1,488 | 1,299 | 568 | 365 | 106 | 32 | 16 | - | - | - | - | - | - | - | 3,874 | 2,386 | 384.434 |
| LIMERICK 1, 2 | BWR | 2,082 | 970 | 366 | 167 | 19 | 2 | 1 | - | - | - | - | - | - | - | 3,607 | 1,525 | 167.797 |
| MCGUIRE 1, 2 | PWR | 2,188 | 977 | 215 | 31 | 2 | - | - | - | - | - | - | - | - | - | 3,413 | 1,225 | 81.321 |
| MILLSTONE 2, 3 | PWR | 2,048 | 429 | 195 | 82 | 11 | 1 | - | - | - | - | - | - | - | - | 2,766 | 718 | 81.589 |
| MONTICELLO | BWR | 1,429 | 388 | 89 | 42 | 8 | 5 | 2 | - | - | - | - | - | - | - | 1,963 | 534 | 56.116 |
| NINE MILE POINT 1, 2 | BWR | 1,849 | 820 | 392 | 254 | 127 | 72 | 38 | - | - | - | - | - | - | - | 3,552 | 1,703 | 375.424 |
| NORTH ANNA 1, 2 | PWR | 3,503 | 562 | 267 | 110 | 51 | 17 | 25 | - | - | - | - | - | - | - | 4,535 | 1,032 | 182.289 |

NOTE: Totals corrected for transients on page B-3.
* Dose values exactly equal to the values separating ranges are reported in the next higher range.

## APPENDIX B
### Annual Whole-Body Doses at Licensed Nuclear Power Facilities
### 2010 (continued)

| PLANT NAME | TYPE | No Meas. Exposure | Meas. <0.10 | 0.10-0.25 | 0.25-0.50 | 0.50-0.75 | 0.75-1.00 | 1.00-2.00 | 2.00-3.00 | 3.00-4.00 | 4.00-5.00 | 5.00-6.00 | 6.00-7.00 | 7.00-12.00 | >12.0 | Total Number Monitored | Number with Meas. Dose | Total Collective TEDE per Site (person-rem) |
|---|---|---|---|---|---|---|---|---|---|---|---|---|---|---|---|---|---|---|
| OCONEE 1, 2, 3 | PWR | 2,893 | 1,295 | 499 | 128 | 29 | 2 | - | - | - | - | - | - | - | - | 4,846 | 1,953 | 193.088 |
| OYSTER CREEK | BWR | 1,470 | 1,056 | 362 | 166 | 53 | 12 | 6 | - | - | - | - | - | - | - | 3,125 | 1,655 | 206.284 |
| PALISADES | PWR | 1,010 | 368 | 244 | 164 | 60 | 46 | 26 | - | - | - | - | - | - | - | 1,918 | 908 | 219.873 |
| PALO VERDE 1, 2, 3 | PWR | 2,703 | 1,322 | 244 | 75 | 12 | 2 | - | - | - | - | - | - | - | - | 4,358 | 1,655 | 112.612 |
| PEACH BOTTOM 2, 3 | BWR | 1,734 | 1,038 | 425 | 189 | 49 | 11 | 4 | - | - | - | - | - | - | - | 3,450 | 1,716 | 219.372 |
| PERRY | BWR | 968 | 153 | 92 | 32 | 1 | - | - | - | - | - | - | - | - | - | 1,246 | 278 | 32.186 |
| PILGRIM 1 | BWR | 861 | 213 | 67 | 21 | 2 | - | - | - | - | - | - | - | - | - | 1,164 | 303 | 25.739 |
| POINT BEACH 1, 2 | PWR | 1,562 | 528 | 242 | 88 | 11 | - | - | - | - | - | - | - | - | - | 2,431 | 869 | 95.695 |
| PRAIRIE ISLAND 1, 2 | PWR | 1,264 | 488 | 128 | 41 | 4 | - | - | - | - | - | - | - | - | - | 1,925 | 661 | 54.933 |
| QUAD CITIES 1, 2 | BWR | 1,609 | 1,511 | 486 | 233 | 27 | 9 | 1 | - | - | - | - | - | - | - | 3,876 | 2,267 | 241.444 |
| RIVER BEND 1 | BWR | 582 | 769 | 95 | 18 | 6 | - | - | - | - | - | - | - | - | - | 1,470 | 888 | 40.356 |
| ROBINSON 2 | PWR | 1,646 | 712 | 217 | 61 | 6 | - | - | - | - | - | - | - | - | - | 2,642 | 996 | 85.917 |
| SALEM 1, 2 | PWR | 521 | 737 | 128 | 72 | 20 | 5 | 2 | - | - | - | - | - | - | - | 1,485 | 964 | 77.828 |
| SAN ONOFRE 2, 3 | PWR | 3,333 | 979 | 421 | 208 | 29 | 5 | - | - | - | - | - | - | - | - | 4,975 | 1,642 | 199.399 |
| SEABROOK | PWR | 935 | 326 | 9 | - | - | - | - | - | - | - | - | - | - | - | 1,270 | 335 | 4.488 |
| SEQUOYAH 1, 2 | PWR | 1,860 | 666 | 113 | 43 | 6 | - | - | - | - | - | - | - | - | - | 2,688 | 828 | 56.956 |
| SOUTH TEXAS 1, 2 | PWR | 1,908 | 590 | 213 | 54 | 10 | - | - | - | - | - | - | - | - | - | 2,775 | 867 | 79.159 |
| ST. LUCIE 1, 2 | PWR | 2,248 | 768 | 344 | 166 | 69 | 9 | 1 | - | - | - | - | - | - | - | 3,605 | 1,357 | 197.359 |
| SUMMER 1 | PWR | 1,435 | 103 | 1 | - | - | - | - | - | - | - | - | - | - | - | 1,539 | 104 | 2.129 |
| SURRY 1, 2 | PWR | 3,660 | 562 | 287 | 87 | 17 | 4 | 1 | - | - | - | - | - | - | - | 4,618 | 958 | 111.129 |
| SUSQUEHANNA 1, 2 | BWR | 1,852 | 1,354 | 452 | 120 | 15 | 8 | 1 | - | - | - | - | - | - | - | 3,802 | 1,950 | 176.161 |
| THREE MILE ISLAND 1 | PWR | 1,750 | 700 | 81 | 9 | - | - | - | - | - | - | - | - | - | - | 2,540 | 790 | 38.994 |
| TURKEY POINT 3, 4 | PWR | 2,314 | 723 | 233 | 66 | 3 | - | - | - | - | - | - | - | - | - | 3,339 | 1,025 | 86.749 |
| VERMONT YANKEE | BWR | 960 | 515 | 300 | 161 | 47 | 28 | 20 | - | - | - | - | - | - | - | 2,031 | 1,071 | 206.321 |
| VOGTLE 1, 2 | PWR | 1,452 | 614 | 242 | 59 | 6 | 3 | - | - | - | - | - | - | - | - | 2,376 | 924 | 89.182 |
| WATERFORD 3 | PWR | 1,169 | 208 | 8 | - | - | - | - | - | - | - | - | - | - | - | 1,385 | 216 | 4.913 |
| WATTS BAR 1 | PWR | 4,830 | 114 | 15 | - | - | - | - | - | - | - | - | - | - | - | 4,959 | 129 | 6.193 |
| WOLF CREEK 1 | PWR | 775 | 449 | 14 | - | - | - | - | - | - | - | - | - | - | - | 1,238 | 463 | 10.516 |
| **Totals BWRs** | BWRs | 31,960 | 23,284 | 8,187 | 4,019 | 1,143 | 385 | 195 | 1 | - | - | - | - | - | - | 69,174 | 37,214 | 4,807.656 |
| **Totals PWRs** | PWRs | 72,678 | 26,287 | 7,855 | 2,637 | 658 | 217 | 138 | 4 | - | - | - | - | - | - | 110,474 | 37,796 | 3,823.728 |
| **Total LWRs** | LWRs | 104,638 | 49,571 | 16,042 | 6,656 | 1,801 | 602 | 333 | 5 | - | - | - | - | - | - | 179,648 | 75,010 | 8,631.384 |
| **Corrected for Transients** | LWRs | 74,218 | 33,874 | 11,670 | 6,356 | 2,231 | 946 | 832 | 42 | 3 | - | - | - | - | - | 130,172 | 55,954 | 8,631.384 |

*Number of Individuals with Whole Body Doses in the Ranges (rem)*

* Dose values exactly equal to the values separating ranges are reported in the next higher range.

## APPENDIX B

### Annual Whole-Body Doses at Licensed Nuclear Power Facilities
### 2010 (continued)

| PLANT NAME | TYPE | No Meas. Exposure | Meas. <0.10 | 0.10- 0.25 | 0.25- 0.50 | 0.50- 0.75 | 0.75- 1.00 | 1.00- 2.00 | 2.00- 3.00 | 3.00- 4.00 | 4.00- 5.00 | 5.00- 6.00 | 6.00- 7.00 | 7.00- 12.00 | >12.0 | Total Number Monitored | Number with Meas. Dose | Total Collective TEDE per Site (person-rem) |
|---|---|---|---|---|---|---|---|---|---|---|---|---|---|---|---|---|---|---|
| REACTORS NOT YET IN COMMERCIAL OPERATION | | | | | | | | | | | | | | | | | | |
| WATTS BAR 2 | PWR | Reported with Watts Bar 1 | | | | | | | | | | | | | | | | - |
| REACTORS NO LONGER IN COMMERCIAL OPERATION | | | | | | | | | | | | | | | | | | |
| BIG ROCK POINT | BWR | 24 | - | - | - | - | - | - | - | - | - | - | - | - | - | 24 | - | - |
| FERMI 1 | FBR | 61 | 30 | 5 | 10 | 3 | - | - | - | - | - | - | - | - | - | 109 | 48 | 7.794 |
| HADDAM NECK | PWR | 26 | 2 | - | - | - | - | - | - | - | - | - | - | - | - | 28 | 2 | 0.024 |
| HUMBOLDT BAY | BWR | 272 | 111 | 23 | 2 | - | - | - | - | - | - | - | - | - | - | 408 | 136 | 7.691 |
| INDIAN POINT 1 | PWR | 47 | 156 | 1 | - | - | - | - | - | - | - | - | - | - | - | 204 | 157 | 0.833 |
| LACROSSE | BWR | 32 | 67 | 11 | - | - | - | - | - | - | - | - | - | - | - | 110 | 78 | 2.971 |
| MAINE YANKEE | PWR | 27 | 1 | - | - | - | - | - | - | - | - | - | - | - | - | 28 | 1 | 0.084 |
| YANKEE ROWE | PWR | 26 | 3 | - | - | - | - | - | - | - | - | - | - | - | - | 29 | 3 | 0.083 |
| ZION 1,2 | PWR | 214 | 17 | - | - | - | - | - | - | - | - | - | - | - | - | 231 | 17 | 0.562 |
| **Total Units Reporting**\*\* | **10** | **729** | **387** | **40** | **12** | **3** | - | - | - | - | - | - | - | - | - | **1,171** | **442** | **20.042** |
| REACTORS NO LONGER IN COMMERCIAL OPERATION, REPORTED WITH OTHER UNITS | | | | | | | | | | | | | | | | | | |
| DRESDEN 1 | BWR | Reported with Dresden 2, 3 | | | | | | | | | | | | | | | | |
| MILLSTONE 1 | BWR | Reported with Millstone Units 2 & 3; estimated dose from Unit 1 is 0.142 person-rem. | | | | | | | | | | | | | | | | |
| SAN ONOFRE 1 | PWR | Reported with San Onofre 2, 3 | | | | | | | | | | | | | | | | |
| THREE MILE ISLAND 2 | PWR | Reported with Three Mile Island 1; estimated dose from Unit 2 is 0.359 person-rem. | | | | | | | | | | | | | | | | |
| REACTORS NO LONGER IN COMMERCIAL OPERATION, DECOMMISSIONED | | | | | | | | | | | | | | | | | | |
| PEACH BOTTOM 1 | HTGR | | | | | | | | | | | | | | | | | |
| RANCHO SECO | PWR | Reported as ISFSI (See Appendix A) | | | | | | | | | | | | | | | | |
| TROJAN | PWR | | | | | | | | | | | | | | | | | |

Note: Totals corrected for transients on page B-3.
\* Dose values exactly equal to the values separating ranges are reported in the next higher range.
\*\* These numbers are for he reactors no longer in commercial operation that report their doses separately (i.e., do not report their doses with other units).

Appendix C*

# PERSONNEL, DOSE, AND POWER GENERATION SUMMARY

# 1969–2010

*A discussion of the methods used to collect and calculate the information contained in this appendix is given in sections 3.1 and 4.2.

| Reporting Organization | Year | Megawatt Years (MW-yr) | Unit Availability Factor | Total Personnel with Measurable Doses | Collective Dose per Site (person-rem) | Average Measurable Dose (rem) | Collective Dose/ MW-yr |
|---|---|---|---|---|---|---|---|
| **ARKANSAS 1, 2** | 1975 | 588.0 | 76.5 | 147 | 21 | 0.14 | 0.04 |
| Docket 50-313, 50-368; | 1976 | 464.6 | 56.6 | 476 | 289 | 0.61 | 0.62 |
| DPR-51; NPF-6 | 1977 | 610.3 | 76.8 | 601 | 256 | 0.43 | 0.42 |
| 1st commercial operation | 1978 | 627.2 | 77.5 | 722 | 189 | 0.26 | 0.30 |
| 12/74, 3/80 | 1979 | 397.0 | 55.3 | 1,321 | 369 | 0.28 | 0.93 |
| Type - PWRs | 1980 | 452.8 | 63.7 | 1,233 | 342 | 0.28 | 0.76 |
| Capacity - 836, 988 MWe | 1981 | 1,104.7 | 68.3 | 2,225 | 1,102 | 0.50 | 1.00 |
| | 1982 | 905.4 | 58.6 | 1,608 | 803 | 0.50 | 0.89 |
| | 1983 | 915.0 | 54.7 | 2,109 | 1,397 | 0.66 | 1.53 |
| | 1984 | 1,289.1 | 77.4 | 1,742 | 806 | 0.46 | 0.63 |
| | 1985 | 1,192.3 | 73.6 | 1,262 | 286 | 0.23 | 0.24 |
| | 1986 | 1,070.3 | 66.9 | 2,135 | 1,141 | 0.53 | 1.07 |
| | 1987 | 1,366.1 | 88.9 | 1,123 | 382 | 0.34 | 0.28 |
| | 1988 | 1,070.3 | 69.4 | 2,421 | 1,387 | 0.57 | 1.30 |
| | 1989 | 1,066.3 | 72.0 | 2,063 | 711 | 0.34 | 0.67 |
| | 1990 | 1,351.9 | 84.2 | 2,493 | 762 | 0.31 | 0.56 |
| | 1991 | 1,515.8 | 88.4 | 2,064 | 351 | 0.17 | 0.23 |
| | 1992 | 1,352.1 | 77.4 | 3,114 | 876 | 0.28 | 0.65 |
| | 1993 | 1,606.0 | 91.3 | 1,981 | 268 | 0.14 | 0.17 |
| | 1994 | 1,662.8 | 93.6 | 1,361 | 172 | 0.13 | 0.10 |
| | 1995 | 1,397.0 | 82.7 | 2,259 | 386 | 0.17 | 0.28 |
| | 1996 | 1,596.0 | 89.5 | 1,441 | 203 | 0.14 | 0.13 |
| | 1997 | 1,621.9 | 95.9 | 1,195 | 119 | 0.10 | 0.07 |
| | 1998 | 1,494.6 | 88.1 | 1,249 | 166.599 | 0.13 | 0.11 |
| | 1999 | 1,477.3 | 86.9 | 1,463 | 183.997 | 0.13 | 0.12 |
| | 2000 | 1,329.2 | 79.5 | 1,977 | 242.326 | 0.12 | 0.18 |
| | 2001 | 1,684.0 | 95.8 | 1,082 | 106.040 | 0.10 | 0.06 |
| | 2002 | 1,659.0 | 91.8 | 1,581 | 265.337 | 0.17 | 0.16 |
| | 2003 | 1,675.8 | 93.1 | 973 | 99.003 | 0.10 | 0.06 |
| | 2004 | 1,759.5 | 95.0 | 1,227 | 106.172 | 0.09 | 0.06 |
| | 2005 | 1,560.0 | 84.5 | 2,335 | 475.784 | 0.20 | 0.31 |
| | 2006 | 1,739.8 | 95.0 | 1,184 | 143.296 | 0.12 | 0.08 |
| | 2007 | 1,769.3 | 96.0 | 1,387 | 105.310 | 0.08 | 0.06 |
| | 2008 | 1,614.8 | 89.7 | 1,791 | 196.047 | 0.11 | 0.12 |
| | 2009 | 1,733.7 | 95.5 | 1,139 | 102.732 | 0.09 | 0.06 |
| | 2010 | 1,716.6 | 93.7 | 1,388 | 99.376 | 0.07 | 0.06 |
| **BEAVER VALLEY 1, 2** | 1977 | 355.6 | 57.0 | 331 | 87 | 0.26 | 0.24 |
| Docket 50-334, 50-412; | 1978 | 304.2 | 40.8 | 646 | 190 | 0.29 | 0.62 |
| DPR-66; NPF-73 | 1979 | 221.0 | 40.0 | 704 | 132 | 0.19 | 0.60 |
| 1st commercial operation | 1980 | 39.8 | 6.8 | 1,817 | 553 | 0.30 | 13.89 |
| 10/76, 11/87 | 1981 | 573.4 | 73.6 | 1,237 | 229 | 0.19 | 0.40 |
| Type - PWRs | 1982 | 326.7 | 41.6 | 1,755 | 599 | 0.34 | 1.83 |
| Capacity - 892, 885 MWe | 1983 | 561.2 | 68.2 | 1,485 | 772 | 0.52 | 1.38 |
| | 1984 | 576.7 | 71.8 | 1,393 | 504 | 0.36 | 0.87 |
| | 1985 | 717.7 | 91.9 | 619 | 60 | 0.10 | 0.08 |
| | 1986 | 581.3 | 70.7 | 1,575 | 627 | 0.40 | 1.08 |
| | 1987 | 684.1 | 83.8 | 1,282 | 210 | 0.16 | 0.31 |
| | 1988 | 1,386.1 | 87.4 | 1,764 | 530 | 0.30 | 0.38 |
| | 1989 | 1,017.4 | 69.6 | 2,349 | 1,378 | 0.59 | 1.35 |
| | 1990 | 1,271.0 | 85.3 | 1,675 | 348 | 0.21 | 0.27 |
| | 1991 | 1,267.5 | 78.6 | 1,689 | 495 | 0.29 | 0.39 |
| | 1992 | 1,441.9 | 89.1 | 1,414 | 289 | 0.20 | 0.20 |
| | 1993 | 1,157.9 | 73.1 | 2,087 | 621 | 0.30 | 0.54 |
| | 1994 | 1,514.6 | 88.6 | 487 | 44 | 0.09 | 0.03 |
| | 1995 | 1,389.2 | 83.1 | 1,536 | 453 | 0.29 | 0.33 |
| | 1996 | 1,269.0 | 76.5 | 1,688 | 449 | 0.27 | 0.35 |
| | 1997 | 1,159.3 | 72.1 | 1,391 | 306 | 0.22 | 0.26 |
| | 1998 | 523.1 | 33.5 | 700 | 59.311 | 0.08 | 0.11 |
| | 1999 | 1,353.7 | 85.9 | 841 | 99.461 | 0.12 | 0.07 |
| | 2000 | 1,378.7 | 87.3 | 1,730 | 337.867 | 0.20 | 0.24 |
| | 2001 | 1,500.8 | 92.3 | 1,202 | 184.361 | 0.15 | 0.12 |
| | 2002 | 1,548.0 | 95.4 | 1,048 | 90.479 | 0.09 | 0.06 |

| Reporting Organization | Year | Megawatt Years (MW-yr) | Unit Availability Factor | Total Personnel with Measurable Doses | Collective Dose per Site (person-rem) | Average Measurable Dose (rem) | Collective Dose/ MW-yr |
|---|---|---|---|---|---|---|---|
| **BEAVER VALLEY 1, 2** | 2003 | 1,437.0 | 88.4 | 1,623 | 277.168 | 0.17 | 0.19 |
| (continued) | 2004 | 1,593.1 | 96.3 | 1,270 | 156.509 | 0.12 | 0.10 |
| | 2005 | 1,590.4 | 96.7 | 978 | 79.055 | 0.08 | 0.05 |
| | 2006 | 1,385.6 | 84.0 | 2,174 | 370.146 | 0.17 | 0.27 |
| | 2007 | 1,664.1 | 96.0 | 955 | 86.595 | 0.09 | 0.05 |
| | 2008 | 1,670.2 | 94.4 | 991 | 83.394 | 0.08 | 0.05 |
| | 2009 | 1,599.3 | 89.6 | 1,504 | 224.516 | 0.15 | 0.14 |
| | 2010 | 1,714.2 | 95.6 | 750 | 49.983 | 0.07 | 0.03 |
| **BIG ROCK POINT**[1] | 1969 | 48.1 | | 165 | 136 | 0.82 | 2.83 |
| Docket 50-155; | 1970 | 43.5 | | 290 | 194 | 0.67 | 4.46 |
| DPR-6 | 1971 | 44.4 | | 260 | 184 | 0.71 | 4.14 |
| 1st commercial operation 3/63 | 1972 | 43.5 | | 195 | 181 | 0.93 | 4.16 |
| Type - BWR | 1973 | 50.9 | | 241 | 285 | 1.18 | 5.60 |
| Capacity - (67) MWe | 1974 | 40.7 | 70.3 | 281 | 276 | 0.98 | 6.78 |
| | 1975 | 35.1 | 59.8 | 300 | 180 | 0.60 | 5.13 |
| | 1976 | 29.5 | 50.1 | 488 | 289 | 0.59 | 9.80 |
| | 1977 | 43.6 | 73.4 | 465 | 334 | 0.72 | 7.66 |
| | 1978 | 48.5 | 77.9 | 285 | 175 | 0.61 | 3.61 |
| | 1979 | 13.0 | 23.5 | 623 | 455 | 0.73 | 35.00 |
| | 1980 | 48.9 | 79.0 | 599 | 354 | 0.59 | 7.24 |
| | 1981 | 56.9 | 90.6 | 479 | 160 | 0.33 | 2.81 |
| | 1982 | 43.6 | 70.8 | 521 | 328 | 0.63 | 7.52 |
| | 1983 | 42.3 | 71.0 | 493 | 263 | 0.53 | 6.22 |
| | 1984 | 50.3 | 78.6 | 297 | 155 | 0.52 | 3.08 |
| | 1985 | 43.8 | 73.5 | 435 | 291 | 0.67 | 6.64 |
| | 1986 | 61.0 | 95.5 | 202 | 84 | 0.42 | 1.38 |
| | 1987 | 45.3 | 71.0 | 251 | 222 | 0.88 | 4.90 |
| | 1988 | 46.1 | 72.8 | 303 | 170 | 0.56 | 3.69 |
| | 1989 | 50.2 | 79.0 | 418 | 177 | 0.42 | 3.53 |
| | 1990 | 51.3 | 77.2 | 351 | 232 | 0.66 | 4.52 |
| | 1991 | 59.1 | 85.2 | 435 | 226 | 0.52 | 3.82 |
| | 1992 | 32.7 | 54.5 | 496 | 277 | 0.56 | 8.47 |
| | 1993 | 51.2 | 79.4 | 419 | 152 | 0.36 | 2.97 |
| | 1994 | 49.5 | 75.3 | 310 | 119 | 0.38 | 2.40 |
| | 1995 | 62.2 | 95.0 | 205 | 54 | 0.26 | 0.87 |
| | 1996 | 1,265.6 | 76.5 | 1,688 | 449 | 0.27 | 0.36 |
| | 1997 | 22.4 | 54.1 | 258 | 55 | 0.21 | 2.46 |
| | 1998 | 0.0 | 0.0 | 432 | 104.130 | 0.24 | --- |
| | 1999 | 0.0 | 0.0 | 285 | 86.577 | 0.31 | --- |
| | 2000 | 0.0 | 0.0 | 226 | 89.271 | 0.40 | --- |
| | 2001 | 0.0 | 0.0 | 167 | 47.556 | 0.28 | --- |
| | 2002 | 0.0 | 0.0 | 170 | 43.538 | 0.26 | --- |
| | 2003 | 0.0 | 0.0 | 336 | 121.045 | 0.36 | --- |
| | 2004 | 0.0 | 0.0 | 227 | 57.599 | 0.25 | --- |
| | 2005 | 0.0 | 0.0 | 223 | 20.227 | 0.09 | --- |
| | 2006 | 0.0 | 0.0 | 27 | 0.382 | 0.01 | --- |
| | 2007 | 0.0 | 0.0 | 0 | 0.000 | --- | --- |
| | 2008 | 0.0 | 0.0 | 0 | 0.000 | --- | --- |
| | 2009 | 0.0 | 0.0 | 0 | 0.000 | --- | --- |
| | 2010 | 0.0 | 0.0 | 0 | 0.000 | --- | --- |
| **BRAIDWOOD 1, 2** | 1989 | 1,381.8 | 75.4 | 1,460 | 296 | 0.20 | 0.21 |
| Docket 50-456, 50-457; | 1990 | 1,740.2 | 84.1 | 1,081 | 186 | 0.17 | 0.11 |
| NPF-72, NPF-77 | 1991 | 1,377.2 | 68.9 | 1,641 | 550 | 0.34 | 0.40 |
| 1st commercial operation | 1992 | 1,885.9 | 89.0 | 1,059 | 228 | 0.22 | 0.12 |
| 7/88, 10/88 | 1993 | 1,899.3 | 86.9 | 1,043 | 273 | 0.26 | 0.14 |
| Type - PWRs | 1994 | 1,666.1 | 77.2 | 1,237 | 298 | 0.24 | 0.18 |
| Capacity - 1,156, 1,131 MWe | 1995 | 1,914.7 | 85.4 | 1,134 | 236 | 0.21 | 0.12 |

[1] Big Rock Point was shut down in September 1997 and is no longer included in the count of operating reactors. Parentheses indicate plant capacity when plant was operational.

| Reporting Organization | Year | Megawatt Years (MW-yr) | Unit Availability Factor | Total Personnel with Measurable Doses | Collective Dose per Site (person-rem) | Average Measurable Dose (rem) | Collective Dose/ MW-yr |
|---|---|---|---|---|---|---|---|
| **BRAIDWOOD 1, 2** | 1996 | 1,854.9 | 82.1 | 1,356 | 334 | 0.25 | 0.18 |
| (continued) | 1997 | 1,863.3 | 85.4 | 1,693 | 321 | 0.19 | 0.17 |
| | 1998 | 1,979.1 | 88.9 | 1,869 | 259.236 | 0.14 | 0.13 |
| | 1999 | 2,161.6 | 95.8 | 1,153 | 145.976 | 0.13 | 0.07 |
| | 2000 | 2,142.8 | 94.9 | 1,562 | 194.126 | 0.12 | 0.09 |
| | 2001 | 2,186.4 | 95.8 | 881 | 100.570 | 0.11 | 0.05 |
| | 2002 | 2,284.0 | 96.8 | 975 | 90.716 | 0.09 | 0.04 |
| | 2003 | 2,279.9 | 95.6 | 1,572 | 244.860 | 0.16 | 0.11 |
| | 2004 | 2,277.8 | 97.3 | 986 | 94.942 | 0.10 | 0.04 |
| | 2005 | 2,253.7 | 96.6 | 926 | 88.084 | 0.10 | 0.04 |
| | 2006 | 2,234.1 | 95.0 | 1,624 | 199.168 | 0.12 | 0.09 |
| | 2007 | 2,244.0 | 96.0 | 1,258 | 98.040 | 0.08 | 0.04 |
| | 2008 | 2,252.5 | 96.3 | 1,235 | 103.180 | 0.08 | 0.05 |
| | 2009 | 2,195.0 | 93.8 | 1,397 | 142.066 | 0.10 | 0.06 |
| | 2010 | 2,111.9 | 94.0 | 870 | 63.856 | 0.07 | 0.03 |
| **BROWNS FERRY 1[2], 2, 3** | 1975 | 161.7 | 17.8 | 2,743 | 347 | 0.13 | 2.15 |
| Docket 50-259, 50-260, 50-296 | 1976 | 337.6 | 26.9 | 2,530 | 232 | 0.09 | 0.69 |
| DPR-33, DPR-52, DPR-68 | 1977 | 1,327.5 | 73.7 | 1,985 | 876 | 0.44 | 0.66 |
| 1st commercial operation | 1978 | 1,992.1 | 73.5 | 2,479 | 1,776 | 0.72 | 0.89 |
| 8/74, 3/75, 3/77 | 1979 | 2,393.0 | 79.1 | 2,869 | 1,593 | 0.56 | 0.67 |
| Type - BWRs | 1980 | 2,182.1 | 73.6 | 2,838 | 1,768 | 0.62 | 0.81 |
| Capacity - 1,105, 1,104, | 1981 | 2,132.9 | 69.5 | 3,497 | 2,398 | 0.69 | 1.12 |
| 1,105 MWe | 1982 | 2,025.4 | 67.6 | 3,360 | 2,230 | 0.66 | 1.10 |
| | 1983 | 1,641.0 | 54.3 | 3,410 | 3,375 | 0.99 | 2.06 |
| | 1984 | 1,431.9 | 54.2 | 3,172 | 1,954 | 0.62 | 1.36 |
| | 1985 | 368.2 | 11.9 | 2,854 | 1,164 | 0.41 | 3.16 |
| | 1986 | 0.0 | 0.0 | 3,074 | 1,054 | 0.34 | --- |
| | 1987 | 0.0 | 0.0 | 3,184 | 1,186 | 0.37 | --- |
| | 1988 | 0.0 | 0.0 | 3,390 | 1,158 | 0.34 | --- |
| | 1989 | 0.0 | 0.0 | 2,707 | 657 | 0.24 | --- |
| | 1990 | 0.0 | 0.0 | 2,725 | 1,311 | 0.48 | --- |
| | 1991 | 445.0 | 17.7 | 1,831 | 356 | 0.19 | 0.80 |
| | 1992 | 979.9 | 32.2 | 2,670 | 519 | 0.19 | 0.53 |
| | 1993 | 675.1 | 66.8 | 3,594 | 870 | 0.24 | 1.29 |
| | 1994 | 860.2 | 83.4 | 3,362 | 861 | 0.26 | 1.00 |
| | 1995 | 1,165.8 | 98.6 | 2,567 | 413 | 0.16 | 0.35 |
| | 1996 | 1,972.8 | 93.0 | 1,904 | 389 | 0.20 | 0.20 |
| | 1997 | 1,928.8 | 90.2 | 2,268 | 522 | 0.23 | 0.27 |
| | 1998 | 1,961.9 | 87.7 | 1,612 | 367.716 | 0.23 | 0.19 |
| | 1999 | 2,091.0 | 85.1 | 1,741 | 446.941 | 0.26 | 0.21 |
| | 2000 | 2,143.8 | 97.1 | 1,657 | 333.215 | 0.20 | 0.16 |
| | 2001 | 2,074.0 | 90.7 | 1,525 | 293.879 | 0.19 | 0.14 |
| | 2002 | 2,069.0 | 95.4 | 1,977 | 357.573 | 0.18 | 0.17 |
| | 2003 | 2,014.5 | 93.6 | 2,608 | 602.535 | 0.23 | 0.30 |
| | 2004 | 2,104.7 | 95.5 | 3,242 | 672.714 | 0.21 | 0.32 |
| | 2005 | 2,044.2 | 94.3 | 3,743 | 636.282 | 0.17 | 0.31 |
| | 2006 | 2,040.1 | 94.0 | 3,618 | 641.154 | 0.18 | 0.31 |
| | 2007 | 2,420.2 | 90.0 | 3,027 | 554.314 | 0.18 | 0.23 |
| | 2008 | 2,837.4 | 88.5 | 2,633 | 482.127 | 0.18 | 0.17 |
| | 2009 | 2,933.1 | 91.2 | 2,188 | 348.257 | 0.16 | 0.12 |
| | 2010 | 2,828.0 | 92.3 | 2,825 | 556.749 | 0.20 | 0.20 |
| **BRUNSWICK 1, 2** | 1976 | 297.2 | 56.0 | 1,265 | 326 | 0.26 | 1.10 |
| Docket 50-324, 50-325; | 1977 | 291.1 | 55.7 | 1,512 | 1,120 | 0.74 | 3.85 |
| DPR-62, DPR-71 | 1978 | 1,173.1 | 83.7 | 1,458 | 1,004 | 0.69 | 0.86 |
| 1st commercial operation | 1979 | 810.0 | 60.1 | 2,891 | 2,602 | 0.90 | 3.21 |
| 3/77, 11/75 | 1980 | 687.2 | 52.2 | 3,788 | 3,870 | 1.02 | 5.63 |
| Type - BWRs | 1981 | 925.2 | 56.9 | 3,854 | 2,638 | 0.68 | 2.85 |
| Capacity - 938, 920 MWe | 1982 | 540.3 | 50.3 | 4,957 | 3,792 | 0.76 | 7.02 |

[2] All three Brown's Ferry units were placed on administrative hold in 1985. Units 2 & 3 were restarted in 1991 and 1995, respectively. Brown's Ferry Unit 1 was restarted during 2007.

| Reporting Organization | Year | Megawatt Years (MW-yr) | Unit Availability Factor | Total Personnel with Measurable Doses | Collective Dose per Site (person-rem) | Average Measurable Dose (rem) | Collective Dose/ MW-yr |
|---|---|---|---|---|---|---|---|
| **BRUNSWICK 1, 2** | 1983 | 636.7 | 44.3 | 5,602 | 3,475 | 0.62 | 5.46 |
| (continued) | 1984 | 761.3 | 51.5 | 5,046 | 3,260 | 0.65 | 4.28 |
| | 1985 | 822.2 | 58.4 | 4,057 | 2,804 | 0.69 | 3.41 |
| | 1986 | 1,051.3 | 69.1 | 3,370 | 1,909 | 0.57 | 1.82 |
| | 1987 | 1,152.4 | 80.6 | 3,052 | 1,419 | 0.46 | 1.23 |
| | 1988 | 990.8 | 70.1 | 2,648 | 1,747 | 0.66 | 1.76 |
| | 1989 | 990.9 | 65.8 | 3,844 | 1,786 | 0.46 | 1.80 |
| | 1990 | 991.6 | 67.8 | 3,182 | 1,548 | 0.49 | 1.56 |
| | 1991 | 952.8 | 64.5 | 2,586 | 778 | 0.30 | 0.82 |
| | 1992 | 375.9 | 27.9 | 2,690 | 623 | 0.23 | 1.66 |
| | 1993 | 470.0 | 33.8 | 2,921 | 872 | 0.30 | 1.86 |
| | 1994 | 1,268.4 | 83.0 | 3,049 | 999 | 0.33 | 0.79 |
| | 1995 | 1,411.7 | 92.9 | 2,657 | 683 | 0.26 | 0.48 |
| | 1996 | 1,261.1 | 85.9 | 2,784 | 716 | 0.26 | 0.57 |
| | 1997 | 1,474.0 | 94.1 | 2,212 | 411 | 0.19 | 0.28 |
| | 1998 | 1,521.0 | 94.3 | 2,005 | 395.526 | 0.20 | 0.26 |
| | 1999 | 1,494.7 | 92.8 | 1,818 | 418.417 | 0.23 | 0.28 |
| | 2000 | 1,571.2 | 95.6 | 1,648 | 321.785 | 0.20 | 0.20 |
| | 2001 | 1,576.0 | 95.8 | 1,623 | 302.812 | 0.19 | 0.19 |
| | 2002 | 1,568.0 | 94.5 | 1,743 | 275.534 | 0.16 | 0.18 |
| | 2003 | 1,676.9 | 95.6 | 1,794 | 248.622 | 0.14 | 0.15 |
| | 2004 | 1,690.6 | 94.5 | 2,140 | 244.577 | 0.11 | 0.14 |
| | 2005 | 1,654.9 | 92.2 | 1,944 | 305.978 | 0.16 | 0.19 |
| | 2006 | 1,661.2 | 90.0 | 2,103 | 280.465 | 0.13 | 0.17 |
| | 2007 | 1,714.9 | 92.0 | 2,186 | 290.093 | 0.13 | 0.17 |
| | 2008 | 1,694.5 | 91.7 | 2,546 | 354.212 | 0.14 | 0.21 |
| | 2009 | 1,647.9 | 89.6 | 2,683 | 350.347 | 0.13 | 0.21 |
| | 2010 | 1,690.7 | 91.3 | 3,227 | 407.424 | 0.13 | 0.24 |
| **BYRON 1, 2** | 1986 | 894.5 | 88.6 | 1,081 | 76 | 0.07 | 0.08 |
| Docket 50-454, 50-455; | 1987 | 650.9 | 70.9 | 1,826 | 769 | 0.42 | 1.18 |
| NPF-37, NPF-66 | 1988 | 1,534.7 | 86.3 | 1,222 | 459 | 0.38 | 0.30 |
| 1st commercial operation | 1989 | 1,812.6 | 90.2 | 1,109 | 172 | 0.16 | 0.09 |
| 9/85, 8/87 | 1990 | 1,567.3 | 78.8 | 1,396 | 434 | 0.31 | 0.28 |
| Type - PWRs | 1991 | 1,816.3 | 89.9 | 1,077 | 268 | 0.25 | 0.15 |
| Capacity - 1,152, 1,125 MWe | 1992 | 1,888.4 | 90.1 | 1,021 | 199 | 0.19 | 0.11 |
| | 1993 | 1,785.6 | 83.5 | 1,370 | 432 | 0.32 | 0.24 |
| | 1994 | 1,953.3 | 90.7 | 962 | 280 | 0.29 | 0.14 |
| | 1995 | 1,900.6 | 85.5 | 1,107 | 306 | 0.28 | 0.16 |
| | 1996 | 1,758.4 | 79.3 | 1,610 | 455 | 0.28 | 0.26 |
| | 1997 | 1,856.7 | 86.6 | 1,546 | 241 | 0.16 | 0.13 |
| | 1998 | 1,869.8 | 85.9 | 1,809 | 275.221 | 0.15 | 0.15 |
| | 1999 | 2,064.2 | 92.3 | 1,478 | 239.102 | 0.16 | 0.12 |
| | 2000 | 2,196.9 | 97.4 | 959 | 193.871 | 0.20 | 0.09 |
| | 2001 | 2,301.5 | 97.8 | 719 | 59.451 | 0.08 | 0.03 |
| | 2002 | 2,205.0 | 93.8 | 1,287 | 195.013 | 0.15 | 0.09 |
| | 2003 | 2,294.8 | 97.2 | 824 | 87.129 | 0.11 | 0.04 |
| | 2004 | 2,277.4 | 97.7 | 906 | 89.147 | 0.10 | 0.04 |
| | 2005 | 2,175.6 | 94.2 | 1,542 | 199.812 | 0.13 | 0.09 |
| | 2006 | 2,223.3 | 95.0 | 1,163 | 134.497 | 0.12 | 0.06 |
| | 2007 | 2,152.1 | 93.0 | 1,311 | 128.797 | 0.10 | 0.06 |
| | 2008 | 2,203.7 | 94.6 | 1,483 | 140.809 | 0.09 | 0.06 |
| | 2009 | 2,250.9 | 96.7 | 985 | 83.443 | 0.08 | 0.04 |
| | 2010 | 2,266.6 | 97.4 | 922 | 56.425 | 0.06 | 0.02 |
| **CALLAWAY 1** | 1985 | 967.4 | 90.0 | 964 | 36 | 0.04 | 0.04 |
| Docket 50-483; | 1986 | 865.2 | 81.3 | 1,052 | 225 | 0.21 | 0.26 |
| NPF-30 | 1987 | 759.0 | 71.1 | 1,082 | 393 | 0.36 | 0.52 |
| 1st commercial operation 12/84 | 1988 | 1,069.2 | 93.4 | 353 | 27 | 0.08 | 0.03 |
| Type - PWR | 1989 | 1,000.3 | 85.4 | 1,055 | 283 | 0.27 | 0.28 |
| Capacity - 1,190 MWe | 1990 | 960.7 | 84.1 | 1,134 | 442 | 0.39 | 0.46 |
| | 1991 | 1,193.1 | 99.7 | 280 | 21 | 0.07 | 0.02 |
| | 1992 | 967.5 | 83.0 | 1,133 | 336 | 0.30 | 0.35 |

| Reporting Organization | Year | Megawatt Years (MW-yr) | Unit Availability Factor | Total Personnel with Measurable Doses | Collective Dose per Site (person-rem) | Average Measurable Dose (rem) | Collective Dose/ MW-yr |
|---|---|---|---|---|---|---|---|
| **CALLAWAY 1** | 1993 | 1,002.9 | 86.4 | 1,126 | 225 | 0.20 | 0.22 |
| (continued) | 1994 | 1,196.4 | 100.0 | 191 | 14 | 0.07 | 0.01 |
| | 1995 | 989.6 | 84.7 | 1,062 | 187 | 0.18 | 0.19 |
| | 1996 | 1,066.0 | 90.5 | 980 | 248 | 0.25 | 0.23 |
| | 1997 | 1,022.2 | 100.0 | 248 | 12 | 0.05 | 0.01 |
| | 1998 | 972.2 | 91.3 | 929 | 200.729 | 0.22 | 0.21 |
| | 1999 | 981.3 | 88.7 | 1,098 | 320.554 | 0.29 | 0.33 |
| | 2000 | 1,137.5 | 99.8 | 244 | 16.058 | 0.07 | 0.01 |
| | 2001 | 954.5 | 86.7 | 873 | 106.782 | 0.12 | 0.11 |
| | 2002 | 955.0 | 86.2 | 983 | 95.648 | 0.10 | 0.10 |
| | 2003 | 1,104.3 | 96.2 | 252 | 8.297 | 0.03 | 0.01 |
| | 2004 | 892.8 | 78.9 | 1,124 | 120.621 | 0.11 | 0.14 |
| | 2005 | 913.2 | 80.7 | 1,600 | 222.629 | 0.14 | 0.24 |
| | 2006 | 1,152.8 | 95.0 | 225 | 6.308 | 0.03 | 0.01 |
| | 2007 | 1,069.7 | 89.0 | 1,079 | 73.236 | 0.07 | 0.07 |
| | 2008 | 1,067.6 | 89.8 | 729 | 45.738 | 0.06 | 0.04 |
| | 2009 | 1,170.3 | 97.6 | 164 | 4.821 | 0.03 | 0.00 |
| | 2010 | 1,029.9 | 84.8 | 800 | 58.735 | 0.07 | 0.06 |
| **CALVERT CLIFFS 1, 2** | 1976 | 753.4 | 95.2 | 507 | 74 | 0.15 | 0.10 |
| Docket 50-317, 50-318; | 1977 | 583.0 | 72.1 | 2,265 | 547 | 0.24 | 0.94 |
| DPR-53, DPR-69 | 1978 | 1,188.5 | 75.8 | 1,391 | 500 | 0.36 | 0.42 |
| 1st commercial operation | 1979 | 1,161.0 | 74.0 | 1,428 | 805 | 0.56 | 0.69 |
| 5/75, 4/77 | 1980 | 1,309.9 | 84.1 | 1,496 | 677 | 0.45 | 0.52 |
| Type - PWRs | 1981 | 1,379.7 | 83.1 | 1,555 | 607 | 0.39 | 0.44 |
| Capacity - 870, 858 MWe | 1982 | 1,238.3 | 73.7 | 1,805 | 1,057 | 0.59 | 0.85 |
| | 1983 | 1,397.2 | 81.6 | 1,915 | 668 | 0.35 | 0.48 |
| | 1984 | 1,389.4 | 79.3 | 1,369 | 479 | 0.35 | 0.34 |
| | 1985 | 1,189.8 | 68.4 | 1,598 | 694 | 0.43 | 0.58 |
| | 1986 | 1,530.0 | 87.2 | 1,296 | 347 | 0.27 | 0.23 |
| | 1987 | 1,207.3 | 71.8 | 1,384 | 412 | 0.30 | 0.34 |
| | 1988 | 1,397.7 | 81.0 | 1,296 | 291 | 0.22 | 0.21 |
| | 1989 | 333.6 | 20.1 | 1,786 | 346 | 0.19 | 1.04 |
| | 1990 | 161.1 | 11.0 | 2,019 | 304 | 0.15 | 1.89 |
| | 1991 | 1,085.0 | 64.7 | 1,974 | 132 | 0.07 | 0.12 |
| | 1992 | 1,271.2 | 73.9 | 1,979 | 330 | 0.17 | 0.26 |
| | 1993 | 1,462.1 | 83.9 | 1,462 | 405 | 0.28 | 0.28 |
| | 1994 | 1,342.1 | 79.4 | 1,482 | 454 | 0.31 | 0.34 |
| | 1995 | 1,542.8 | 89.9 | 1,203 | 235 | 0.20 | 0.15 |
| | 1996 | 1,438.5 | 82.4 | 1,167 | 239 | 0.20 | 0.17 |
| | 1997 | 1,499.6 | 89.1 | 1,091 | 229 | 0.21 | 0.15 |
| | 1998 | 1,523.1 | 89.3 | 1,042 | 186.887 | 0.18 | 0.12 |
| | 1999 | 1,521.4 | 90.1 | 1,134 | 191.778 | 0.17 | 0.13 |
| | 2000 | 1,575.7 | 92.7 | 912 | 134.689 | 0.15 | 0.09 |
| | 2001 | 1,554.7 | 91.7 | 895 | 166.864 | 0.19 | 0.11 |
| | 2002 | 1,380.0 | 81.7 | 1,582 | 245.075 | 0.16 | 0.18 |
| | 2003 | 1,558.4 | 90.9 | 1,671 | 265.164 | 0.16 | 0.17 |
| | 2004 | 1,653.7 | 95.7 | 1,205 | 143.944 | 0.12 | 0.09 |
| | 2005 | 1,678.1 | 97.2 | 942 | 168.390 | 0.18 | 0.10 |
| | 2006 | 1,581.8 | 92.0 | 1,215 | 203.790 | 0.17 | 0.13 |
| | 2007 | 1,641.6 | 95.0 | 1,191 | 153.335 | 0.13 | 0.09 |
| | 2008 | 1,670.7 | 97.4 | 745 | 74.149 | 0.10 | 0.04 |
| | 2009 | 1,660.9 | 96.6 | 891 | 95.756 | 0.11 | 0.06 |
| | 2010 | 1,597.3 | 93.5 | 834 | 128.581 | 0.15 | 0.08 |
| **CATAWBA 1, 2** | 1986 | 638.9 | 49.9 | 1,724 | 286 | 0.17 | 0.45 |
| Docket 50-413, 50-414; | 1987 | 1,651.2 | 75.9 | 1,865 | 449 | 0.24 | 0.27 |
| NPF-35, NPF-52 | 1988 | 1,675.2 | 77.2 | 2,009 | 556 | 0.28 | 0.33 |
| 1st commercial operation | 1989 | 1,733.6 | 79.5 | 1,660 | 334 | 0.20 | 0.19 |
| 6/85, 8/86 | 1990 | 1,616.3 | 70.8 | 2,174 | 809 | 0.37 | 0.50 |
| Type - PWRs | 1991 | 1,691.5 | 74.6 | 1,871 | 462 | 0.25 | 0.27 |
| Capacity - 1,129, 1,129 MWe | 1992 | 1,962.8 | 83.9 | 1,515 | 414 | 0.27 | 0.21 |
| | 1993 | 1,896.1 | 81.5 | 1,564 | 396 | 0.25 | 0.21 |

| Reporting Organization | Year | Megawatt Years (MW-yr) | Unit Availability Factor | Total Personnel with Measurable Doses | Collective Dose per Site (person-rem) | Average Measurable Dose (rem) | Collective Dose/ MW-yr |
|---|---|---|---|---|---|---|---|
| **CATAWBA 1, 2** | 1994 | 2,105.2 | 90.2 | 1,268 | 207 | 0.16 | 0.10 |
| (continued) | 1995 | 2,011.9 | 85.3 | 1,892 | 462 | 0.24 | 0.23 |
| | 1996 | 1,879.1 | 80.5 | 1,588 | 302 | 0.19 | 0.16 |
| | 1997 | 2,028.2 | 89.3 | 1,561 | 266 | 0.17 | 0.13 |
| | 1998 | 2,006.4 | 89.6 | 1,123 | 162.068 | 0.14 | 0.08 |
| | 1999 | 2,046.7 | 90.2 | 1,024 | 118.662 | 0.12 | 0.06 |
| | 2000 | 2,038.3 | 90.3 | 1,185 | 186.532 | 0.16 | 0.09 |
| | 2001 | 2,119.9 | 92.9 | 960 | 116.241 | 0.12 | 0.05 |
| | 2002 | 2,238.0 | 97.2 | 884 | 81.325 | 0.09 | 0.04 |
| | 2003 | 1,991.8 | 89.2 | 1,409 | 210.617 | 0.15 | 0.11 |
| | 2004 | 2,111.4 | 93.0 | 1,123 | 122.831 | 0.11 | 0.06 |
| | 2005 | 2,194.5 | 96.0 | 1,019 | 83.679 | 0.08 | 0.04 |
| | 2006 | 1,928.6 | 85.0 | 1,792 | 212.570 | 0.12 | 0.11 |
| | 2007 | 2,102.5 | 92.0 | 1,399 | 144.218 | 0.10 | 0.07 |
| | 2008 | 2,160.3 | 93.5 | 1,110 | 85.080 | 0.08 | 0.04 |
| | 2009 | 2,044.8 | 89.1 | 1,385 | 169.409 | 0.12 | 0.08 |
| | 2010 | 2,164.8 | 94.8 | 1,045 | 97.010 | 0.09 | 0.04 |
| **CLINTON** | 1988 | 701.3 | 84.2 | 769 | 130 | 0.17 | 0.19 |
| Docket 50-461; | 1989 | 348.3 | 48.5 | 1,196 | 372 | 0.31 | 1.07 |
| NPF-62 | 1990 | 435.8 | 55.1 | 1,390 | 553 | 0.40 | 1.27 |
| 1st commercial operation 11/87 | 1991 | 722.7 | 80.8 | 1,010 | 233 | 0.23 | 0.32 |
| Type - BWR | 1992 | 589.7 | 68.6 | 1,195 | 431 | 0.36 | 0.73 |
| Capacity - 1,022 MWe | 1993 | 701.5 | 79.6 | 1,253 | 498 | 0.40 | 0.71 |
| | 1994 | 883.3 | 94.8 | 409 | 63 | 0.15 | 0.07 |
| | 1995 | 731.1 | 83.0 | 1,182 | 316 | 0.27 | 0.43 |
| | 1996 | 634.7 | 66.7 | 1,154 | 350 | 0.30 | 0.55 |
| | 1997 | 0.0 | 0.0 | 738 | 172 | 0.23 | --- |
| | 1998 | 0.0 | 0.0 | 866 | 144.140 | 0.17 | --- |
| | 1999 | 537.0 | 63.5 | 637 | 87.489 | 0.14 | 0.16 |
| | 2000 | 784.2 | 87.8 | 1,248 | 253.382 | 0.20 | 0.32 |
| | 2001 | 896.8 | 98.5 | 329 | 33.770 | 0.10 | 0.04 |
| | 2002 | 872.0 | 90.5 | 1,418 | 208.094 | 0.15 | 0.24 |
| | 2003 | 990.5 | 99.1 | 372 | 57.118 | 0.15 | 0.06 |
| | 2004 | 910.8 | 92.6 | 1,622 | 282.833 | 0.17 | 0.31 |
| | 2005 | 989.1 | 97.4 | 298 | 36.019 | 0.12 | 0.04 |
| | 2006 | 939.9 | 92.0 | 1,649 | 295.720 | 0.18 | 0.32 |
| | 2007 | 1,049.2 | 100.0 | 310 | 30.618 | 0.10 | 0.03 |
| | 2008 | 973.0 | 93.3 | 1,381 | 205.086 | 0.15 | 0.21 |
| | 2009 | 1,014.6 | 96.6 | 435 | 48.009 | 0.11 | 0.05 |
| | 2010 | 983.1 | 93.5 | 1,540 | 219.954 | 0.14 | 0.22 |
| **COLUMBIA GENERATING**[3] | 1985 | 616.0 | 87.6 | 755 | 119 | 0.16 | 0.19 |
| Docket 50-397; | 1986 | 616.0 | 74.4 | 1,013 | 222 | 0.22 | 0.36 |
| NPF-21 | 1987 | 639.0 | 70.8 | 1,201 | 406 | 0.34 | 0.64 |
| 1st commercial operation 12/84 | 1988 | 707.7 | 71.8 | 1,050 | 353 | 0.34 | 0.50 |
| Type - BWR | 1989 | 727.2 | 78.3 | 1,299 | 492 | 0.38 | 0.68 |
| Capacity - 1,107 MWe | 1990 | 684.7 | 67.5 | 1,348 | 536 | 0.40 | 0.78 |
| | 1991 | 508.5 | 50.3 | 1,088 | 387 | 0.36 | 0.76 |
| | 1992 | 682.3 | 65.6 | 1,489 | 612 | 0.41 | 0.90 |
| | 1993 | 849.6 | 79.5 | 1,385 | 469 | 0.34 | 0.55 |
| | 1994 | 803.8 | 75.2 | 1,870 | 866 | 0.46 | 1.08 |
| | 1995 | 824.7 | 83.8 | 1,694 | 456 | 0.27 | 0.55 |
| | 1996 | 662.9 | 82.2 | 1,453 | 373 | 0.26 | 0.56 |
| | 1997 | 697.0 | 72.7 | 1,218 | 251 | 0.21 | 0.36 |
| | 1998 | 789.5 | 75.3 | 1,220 | 286.020 | 0.23 | 0.36 |
| | 1999 | 694.7 | 70.0 | 1,022 | 155.109 | 0.15 | 0.22 |
| | 2000 | 979.6 | 96.3 | 706 | 53.152 | 0.08 | 0.05 |
| | 2001 | 939.3 | 88.1 | 1,515 | 226.675 | 0.15 | 0.24 |
| | 2002 | 1,023.0 | 97.5 | 647 | 46.650 | 0.07 | 0.05 |
| | 2003 | 866.9 | 81.8 | 1,618 | 205.225 | 0.13 | 0.24 |

[3] Energy Northwest has changed the name of Washington Nuclear 2 to Columbia Generating Station.

| Reporting Organization | Year | Megawatt Years (MW-yr) | Unit Availability Factor | Total Personnel with Measurable Doses | Collective Dose per Site (person-rem) | Average Measurable Dose (rem) | Collective Dose/ MW-yr |
|---|---|---|---|---|---|---|---|
| **COLUMBIA GENERATING**[3] | 2004 | 1,022.5 | 94.6 | 716 | 66.130 | 0.09 | 0.06 |
| (continued) | 2005 | 938.3 | 87.3 | 1,718 | 325.025 | 0.19 | 0.35 |
| | 2006 | 1,064.9 | 98.0 | 623 | 55.817 | 0.09 | 0.05 |
| | 2007 | 925.6 | 87.0 | 2,147 | 306.443 | 0.14 | 0.33 |
| | 2008 | 1,055.3 | 98.3 | 715 | 54.957 | 0.08 | 0.05 |
| | 2009 | 757.2 | 76.3 | 1,958 | 305.163 | 0.16 | 0.40 |
| | 2010 | 1,054.9 | 100.0 | 733 | 54.712 | 0.07 | 0.05 |
| **COMANCHE PEAK 1, 2** | 1991 | 644.4 | 82.2 | 985 | 148 | 0.15 | 0.23 |
| Docket 50-445, 50-446; | 1992 | 830.8 | 84.0 | 1,128 | 188 | 0.17 | 0.23 |
| NPF-87, NPF-89 | 1993 | 853.8 | 81.2 | 945 | 109 | 0.12 | 0.13 |
| 1st commercial operation | 1994 | 1,750.0 | 93.7 | 970 | 90 | 0.09 | 0.05 |
| 8/90, 8/93 | 1995 | 2,022.6 | 92.5 | 951 | 179 | 0.19 | 0.09 |
| Type - PWR | 1996 | 1,804.8 | 81.4 | 1,462 | 288 | 0.20 | 0.16 |
| Capacity - 1,150, 1,150 MWe | 1997 | 2,002.4 | 93.4 | 870 | 146 | 0.17 | 0.07 |
| | 1998 | 2,037.8 | 94.9 | 967 | 232.026 | 0.24 | 0.11 |
| | 1999 | 1,981.5 | 90.9 | 1,316 | 251.276 | 0.19 | 0.13 |
| | 2000 | 2,104.7 | 95.3 | 759 | 77.679 | 0.10 | 0.04 |
| | 2001 | 2,085.9 | 94.7 | 853 | 114.968 | 0.13 | 0.06 |
| | 2002 | 1,887.0 | 86.9 | 1,106 | 225.317 | 0.20 | 0.12 |
| | 2003 | 2,020.6 | 91.6 | 639 | 66.313 | 0.10 | 0.03 |
| | 2004 | 2,169.5 | 95.1 | 864 | 135.388 | 0.16 | 0.06 |
| | 2005 | 2,099.6 | 91.5 | 1,365 | 242.481 | 0.18 | 0.12 |
| | 2006 | 2,271.3 | 97.0 | 686 | 59.959 | 0.09 | 0.03 |
| | 2007 | 2,151.3 | 93.0 | 1,616 | 219.799 | 0.14 | 0.10 |
| | 2008 | 2,189.7 | 94.3 | 1,037 | 168.836 | 0.16 | 0.08 |
| | 2009 | 2,299.3 | 96.7 | 938 | 51.420 | 0.05 | 0.02 |
| | 2010 | 2,316.8 | 96.3 | 1,037 | 70.807 | 0.07 | 0.03 |
| **COOK  1, 2** | 1976 | 807.4 | 83.1 | 395 | 116 | 0.29 | 0.14 |
| Docket 50-315, 50-316; | 1977 | 573.0 | 76.1 | 802 | 300 | 0.37 | 0.52 |
| DPR-58, DPR-74 | 1978 | 744.8 | 73.6 | 778 | 336 | 0.43 | 0.45 |
| 1st commercial operation | 1979 | 1,373.0 | 65.3 | 1,445 | 718 | 0.50 | 0.52 |
| 8/75, 7/78 | 1980 | 1,552.4 | 74.1 | 1,345 | 493 | 0.37 | 0.32 |
| Type - PWRs | 1981 | 1,557.3 | 73.4 | 1,341 | 656 | 0.49 | 0.42 |
| Capacity - 1,030, 1,077 MWe | 1982 | 1,461.6 | 69.8 | 1,527 | 699 | 0.46 | 0.48 |
| | 1983 | 1,456.5 | 71.2 | 1,418 | 658 | 0.46 | 0.45 |
| | 1984 | 1,526.0 | 75.3 | 1,559 | 762 | 0.49 | 0.50 |
| | 1985 | 925.4 | 47.6 | 1,984 | 945 | 0.48 | 1.02 |
| | 1986 | 1,307.1 | 73.4 | 1,774 | 745 | 0.42 | 0.57 |
| | 1987 | 1,199.5 | 70.2 | 1,696 | 666 | 0.39 | 0.56 |
| | 1988 | 1,160.4 | 63.5 | 2,266 | 867 | 0.38 | 0.75 |
| | 1989 | 1,433.1 | 72.8 | 1,575 | 493 | 0.31 | 0.34 |
| | 1990 | 1,318.5 | 67.9 | 1,851 | 580 | 0.31 | 0.44 |
| | 1991 | 1,837.4 | 90.2 | 815 | 69 | 0.08 | 0.04 |
| | 1992 | 760.9 | 50.8 | 1,954 | 492 | 0.25 | 0.65 |
| | 1993 | 1,927.7 | 98.5 | 587 | 44 | 0.07 | 0.02 |
| | 1994 | 1,105.2 | 65.2 | 1,748 | 479 | 0.27 | 0.43 |
| | 1995 | 1,656.0 | 82.1 | 1,310 | 203 | 0.15 | 0.12 |
| | 1996 | 1,938.9 | 92.7 | 1,114 | 214 | 0.19 | 0.11 |
| | 1997 | 1,189.7 | 59.7 | 1,864 | 550 | 0.30 | 0.46 |
| | 1998 | 0.0 | 0.0 | 1,155 | 104.638 | 0.09 | --- |
| | 1999 | 0.0 | 0.0 | 1,662 | 171.479 | 0.10 | --- |
| | 2000 | 560.1 | 28.1 | 2,506 | 337.584 | 0.14 | 0.60 |
| | 2001 | 1,794.3 | 89.2 | 423 | 27.290 | 0.06 | 0.02 |
| | 2002 | 1,756.0 | 87.3 | 1,624 | 278.001 | 0.17 | 0.16 |
| | 2003 | 1,557.6 | 75.7 | 1,408 | 209.526 | 0.15 | 0.13 |
| | 2004 | 1,909.2 | 91.4 | 1,015 | 156.213 | 0.15 | 0.08 |
| | 2005 | 1,989.0 | 95.0 | 852 | 91.192 | 0.11 | 0.05 |
| | 2006 | 1,790.5 | 86.0 | 1,780 | 312.214 | 0.18 | 0.17 |
| | 2007 | 1,983.7 | 93.0 | 1,310 | 238.829 | 0.18 | 0.12 |

[3] Energy Northwest has changed the name of Washington Nuclear 2 to Columbia Generating Station.

| Reporting Organization | Year | Megawatt Years (MW-yr) | Unit Availability Factor | Total Personnel with Measurable Doses | Collective Dose per Site (person-rem) | Average Measurable Dose (rem) | Collective Dose/ MW-yr |
|---|---|---|---|---|---|---|---|
| **COOK 1, 2** | 2008 | 1,711.8 | 80.8 | 971 | 76.460 | 0.08 | 0.04 |
| (continued) | 2009 | 950.5 | 45.3 | 693 | 40.007 | 0.06 | 0.04 |
|  | 2010 | 1,786.1 | 86.7 | 1,116 | 83.276 | 0.07 | 0.05 |
| **COOPER STATION** | 1975 | 456.4 | 83.6 | 579 | 117 | 0.20 | 0.26 |
| Docket 50-298; | 1976 | 433.3 | 75.5 | 763 | 350 | 0.46 | 0.81 |
| DPR-46 | 1977 | 538.2 | 86.2 | 315 | 198 | 0.63 | 0.37 |
| 1st commercial operation 7/74 | 1978 | 576.0 | 91.0 | 297 | 158 | 0.53 | 0.27 |
| Type - BWR | 1979 | 591.0 | 87.6 | 426 | 221 | 0.52 | 0.37 |
| Capacity - 769 MWe | 1980 | 448.3 | 71.2 | 785 | 859 | 1.09 | 1.92 |
|  | 1981 | 457.1 | 71.2 | 935 | 579 | 0.62 | 1.27 |
|  | 1982 | 622.3 | 84.6 | 743 | 542 | 0.73 | 0.87 |
|  | 1983 | 396.6 | 63.3 | 1,383 | 1,293 | 0.93 | 3.26 |
|  | 1984 | 411.9 | 67.2 | 1,598 | 799 | 0.50 | 1.94 |
|  | 1985 | 127.3 | 21.5 | 1,980 | 1,333 | 0.67 | 10.47 |
|  | 1986 | 480.0 | 74.7 | 895 | 320 | 0.36 | 0.67 |
|  | 1987 | 652.3 | 96.2 | 549 | 103 | 0.19 | 0.16 |
|  | 1988 | 493.4 | 67.9 | 942 | 251 | 0.27 | 0.51 |
|  | 1989 | 564.3 | 76.2 | 1,202 | 343 | 0.29 | 0.61 |
|  | 1990 | 602.0 | 79.4 | 1,174 | 379 | 0.32 | 0.63 |
|  | 1991 | 566.3 | 78.8 | 1,099 | 405 | 0.37 | 0.72 |
|  | 1992 | 731.0 | 96.4 | 463 | 84 | 0.18 | 0.11 |
|  | 1993 | 436.1 | 58.8 | 1,130 | 391 | 0.35 | 0.90 |
|  | 1994 | 262.2 | 35.1 | 333 | 79 | 0.24 | 0.30 |
|  | 1995 | 486.5 | 66.8 | 1,095 | 228 | 0.21 | 0.47 |
|  | 1996 | 742.1 | 97.9 | 468 | 48 | 0.10 | 0.06 |
|  | 1997 | 622.8 | 84.4 | 1,125 | 174 | 0.16 | 0.28 |
|  | 1998 | 555.9 | 75.9 | 977 | 181.858 | 0.19 | 0.33 |
|  | 1999 | 743.2 | 98.1 | 318 | 47.815 | 0.15 | 0.06 |
|  | 2000 | 539.2 | 74.2 | 963 | 199.589 | 0.21 | 0.37 |
|  | 2001 | 592.7 | 80.9 | 1,309 | 168.665 | 0.13 | 0.28 |
|  | 2002 | 719.0 | 98.6 | 362 | 38.739 | 0.11 | 0.05 |
|  | 2003 | 511.4 | 74.1 | 882 | 135.249 | 0.15 | 0.26 |
|  | 2004 | 702.6 | 94.7 | 481 | 47.064 | 0.10 | 0.07 |
|  | 2005 | 670.8 | 89.4 | 1,266 | 275.652 | 0.22 | 0.41 |
|  | 2006 | 674.7 | 90.0 | 1,265 | 270.135 | 0.21 | 0.40 |
|  | 2007 | 761.6 | 99.0 | 730 | 49.902 | 0.07 | 0.07 |
|  | 2008 | 679.0 | 89.9 | 1,715 | 359.926 | 0.21 | 0.53 |
|  | 2009 | 654.6 | 86.6 | 1,638 | 254.032 | 0.16 | 0.39 |
|  | 2010 | 775.4 | 100.0 | 773 | 61.303 | 0.08 | 0.08 |
| **CRYSTAL RIVER 3** | 1978 | 311.5 | 41.4 | 643 | 321 | 0.50 | 1.03 |
| Docket 50-302; | 1979 | 453.0 | 58.9 | 1,150 | 495 | 0.43 | 1.09 |
| DPR-72 | 1980 | 404.1 | 53.2 | 1,053 | 625 | 0.59 | 1.55 |
| 1st commercial operation 3/77 | 1981 | 490.4 | 62.2 | 1,120 | 408 | 0.36 | 0.83 |
| Type - PWR | 1982 | 589.8 | 76.0 | 780 | 177 | 0.23 | 0.30 |
| Capacity - 860 MWe | 1983 | 452.1 | 58.8 | 1,720 | 552 | 0.32 | 1.22 |
|  | 1984 | 774.2 | 94.5 | 549 | 49 | 0.09 | 0.06 |
|  | 1985 | 344.2 | 47.6 | 1,976 | 689 | 0.35 | 2.00 |
|  | 1986 | 319.5 | 41.8 | 1,057 | 472 | 0.45 | 1.48 |
|  | 1987 | 436.0 | 60.9 | 1,384 | 488 | 0.35 | 1.12 |
|  | 1988 | 690.2 | 84.0 | 569 | 64 | 0.11 | 0.09 |
|  | 1989 | 352.8 | 48.8 | 880 | 234 | 0.27 | 0.66 |
|  | 1990 | 497.8 | 63.8 | 1,441 | 476 | 0.33 | 0.96 |
|  | 1991 | 654.6 | 82.0 | 821 | 116 | 0.14 | 0.18 |
|  | 1992 | 632.1 | 76.1 | 1,403 | 424 | 0.30 | 0.67 |
|  | 1993 | 722.4 | 85.0 | 683 | 60 | 0.09 | 0.08 |
|  | 1994 | 711.9 | 84.3 | 1,079 | 228 | 0.21 | 0.32 |
|  | 1995 | 866.3 | 100.0 | 209 | 8 | 0.04 | 0.01 |
|  | 1996 | 290.8 | 37.7 | 1,192 | 353 | 0.30 | 1.21 |
|  | 1997 | 0.0 | 0.0 | 973 | 179 | 0.18 | --- |
|  | 1998 | 739.9 | 90.3 | 313 | 19.298 | 0.06 | 0.03 |
|  | 1999 | 727.5 | 87.8 | 1,324 | 251.077 | 0.19 | 0.35 |

| Reporting Organization | Year | Megawatt Years (MW-yr) | Unit Availability Factor | Total Personnel with Measurable Doses | Collective Dose per Site (person-rem) | Average Measurable Dose (rem) | Collective Dose/ MW-yr |
|---|---|---|---|---|---|---|---|
| **CRYSTAL RIVER 3** | 2000 | 819.4 | 97.6 | 257 | 14.649 | 0.06 | 0.02 |
| (continued) | 2001 | 741.6 | 89.2 | 902 | 147.946 | 0.16 | 0.20 |
|  | 2002 | 831.0 | 99.4 | 128 | 5.039 | 0.04 | 0.01 |
|  | 2003 | 749.0 | 90.8 | 961 | 126.554 | 0.13 | 0.17 |
|  | 2004 | 831.4 | 98.1 | 131 | 4.044 | 0.03 | 0.0 |
|  | 2005 | 723.0 | 88.5 | 939 | 122.608 | 0.13 | 0.17 |
|  | 2006 | 793.8 | 95.0 | 138 | 4.474 | 0.03 | 0.01 |
|  | 2007 | 761.7 | 91.0 | 1,135 | 184.554 | 0.16 | 0.24 |
|  | 2008 | 796.9 | 93.7 | 282 | 16.110 | 0.06 | 0.02 |
|  | 2009 | 615.0 | 72.5 | 1,705 | 222.344 | 0.13 | 0.36 |
|  | 2010 | - | 0.0 | 666 | 31.922 | 0.05 | - |
| **DAVIS-BESSE 1** | 1978 | 326.4 | 48.7 | 421 | 48 | 0.11 | 0.15 |
| Docket 50-346; | 1979 | 381.0 | 67.0 | 304 | 30 | 0.10 | 0.08 |
| NPF-3 | 1980 | 256.4 | 36.2 | 1,283 | 154 | 0.12 | 0.60 |
| 1st commercial operation 7/78 | 1981 | 531.4 | 67.4 | 578 | 58 | 0.10 | 0.11 |
| Type - PWR | 1982 | 390.8 | 51.5 | 1,350 | 164 | 0.12 | 0.42 |
| Capacity - 894 MWe | 1983 | 592.1 | 73.0 | 718 | 80 | 0.11 | 0.14 |
|  | 1984 | 518.5 | 62.5 | 1,088 | 177 | 0.16 | 0.34 |
|  | 1985 | 238.3 | 31.2 | 718 | 71 | 0.10 | 0.30 |
|  | 1986 | 3.3 | 1.3 | 981 | 124 | 0.13 | 37.58 |
|  | 1987 | 618.0 | 89.6 | 625 | 47 | 0.08 | 0.08 |
|  | 1988 | 144.1 | 27.1 | 1,183 | 307 | 0.26 | 2.13 |
|  | 1989 | 880.0 | 98.6 | 404 | 38 | 0.09 | 0.04 |
|  | 1990 | 500.0 | 56.7 | 1,377 | 489 | 0.36 | 0.98 |
|  | 1991 | 703.6 | 81.8 | 1,000 | 216 | 0.22 | 0.31 |
|  | 1992 | 915.2 | 100.0 | 287 | 19 | 0.07 | 0.02 |
|  | 1993 | 729.5 | 83.4 | 1,244 | 348 | 0.28 | 0.48 |
|  | 1994 | 768.4 | 88.0 | 861 | 144 | 0.17 | 0.19 |
|  | 1995 | 920.4 | 100.0 | 256 | 7 | 0.03 | 0.01 |
|  | 1996 | 775.8 | 85.3 | 949 | 167 | 0.18 | 0.22 |
|  | 1997 | 820.0 | 94.0 | 213 | 10 | 0.05 | 0.01 |
|  | 1998 | 699.8 | 83.2 | 980 | 155.269 | 0.16 | 0.22 |
|  | 1999 | 841.3 | 95.6 | 397 | 27.951 | 0.07 | 0.03 |
|  | 2000 | 770.8 | 87.3 | 1,109 | 168.044 | 0.15 | 0.22 |
|  | 2001 | 875.6 | 100.0 | 119 | 5.505 | 0.05 | 0.01 |
|  | 2002 | 106.0 | 12.6 | 1,983 | 402.766 | 0.20 | 3.81 |
|  | 2003 | 0.0 | 0.0 | 1,047 | 219.696 | 0.21 | --- |
|  | 2004 | 657.8 | 77.6 | 161 | 6.594 | 0.04 | 0.01 |
|  | 2005 | 817.1 | 93.3 | 577 | 51.332 | 0.09 | 0.06 |
|  | 2006 | 727.8 | 84.0 | 1,331 | 204.201 | 0.15 | 0.28 |
|  | 2007 | 879.7 | 100.0 | 189 | 7.088 | 0.04 | 0.01 |
|  | 2008 | 777.5 | 89.4 | 985 | 106.603 | 0.11 | 0.14 |
|  | 2009 | 868.7 | 95.7 | 115 | 3.621 | 0.03 | 0.00 |
|  | 2010 | 598.0 | 67.1 | 1,649 | 464.095 | 0.28 | 0.78 |
| **DIABLO CANYON 1, 2** | 1986 | 641.5 | 80.6 | 1,260 | 304 | 0.24 | 0.47 |
| Docket 50-275, 50-323; | 1987 | 1,688.6 | 83.0 | 1,170 | 336 | 0.29 | 0.20 |
| DPR-80, DPR-82 | 1988 | 1,386.1 | 67.6 | 1,826 | 877 | 0.48 | 0.63 |
| 1st commercial operation | 1989 | 1,899.0 | 87.5 | 1,646 | 465 | 0.28 | 0.24 |
| 5/85, 3/86 | 1990 | 1,952.6 | 91.0 | 1,441 | 323 | 0.22 | 0.17 |
| Type - PWRs | 1991 | 1,809.6 | 83.8 | 2,040 | 546 | 0.27 | 0.30 |
| Capacity - 1,122, 1,118 MWe | 1992 | 1,995.7 | 90.9 | 1,850 | 459 | 0.25 | 0.23 |
|  | 1993 | 2,008.6 | 91.4 | 1,508 | 281 | 0.19 | 0.14 |
|  | 1994 | 1,832.6 | 83.3 | 2,317 | 590 | 0.25 | 0.32 |
|  | 1995 | 1,950.3 | 90.0 | 1,615 | 286 | 0.18 | 0.15 |
|  | 1996 | 2,003.6 | 90.7 | 1,462 | 176 | 0.12 | 0.09 |
|  | 1997 | 1,948.7 | 92.7 | 1,331 | 219 | 0.17 | 0.11 |
|  | 1998 | 1,955.1 | 92.8 | 1,313 | 173.238 | 0.13 | 0.09 |
|  | 1999 | 1,902.8 | 90.1 | 1,566 | 448.634 | 0.29 | 0.24 |
|  | 2000 | 1,940.1 | 92.0 | 1,057 | 180.792 | 0.17 | 0.09 |
|  | 2001 | 2,067.7 | 96.4 | 1,074 | 117.804 | 0.11 | 0.06 |
|  | 2002 | 1,860.0 | 88.4 | 1,016 | 148.690 | 0.15 | 0.08 |

| Reporting Organization | Year | Megawatt Years (MW-yr) | Unit Availability Factor | Total Personnel with Measurable Doses | Collective Dose per Site (person-rem) | Average Measurable Dose (rem) | Collective Dose/ MW-yr |
|---|---|---|---|---|---|---|---|
| **DIABLO CANYON 1, 2** | 2003 | 1,970.7 | 91.6 | 1,004 | 135.482 | 0.13 | 0.07 |
| (continued) | 2004 | 1,736.3 | 83.5 | 1,230 | 254.367 | 0.21 | 0.15 |
| | 2005 | 2,022.4 | 94.8 | 955 | 124.469 | 0.13 | 0.06 |
| | 2006 | 2,109.0 | 94.0 | 1,086 | 82.248 | 0.08 | 0.04 |
| | 2007 | 2,131.4 | 95.0 | 1,269 | 111.866 | 0.09 | 0.05 |
| | 2008 | 1,952.1 | 87.7 | 2,121 | 235.034 | 0.11 | 0.12 |
| | 2009 | 1,873.0 | 85.3 | 2,534 | 337.831 | 0.13 | 0.18 |
| | 2010 | 2,115.2 | 94.7 | 1,367 | 125.457 | 0.09 | 0.06 |
| **DRESDEN 1[4], 2, 3** | 1969 | 99.7 | | | 286 | | 2.87 |
| Docket 50-010, 50-237, 50-249; | 1970 | 163.1 | | | 143 | | 0.88 |
| DPR-2, DPR-19, DPR-25 | 1971 | 394.5 | | | 715 | | 1.81 |
| 1st commercial operation 7/60, | 1972 | 1,243.7 | | | 728 | | 0.59 |
| 6/70, 11/71 | 1973 | 1,112.2 | | 1,341 | 939 | 0.70 | 0.84 |
| Type - BWRs | 1974 | 842.5 | 54.9 | 1,594 | 1,662 | 1.04 | 1.97 |
| Capacity - (197), 850, 850 MWe | 1975 | 708.1 | 54.6 | 2,310 | 3,423 | 1.48 | 4.83 |
| | 1976 | 1,127.2 | 80.8 | 1,746 | 1,680 | 0.96 | 1.49 |
| | 1977 | 1,132.9 | 77.0 | 1,862 | 1,694 | 0.91 | 1.50 |
| | 1978 | 1,242.2 | 79.5 | 1,946 | 1,529 | 0.79 | 1.23 |
| | 1979 | 1,013.0 | 74.7 | 2,407 | 1,800 | 0.75 | 1.78 |
| | 1980 | 1,074.4 | 55.0 | 2,717 | 2,105 | 0.77 | 1.96 |
| | 1981 | 1,035.7 | 51.5 | 2,331 | 2,802 | 1.20 | 2.71 |
| | 1982 | 1,085.3 | 77.9 | 2,572 | 2,923 | 1.14 | 2.69 |
| | 1983 | 913.6 | 65.6 | 2,854 | 3,582 | 1.26 | 3.92 |
| | 1984 | 789.8 | 55.3 | 2,261 | 1,774 | 0.78 | 2.25 |
| | 1985 | 903.0 | 64.5 | 2,817 | 1,686 | 0.60 | 1.87 |
| | 1986 | 740.5 | 52.6 | 3,111 | 2,668 | 0.86 | 3.60 |
| | 1987 | 933.9 | 74.0 | 2,052 | 1,145 | 0.56 | 1.23 |
| | 1988 | 1,014.7 | 75.8 | 2,414 | 1,409 | 0.58 | 1.39 |
| | 1989 | 1,184.2 | 83.1 | 2,259 | 1,131 | 0.50 | 0.96 |
| | 1990 | 1,107.8 | 76.6 | 2,235 | 1,400 | 0.63 | 1.26 |
| | 1991 | 675.2 | 60.7 | 2,044 | 1,005 | 0.49 | 1.49 |
| | 1992 | 872.4 | 75.4 | 1,812 | 619 | 0.34 | 0.71 |
| | 1993 | 960.1 | 68.5 | 2,751 | 1,655 | 0.60 | 1.72 |
| | 1994 | 690.2 | 51.7 | 2,336 | 833 | 0.36 | 1.21 |
| | 1995 | 643.1 | 49.8 | 2,482 | 875 | 0.35 | 1.36 |
| | 1996 | 612.6 | 47.7 | 1,788 | 456 | 0.26 | 0.74 |
| | 1997 | 1,096.2 | 79.5 | 2,747 | 467 | 0.17 | 0.43 |
| | 1998 | 1,354.7 | 90.6 | 2,311 | 426.918 | 0.18 | 0.32 |
| | 1999 | 1,410.9 | 92.5 | 3,243 | 591.443 | 0.18 | 0.42 |
| | 2000 | 1,506.4 | 97.3 | 2,341 | 261.684 | 0.11 | 0.17 |
| | 2001 | 1,427.4 | 94.5 | 2,769 | 400.702 | 0.14 | 0.28 |
| | 2002 | 1,547.0 | 95.7 | 2,819 | 355.011 | 0.13 | 0.23 |
| | 2003 | 1,555.9 | 93.5 | 2,098 | 356.572 | 0.17 | 0.23 |
| | 2004 | 1,405.5 | 84.8 | 2,044 | 381.054 | 0.19 | 0.27 |
| | 2005 | 1,550.8 | 92.0 | 2,006 | 258.799 | 0.13 | 0.17 |
| | 2006 | 1,649.0 | 96.0 | 2,042 | 289.167 | 0.14 | 0.18 |
| | 2007 | 1,658.8 | 97.0 | 2,310 | 275.697 | 0.12 | 0.17 |
| | 2008 | 1,638.0 | 95.9 | 2,307 | 198.153 | 0.09 | 0.12 |
| | 2009 | 1,628.7 | 95.4 | 1,932 | 231.688 | 0.12 | 0.14 |
| | 2010 | 1,665.9 | 96.3 | 2,152 | 213.825 | 0.10 | 0.13 |
| **DUANE ARNOLD** | 1976 | 305.2 | 78.0 | 350 | 105 | 0.30 | 0.34 |
| Docket 50-331; | 1977 | 353.6 | 78.9 | 538 | 299 | 0.56 | 0.85 |
| DPR-49 | 1978 | 149.2 | 33.2 | 1,112 | 974 | 0.88 | 6.53 |
| 1st commercial operation 2/75 | 1979 | 352.0 | 78.0 | 757 | 275 | 0.36 | 0.78 |
| Type - BWR | 1980 | 339.1 | 73.3 | 1,108 | 671 | 0.61 | 1.98 |
| Capacity - 602 MWe | 1981 | 277.7 | 69.8 | 1,286 | 790 | 0.61 | 2.84 |
| | 1982 | 278.5 | 74.7 | 524 | 229 | 0.44 | 0.82 |

[4] Dresden 1 has been shut down since 1978, and in 1985, it was decided that it would not be put in commercial operation again. Therefore, it is no longer included in the count of operating reactors. Parentheses indicate plant capacity when plant was operational.

| Reporting Organization | Year | Megawatt Years (MW-yr) | Unit Availability Factor | Total Personnel with Measurable Doses | Collective Dose per Site (person-rem) | Average Measurable Dose (rem) | Collective Dose/ MW-yr |
|---|---|---|---|---|---|---|---|
| **DUANE ARNOLD** | 1983 | 283.0 | 62.9 | 1,468 | 1,135 | 0.77 | 4.01 |
| (continued) | 1984 | 329.4 | 72.9 | 611 | 189 | 0.31 | 0.57 |
| | 1985 | 236.2 | 53.8 | 1,414 | 1,112 | 0.79 | 4.71 |
| | 1986 | 365.5 | 82.0 | 476 | 187 | 0.39 | 0.51 |
| | 1987 | 308.4 | 64.7 | 1,094 | 667 | 0.61 | 2.16 |
| | 1988 | 386.5 | 75.2 | 1,136 | 614 | 0.54 | 1.59 |
| | 1989 | 388.5 | 79.0 | 425 | 194 | 0.46 | 0.50 |
| | 1990 | 367.4 | 75.8 | 1,460 | 861 | 0.59 | 2.34 |
| | 1991 | 503.7 | 94.5 | 336 | 202 | 0.60 | 0.40 |
| | 1992 | 416.5 | 81.9 | 1,043 | 502 | 0.48 | 1.21 |
| | 1993 | 393.4 | 79.5 | 1,043 | 407 | 0.39 | 1.03 |
| | 1994 | 498.6 | 94.0 | 493 | 120 | 0.24 | 0.24 |
| | 1995 | 452.5 | 83.8 | 1,129 | 357 | 0.32 | 0.79 |
| | 1996 | 476.8 | 90.7 | 1,093 | 270 | 0.25 | 0.57 |
| | 1997 | 474.4 | 94.4 | 352 | 63 | 0.18 | 0.13 |
| | 1998 | 438.3 | 86.6 | 1,019 | 236.693 | 0.23 | 0.54 |
| | 1999 | 416.6 | 84.3 | 834 | 201.196 | 0.24 | 0.48 |
| | 2000 | 507.3 | 98.4 | 317 | 44.181 | 0.14 | 0.09 |
| | 2001 | 439.5 | 86.8 | 898 | 137.564 | 0.15 | 0.31 |
| | 2002 | 522.0 | 94.4 | 319 | 35.061 | 0.11 | 0.07 |
| | 2003 | 455.2 | 84.8 | 829 | 124.402 | 0.15 | 0.27 |
| | 2004 | 561.2 | 98.3 | 220 | 18.993 | 0.09 | 0.03 |
| | 2005 | 517.4 | 90.5 | 879 | 139.622 | 0.16 | 0.27 |
| | 2006 | 581.7 | 99.0 | 254 | 29.392 | 0.12 | 0.05 |
| | 2007 | 515.8 | 88.0 | 1,062 | 183.609 | 0.17 | 0.36 |
| | 2008 | 601.4 | 100.0 | 276 | 24.187 | 0.09 | 0.04 |
| | 2009 | 534.1 | 91.3 | 960 | 140.206 | 0.15 | 0.26 |
| | 2010 | 508.1 | 86.9 | 1,093 | 200.601 | 0.18 | 0.39 |
| **FARLEY 1, 2** | 1978 | 713.8 | 86.5 | 527 | 108 | 0.20 | 0.15 |
| Docket 50-348, 50-364; | 1979 | 211.0 | 28.6 | 1,227 | 643 | 0.52 | 3.05 |
| NPF-2, NPF-8 | 1980 | 557.3 | 69.3 | 1,330 | 435 | 0.33 | 0.78 |
| 1st commercial operation | 1981 | 310.2 | 41.4 | 1,331 | 512 | 0.38 | 1.65 |
| 12/77, 7/81 | 1982 | 1,271.5 | 79.2 | 1,453 | 484 | 0.33 | 0.38 |
| Type - PWRs | 1983 | 1,356.5 | 83.0 | 1,938 | 1,021 | 0.53 | 0.75 |
| Capacity - 851, 860 MWe | 1984 | 1,447.0 | 86.6 | 2,046 | 902 | 0.44 | 0.62 |
| | 1985 | 1,368.2 | 81.1 | 2,551 | 799 | 0.31 | 0.58 |
| | 1986 | 1,409.4 | 83.8 | 2,314 | 858 | 0.37 | 0.61 |
| | 1987 | 1,369.7 | 84.7 | 1,871 | 598 | 0.32 | 0.44 |
| | 1988 | 1,567.7 | 92.3 | 1,840 | 552 | 0.30 | 0.35 |
| | 1989 | 1,402.9 | 84.6 | 2,206 | 749 | 0.34 | 0.53 |
| | 1990 | 1,464.0 | 86.7 | 1,700 | 457 | 0.27 | 0.31 |
| | 1991 | 1,464.0 | 88.1 | 1,645 | 648 | 0.39 | 0.44 |
| | 1992 | 1,331.7 | 81.8 | 2,018 | 805 | 0.40 | 0.60 |
| | 1993 | 1,455.5 | 88.3 | 1,284 | 333 | 0.26 | 0.23 |
| | 1994 | 1,587.2 | 93.0 | 1,035 | 250 | 0.24 | 0.16 |
| | 1995 | 1,311.2 | 83.8 | 1,574 | 460 | 0.29 | 0.35 |
| | 1996 | 1,549.2 | 90.9 | 1,150 | 232 | 0.20 | 0.15 |
| | 1997 | 1,449.7 | 89.0 | 1,105 | 278 | 0.25 | 0.19 |
| | 1998 | 1,313.9 | 80.9 | 1,380 | 431.821 | 0.31 | 0.33 |
| | 1999 | 1,436.0 | 91.4 | 1,102 | 190.463 | 0.17 | 0.13 |
| | 2000 | 1,430.1 | 88.6 | 1,683 | 359.855 | 0.21 | 0.25 |
| | 2001 | 1,384.3 | 84.4 | 1,810 | 320.509 | 0.18 | 0.23 |
| | 2002 | 1,558.0 | 93.5 | 772 | 96.431 | 0.13 | 0.06 |
| | 2003 | 1,592.6 | 95.3 | 788 | 111.016 | 0.14 | 0.07 |
| | 2004 | 1,496.8 | 89.4 | 1,141 | 107.227 | 0.09 | 0.07 |
| | 2005 | 1,564.2 | 93.3 | 810 | 67.826 | 0.08 | 0.04 |
| | 2006 | 1,602.7 | 94.0 | 747 | 66.189 | 0.09 | 0.04 |
| | 2007 | 1,495.8 | 88.0 | 1,226 | 139.716 | 0.11 | 0.09 |
| | 2008 | 1,602.6 | 94.4 | 669 | 40.833 | 0.06 | 0.03 |
| | 2009 | 1,595.2 | 94.1 | 657 | 41.851 | 0.06 | 0.03 |
| | 2010 | 1,503.4 | 89.0 | 1,321 | 121.313 | 0.09 | 0.08 |

| Reporting Organization | Year | Megawatt Years (MW-yr) | Unit Availability Factor | Total Personnel with Measurable Doses | Collective Dose per Site (person-rem) | Average Measurable Dose (rem) | Collective Dose/ MW-yr |
|---|---|---|---|---|---|---|---|
| **FERMI 2** | 1989 | 624.0 | 68.5 | 1,270 | 255 | 0.20 | 0.41 |
| Docket 50-341; | 1990 | 848.2 | 84.7 | 462 | 83 | 0.18 | 0.10 |
| NPF-43 | 1991 | 739.0 | 77.0 | 1,223 | 228 | 0.19 | 0.31 |
| 1st commercial operation 1/88 | 1992 | 874.3 | 81.3 | 1,213 | 245 | 0.20 | 0.28 |
| Type - BWR | 1993 | 984.3 | 92.9 | 360 | 35 | 0.10 | 0.04 |
| Capacity - 1,058 MWe | 1994 | 0.0 | 2.2 | 1,130 | 213 | 0.19 | --- |
| | 1995 | 618.3 | 86.9 | 390 | 28 | 0.07 | 0.05 |
| | 1996 | 577.5 | 69.1 | 1,402 | 157 | 0.11 | 0.27 |
| | 1997 | 637.0 | 66.6 | 623 | 49 | 0.08 | 0.08 |
| | 1998 | 815.8 | 79.9 | 1,362 | 207.593 | 0.15 | 0.25 |
| | 1999 | 1,082.7 | 99.5 | 461 | 36.152 | 0.08 | 0.03 |
| | 2000 | 939.6 | 87.6 | 1,266 | 145.964 | 0.12 | 0.15 |
| | 2001 | 975.0 | 90.9 | 1,202 | 168.689 | 0.14 | 0.17 |
| | 2002 | 1,059.0 | 98.7 | 463 | 38.235 | 0.08 | 0.04 |
| | 2003 | 925.3 | 86.9 | 1,207 | 168.138 | 0.14 | 0.18 |
| | 2004 | 962.3 | 90.0 | 1,302 | 145.090 | 0.11 | 0.15 |
| | 2005 | 998.1 | 91.7 | 538 | 61.626 | 0.11 | 0.06 |
| | 2006 | 855.9 | 83.0 | 1,430 | 181.300 | 0.13 | 0.21 |
| | 2007 | 950.2 | 87.0 | 1,484 | 194.039 | 0.13 | 0.20 |
| | 2008 | 1,094.5 | 99.5 | 460 | 35.186 | 0.08 | 0.03 |
| | 2009 | 847.8 | 79.3 | 1,497 | 148.846 | 0.10 | 0.18 |
| | 2010 | 885.0 | 86.4 | 1,625 | 146.490 | 0.09 | 0.17 |
| **FITZPATRICK** | 1976 | 489.0 | 71.6 | 600 | 202 | 0.34 | 0.41 |
| Docket 50-333; | 1977 | 460.5 | 68.4 | 1,380 | 1,080 | 0.78 | 2.35 |
| DPR-59 | 1978 | 497.0 | 72.1 | 904 | 909 | 1.01 | 1.83 |
| 1st commercial operation 7/75 | 1979 | 349.0 | 50.8 | 850 | 859 | 1.01 | 2.46 |
| Type - BWR | 1980 | 509.5 | 70.3 | 2,056 | 2,040 | 0.99 | 4.00 |
| Capacity - 813 MWe | 1981 | 562.9 | 74.7 | 2,490 | 1,425 | 0.57 | 2.53 |
| | 1982 | 583.6 | 75.0 | 2,322 | 1,190 | 0.51 | 2.04 |
| | 1983 | 546.2 | 70.6 | 1,715 | 1,090 | 0.64 | 2.00 |
| | 1984 | 576.2 | 76.8 | 1,610 | 971 | 0.60 | 1.69 |
| | 1985 | 492.3 | 63.7 | 1,845 | 1,051 | 0.57 | 2.13 |
| | 1986 | 711.2 | 90.6 | 1,185 | 411 | 0.35 | 0.58 |
| | 1987 | 496.2 | 70.3 | 1,578 | 940 | 0.60 | 1.89 |
| | 1988 | 514.0 | 69.0 | 1,553 | 786 | 0.51 | 1.53 |
| | 1989 | 727.5 | 92.3 | 1,027 | 377 | 0.37 | 0.52 |
| | 1990 | 543.8 | 72.6 | 1,536 | 884 | 0.58 | 1.63 |
| | 1991 | 399.7 | 53.4 | 1,269 | 333 | 0.26 | 0.83 |
| | 1992 | 0.0 | 0.0 | 2,374 | 674 | 0.28 | --- |
| | 1993 | 559.6 | 81.7 | 1,427 | 232 | 0.16 | 0.41 |
| | 1994 | 588.4 | 83.2 | 1,595 | 322 | 0.20 | 0.55 |
| | 1995 | 569.8 | 74.5 | 1,249 | 327 | 0.26 | 0.57 |
| | 1996 | 623.3 | 83.1 | 1,384 | 357 | 0.26 | 0.57 |
| | 1997 | 756.2 | 95.9 | 662 | 91 | 0.14 | 0.12 |
| | 1998 | 562.8 | 78.0 | 1,781 | 357.826 | 0.20 | 0.64 |
| | 1999 | 749.7 | 95.5 | 558 | 68.409 | 0.12 | 0.09 |
| | 2000 | 685.9 | 88.4 | 1,267 | 300.997 | 0.24 | 0.44 |
| | 2001 | 807.2 | 98.9 | 665 | 63.229 | 0.10 | 0.08 |
| | 2002 | 751.0 | 93.3 | 1,234 | 230.523 | 0.19 | 0.31 |
| | 2003 | 793.0 | 97.9 | 298 | 51.156 | 0.17 | 0.06 |
| | 2004 | 735.0 | 92.1 | 1,091 | 186.055 | 0.17 | 0.25 |
| | 2005 | 802.9 | 96.3 | 382 | 62.697 | 0.16 | 0.08 |
| | 2006 | 771.5 | 93.0 | 1,527 | 234.425 | 0.15 | 0.30 |
| | 2007 | 790.1 | 96.0 | 526 | 58.741 | 0.11 | 0.07 |
| | 2008 | 761.7 | 92.9 | 1,430 | 184.772 | 0.13 | 0.24 |
| | 2009 | 844.5 | 100.0 | 487 | 35.119 | 0.07 | 0.04 |
| | 2010 | 726.2 | 91.3 | 1,429 | 219.887 | 0.15 | 0.30 |

| Reporting Organization | Year | Megawatt Years (MW-yr) | Unit Availability Factor | Total Personnel with Measurable Doses | Collective Dose per Site (person-rem) | Average Measurable Dose (rem) | Collective Dose/ MW-yr |
|---|---|---|---|---|---|---|---|
| **FORT CALHOUN** | 1975 | 252.3 | 67.4 | 469 | 294 | 0.63 | 1.17 |
| Docket 50-285; | 1976 | 265.9 | 69.5 | 516 | 313 | 0.61 | 1.18 |
| DPR-40 | 1977 | 351.8 | 79.4 | 535 | 297 | 0.56 | 0.84 |
| 1st commercial operation 6/74 | 1978 | 342.3 | 75.1 | 596 | 410 | 0.69 | 1.20 |
| Type - PWR | 1979 | 440.0 | 95.7 | 451 | 126 | 0.28 | 0.29 |
| Capacity - 482 MWe | 1980 | 242.3 | 60.4 | 891 | 668 | 0.75 | 2.76 |
| | 1981 | 260.9 | 72.3 | 822 | 458 | 0.56 | 1.76 |
| | 1982 | 418.0 | 89.7 | 604 | 217 | 0.36 | 0.52 |
| | 1983 | 330.4 | 73.1 | 860 | 433 | 0.50 | 1.31 |
| | 1984 | 279.2 | 59.9 | 913 | 563 | 0.62 | 2.02 |
| | 1985 | 367.0 | 73.7 | 982 | 373 | 0.38 | 1.02 |
| | 1986 | 431.8 | 94.3 | 756 | 75 | 0.10 | 0.17 |
| | 1987 | 366.0 | 75.4 | 1,247 | 388 | 0.31 | 1.06 |
| | 1988 | 315.5 | 74.1 | 1,594 | 272 | 0.17 | 0.86 |
| | 1989 | 395.7 | 89.2 | 1,210 | 93 | 0.08 | 0.24 |
| | 1990 | 290.0 | 64.2 | 760 | 290 | 0.38 | 1.00 |
| | 1991 | 391.1 | 91.7 | 284 | 57 | 0.20 | 0.15 |
| | 1992 | 303.4 | 65.9 | 802 | 272 | 0.34 | 0.90 |
| | 1993 | 369.7 | 80.8 | 713 | 157 | 0.22 | 0.42 |
| | 1994 | 492.8 | 99.6 | 211 | 23 | 0.11 | 0.05 |
| | 1995 | 402.8 | 83.2 | 627 | 139 | 0.22 | 0.35 |
| | 1996 | 374.9 | 79.5 | 740 | 226 | 0.31 | 0.60 |
| | 1997 | 435.9 | 93.6 | 258 | 41 | 0.16 | 0.09 |
| | 1998 | 387.7 | 82.5 | 788 | 223.847 | 0.28 | 0.58 |
| | 1999 | 409.2 | 89.2 | 676 | 158.843 | 0.24 | 0.39 |
| | 2000 | 443.8 | 93.5 | 249 | 35.215 | 0.14 | 0.08 |
| | 2001 | 401.2 | 88.3 | 770 | 225.891 | 0.29 | 0.56 |
| | 2002 | 434.0 | 92.3 | 742 | 163.806 | 0.22 | 0.38 |
| | 2003 | 399.6 | 87.0 | 914 | 212.422 | 0.23 | 0.53 |
| | 2004 | 463.5 | 97.0 | 215 | 21.574 | 0.10 | 0.05 |
| | 2005 | 332.4 | 72.2 | 1,069 | 272.876 | 0.26 | 0.82 |
| | 2006 | 353.9 | 75.0 | 1,591 | 289.100 | 0.18 | 0.82 |
| | 2007 | 499.9 | 100.0 | 100 | 3.990 | 0.04 | 0.01 |
| | 2008 | 400.4 | 82.2 | 839 | 96.155 | 0.11 | 0.24 |
| | 2009 | 422.7 | 87.0 | 870 | 110.918 | 0.13 | 0.26 |
| | 2010 | 486.5 | 98.5 | 171 | 9.763 | 0.06 | 0.02 |
| **GINNA** | 1971 | 327.8 | | 340 | 430 | 1.26 | 1.31 |
| Docket 50-244; | 1972 | 293.6 | | 677 | 1,032 | 1.52 | 3.51 |
| DPR-18 | 1973 | 409.5 | | 319 | 224 | 0.70 | 0.55 |
| 1st commercial operation 7/70 | 1974 | 253.7 | 62.4 | 884 | 1,225 | 1.39 | 4.83 |
| Type - PWR | 1975 | 365.2 | 76.7 | 685 | 538 | 0.79 | 1.47 |
| Capacity - 560 MWe | 1976 | 248.8 | 58.2 | 758 | 636 | 0.84 | 2.56 |
| | 1977 | 365.6 | 85.5 | 530 | 401 | 0.76 | 1.10 |
| | 1978 | 386.5 | 80.6 | 657 | 450 | 0.68 | 1.16 |
| | 1979 | 355.0 | 72.8 | 878 | 592 | 0.67 | 1.67 |
| | 1980 | 370.5 | 76.0 | 1,073 | 708 | 0.66 | 1.91 |
| | 1981 | 399.0 | 82.1 | 925 | 655 | 0.71 | 1.64 |
| | 1982 | 289.0 | 58.8 | 1,117 | 1,140 | 1.02 | 3.94 |
| | 1983 | 365.0 | 74.6 | 969 | 855 | 0.88 | 2.34 |
| | 1984 | 378.1 | 77.2 | 713 | 395 | 0.55 | 1.04 |
| | 1985 | 436.7 | 87.9 | 845 | 426 | 0.50 | 0.98 |
| | 1986 | 433.3 | 87.4 | 901 | 357 | 0.40 | 0.82 |
| | 1987 | 459.0 | 91.5 | 773 | 344 | 0.45 | 0.75 |
| | 1988 | 423.1 | 87.4 | 897 | 295 | 0.33 | 0.70 |
| | 1989 | 369.2 | 75.9 | 1,254 | 605 | 0.48 | 1.64 |
| | 1990 | 414.3 | 84.4 | 991 | 347 | 0.35 | 0.84 |
| | 1991 | 418.6 | 86.7 | 947 | 328 | 0.35 | 0.78 |
| | 1992 | 417.6 | 86.9 | 832 | 261 | 0.31 | 0.63 |
| | 1993 | 419.6 | 86.3 | 856 | 193 | 0.23 | 0.46 |
| | 1994 | 405.3 | 83.2 | 679 | 138 | 0.20 | 0.34 |
| | 1995 | 437.0 | 89.6 | 738 | 136 | 0.18 | 0.31 |
| | 1996 | 347.9 | 71.1 | 976 | 168 | 0.17 | 0.48 |

| Reporting Organization | Year | Megawatt Years (MW-yr) | Unit Availability Factor | Total Personnel with Measurable Doses | Collective Dose per Site (person-rem) | Average Measurable Dose (rem) | Collective Dose/ MW-yr |
|---|---|---|---|---|---|---|---|
| **GINNA** | 1997 | 444.6 | 91.8 | 533 | 81 | 0.15 | 0.18 |
| (continued) | 1998 | 491.8 | 100.0 | 161 | 14.892 | 0.09 | 0.03 |
| | 1999 | 403.4 | 85.6 | 641 | 175.173 | 0.27 | 0.43 |
| | 2000 | 434.2 | 91.6 | 429 | 76.435 | 0.18 | 0.18 |
| | 2001 | 488.0 | 100.0 | 140 | 10.156 | 0.07 | 0.02 |
| | 2002 | 438.0 | 91.3 | 535 | 80.432 | 0.15 | 0.18 |
| | 2003 | 440.4 | 91.1 | 510 | 74.533 | 0.15 | 0.17 |
| | 2004 | 490.5 | 99.5 | 111 | 7.486 | 0.07 | 0.02 |
| | 2005 | 455.0 | 93.9 | 564 | 72.841 | 0.13 | 0.16 |
| | 2006 | 470.2 | 94.0 | 514 | 44.580 | 0.09 | 0.09 |
| | 2007 | 564.4 | 99.0 | 111 | 4.412 | 0.04 | 0.01 |
| | 2008 | 540.1 | 94.5 | 976 | 101.996 | 0.10 | 0.19 |
| | 2009 | 529.2 | 94.3 | 633 | 41.809 | 0.07 | 0.08 |
| | 2010 | 564.9 | 98.9 | 75 | 3.168 | 0.04 | 0.01 |
| **GRAND GULF** | 1986 | 494.7 | 60.9 | 1,486 | 436 | 0.29 | 0.88 |
| Docket 50-416; | 1987 | 920.7 | 82.2 | 1,358 | 420 | 0.31 | 0.46 |
| NPF-29 | 1988 | 1,136.6 | 96.7 | 692 | 147 | 0.21 | 0.13 |
| 1st commercial operation 7/85 | 1989 | 932.6 | 80.0 | 1,972 | 498 | 0.25 | 0.53 |
| Type - BWR | 1990 | 883.5 | 78.9 | 1,765 | 482 | 0.27 | 0.55 |
| Capacity - 1,266 MWe | 1991 | 1,085.2 | 94.0 | 699 | 94 | 0.13 | 0.09 |
| | 1992 | 969.0 | 83.7 | 2,032 | 484 | 0.24 | 0.50 |
| | 1993 | 936.4 | 81.5 | 1,807 | 332 | 0.18 | 0.35 |
| | 1994 | 1,143.2 | 96.6 | 455 | 56 | 0.12 | 0.05 |
| | 1995 | 952.9 | 80.4 | 1,589 | 342 | 0.22 | 0.36 |
| | 1996 | 1,096.2 | 88.7 | 1,564 | 357 | 0.23 | 0.33 |
| | 1997 | 1,234.9 | 100.0 | 514 | 105 | 0.20 | 0.09 |
| | 1998 | 1,049.2 | 88.9 | 1,410 | 303.695 | 0.22 | 0.29 |
| | 1999 | 962.1 | 81.3 | 1,180 | 226.277 | 0.19 | 0.23 |
| | 2000 | 1,217.5 | 99.4 | 289 | 34.877 | 0.12 | 0.03 |
| | 2001 | 1,129.8 | 93.0 | 1,109 | 185.214 | 0.17 | 0.16 |
| | 2002 | 1,145.0 | 93.6 | 1,060 | 176.396 | 0.17 | 0.15 |
| | 2003 | 1,241.2 | 98.6 | 290 | 31.250 | 0.11 | 0.03 |
| | 2004 | 1,165.2 | 92.2 | 1,243 | 158.112 | 0.13 | 0.14 |
| | 2005 | 1,147.3 | 91.9 | 1,326 | 167.914 | 0.13 | 0.15 |
| | 2006 | 1,233.7 | 98.0 | 1,016 | 59.935 | 0.06 | 0.05 |
| | 2007 | 1,070.5 | 88.0 | 1,750 | 177.884 | 0.10 | 0.17 |
| | 2008 | 1,072.1 | 89.5 | 1,843 | 167.859 | 0.09 | 0.16 |
| | 2009 | 1,255.5 | 100.0 | 521 | 30.721 | 0.06 | 0.02 |
| | 2010 | 1,102.0 | 91.5 | 1,822 | 188.370 | 0.10 | 0.17 |
| **HADDAM NECK**[5] | 1969 | 438.5 | | 138 | 106 | 0.77 | 0.24 |
| Docket 50-213; | 1970 | 424.7 | | 734 | 689 | 0.94 | 1.62 |
| DPR-61 | 1971 | 502.2 | | 289 | 342 | 1.18 | 0.68 |
| 1st commercial operation 1/68 | 1972 | 515.6 | | 355 | 325 | 0.91 | 0.63 |
| Type - PWR | 1973 | 293.1 | | 951 | 697 | 0.73 | 2.38 |
| Capacity - (560) MWe | 1974 | 521.4 | 91.2 | 550 | 201 | 0.37 | 0.39 |
| | 1975 | 494.3 | 89.9 | 795 | 703 | 0.88 | 1.42 |
| | 1976 | 482.9 | 82.5 | 644 | 449 | 0.70 | 0.93 |
| | 1977 | 480.7 | 83.9 | 894 | 641 | 0.72 | 1.33 |
| | 1978 | 563.4 | 98.6 | 216 | 117 | 0.54 | 0.21 |
| | 1979 | 493.0 | 87.5 | 1,226 | 1,162 | 0.95 | 2.36 |
| | 1980 | 426.8 | 75.0 | 1,860 | 1,353 | 0.73 | 3.17 |
| | 1981 | 487.5 | 84.3 | 1,554 | 1,036 | 0.67 | 2.13 |
| | 1982 | 543.9 | 93.4 | 559 | 126 | 0.23 | 0.23 |
| | 1983 | 453.7 | 77.8 | 1,645 | 1,384 | 0.84 | 3.05 |
| | 1984 | 404.0 | 71.7 | 1,430 | 1,216 | 0.85 | 3.01 |
| | 1985 | 556.1 | 98.4 | 384 | 101 | 0.26 | 0.18 |
| | 1986 | 294.8 | 53.6 | 1,945 | 1,567 | 0.81 | 5.32 |
| | 1987 | 304.6 | 54.0 | 1,763 | 750 | 0.43 | 2.46 |
| | 1988 | 397.4 | 70.3 | 735 | 237 | 0.32 | 0.60 |

[5] Haddam Neck (also known as Connecticut Yankee) was shut down on December 4, 1996, and is no longer in the count of operating reactors. Parentheses indicate plant capacity when plant was operational.

| Reporting Organization | Year | Megawatt Years (MW-yr) | Unit Availability Factor | Total Personnel with Measurable Doses | Collective Dose per Site (person-rem) | Average Measurable Dose (rem) | Collective Dose/ MW-yr |
|---|---|---|---|---|---|---|---|
| **HADDAM NECK**[5] | 1989 | 356.4 | 67.2 | 1,455 | 596 | 0.41 | 1.67 |
| (continued) | 1990 | 142.7 | 32.2 | 979 | 421 | 0.43 | 2.95 |
| | 1991 | 444.4 | 76.4 | 1,168 | 590 | 0.51 | 1.33 |
| | 1992 | 465.2 | 80.1 | 797 | 202 | 0.25 | 0.43 |
| | 1993 | 448.6 | 81.6 | 1,004 | 408 | 0.41 | 0.91 |
| | 1994 | 455.6 | 77.7 | 463 | 135 | 0.29 | 0.30 |
| | 1995 | 439.4 | 77.7 | 1,006 | 442 | 0.44 | 1.01 |
| | 1996 | 331.8 | 55.7 | 673 | 175 | 0.26 | 0.53 |
| | 1997 | -1.3 | 0.0 | 219 | 11 | 0.05 | — |
| | 1998 | 0.0 | 0.0 | 423 | 93.743 | 0.22 | — |
| | 1999 | 0.0 | 0.0 | 545 | 108.602 | 0.20 | — |
| | 2000 | 0.0 | 0.0 | 555 | 262.192 | 0.47 | — |
| | 2001 | 0.0 | 0.0 | 361 | 95.348 | 0.26 | — |
| | 2002 | 0.0 | 0.0 | 258 | 51.668 | 0.20 | — |
| | 2003 | 0.0 | 0.0 | 400 | 82.022 | 0.21 | — |
| | 2004 | 0.0 | 0.0 | 564 | 91.981 | 0.16 | — |
| | 2005 | 0.0 | 0.0 | 350 | 36.479 | 0.10 | — |
| | 2006 | 0.0 | 0.0 | 124 | 11.883 | 0.10 | — |
| | 2007 | 0.0 | 0.0 | 0 | 0.000 | — | — |
| | 2008 | 0.0 | 0.0 | 1 | 0.011 | 0.01 | — |
| | 2009 | 0.0 | 0.0 | 1 | 0.010 | 0.01 | — |
| | 2010 | 0.0 | 0.0 | 2 | 0.024 | 0.01 | — |
| **HARRIS 1** | 1988 | 652.9 | 75.0 | 721 | 169 | 0.23 | 0.26 |
| Docket 50-400; | 1989 | 690.6 | 79.5 | 929 | 156 | 0.17 | 0.23 |
| NPF-63 | 1990 | 776.4 | 89.6 | 453 | 85 | 0.19 | 0.11 |
| 1st commercial operation 5/87 | 1991 | 724.8 | 81.5 | 872 | 226 | 0.26 | 0.31 |
| Type - PWR | 1992 | 661.8 | 74.9 | 930 | 213 | 0.23 | 0.32 |
| Capacity - 900 MWe | 1993 | 913.0 | 99.7 | 327 | 31 | 0.09 | 0.03 |
| | 1994 | 740.8 | 82.7 | 1,089 | 222 | 0.20 | 0.30 |
| | 1995 | 731.1 | 83.8 | 1,068 | 174 | 0.16 | 0.24 |
| | 1996 | 860.6 | 95.4 | 444 | 17 | 0.04 | 0.02 |
| | 1997 | 673.6 | 80.4 | 1,131 | 149 | 0.13 | 0.22 |
| | 1998 | 766.2 | 90.4 | 931 | 133.497 | 0.14 | 0.17 |
| | 1999 | 827.0 | 97.9 | 247 | 15.538 | 0.06 | 0.02 |
| | 2000 | 783.0 | 92.5 | 888 | 100.981 | 0.11 | 0.13 |
| | 2001 | 611.2 | 72.4 | 1,586 | 252.241 | 0.16 | 0.41 |
| | 2002 | 892.0 | 99.4 | 145 | 6.674 | 0.05 | 0.01 |
| | 2003 | 823.9 | 93.2 | 786 | 68.463 | 0.09 | 0.08 |
| | 2004 | 797.9 | 88.2 | 747 | 57.103 | 0.08 | 0.07 |
| | 2005 | 902.9 | 99.5 | 164 | 8.483 | 0.05 | 0.01 |
| | 2006 | 802.4 | 89.0 | 917 | 87.225 | 0.10 | 0.11 |
| | 2007 | 845.1 | 94.0 | 870 | 64.808 | 0.07 | 0.08 |
| | 2008 | 890.4 | 97.4 | 192 | 10.356 | 0.05 | 0.01 |
| | 2009 | 845.1 | 92.7 | 742 | 41.401 | 0.06 | 0.05 |
| | 2010 | 808.3 | 89.0 | 1,069 | 82.578 | 0.08 | 0.10 |
| **HATCH 1, 2** | 1976 | 496.3 | 83.8 | 630 | 134 | 0.21 | 0.27 |
| Docket 50-321, 50-366; | 1977 | 446.8 | 66.3 | 1,303 | 465 | 0.36 | 1.04 |
| DPR-57; NPF-5 | 1978 | 513.0 | 72.8 | 1,304 | 248 | 0.19 | 0.48 |
| 1st commercial operation | 1979 | 401.0 | 54.6 | 2,131 | 582 | 0.27 | 1.45 |
| 12/75, 9/79 | 1980 | 1,008.7 | 70.9 | 1,930 | 449 | 0.23 | 0.45 |
| Type - BWRs | 1981 | 870.9 | 64.3 | 2,899 | 1,337 | 0.46 | 1.54 |
| Capacity - 876, 883 MWe | 1982 | 768.0 | 56.6 | 3,418 | 1,460 | 0.43 | 1.90 |
| | 1983 | 934.7 | 68.6 | 3,428 | 1,299 | 0.38 | 1.39 |
| | 1984 | 658.6 | 47.3 | 4,110 | 2,218 | 0.54 | 3.37 |
| | 1985 | 1,211.0 | 79.6 | 2,841 | 818 | 0.29 | 0.68 |
| | 1986 | 872.0 | 64.8 | 3,486 | 1,497 | 0.43 | 1.72 |
| | 1987 | 1,295.4 | 89.7 | 2,202 | 816 | 0.37 | 0.63 |
| | 1988 | 1,001.4 | 70.4 | 2,509 | 1,401 | 0.56 | 1.40 |
| | 1989 | 1,271.1 | 87.1 | 1,350 | 556 | 0.41 | 0.44 |
| | 1990 | 1,268.0 | 83.5 | 2,902 | 1,455 | 0.50 | 1.15 |
| | 1991 | 1,152.4 | 77.4 | 2,508 | 1,161 | 0.46 | 1.01 |

[5] Haddam Neck (also known as Connecticut Yankee) was shut down on December 4, 1996, and is no longer in the count of operating reactors. Parentheses indicate plant capacity when plant was operational.

| Reporting Organization | Year | Megawatt Years (MW-yr) | Unit Availability Factor | Total Personnel with Measurable Doses | Collective Dose per Site (person-rem) | Average Measurable Dose (rem) | Collective Dose/ MW-yr |
|---|---|---|---|---|---|---|---|
| **HATCH 1, 2** | 1992 | 1,293.8 | 88.6 | 1,615 | 550 | 0.34 | 0.43 |
| (continued) | 1993 | 1,189.6 | 85.5 | 1,733 | 669 | 0.39 | 0.56 |
| | 1994 | 1,289.0 | 87.1 | 2,243 | 864 | 0.39 | 0.67 |
| | 1995 | 1,376.3 | 90.6 | 1,458 | 488 | 0.33 | 0.35 |
| | 1996 | 1,519.6 | 94.0 | 1,495 | 441 | 0.29 | 0.29 |
| | 1997 | 1,374.7 | 88.1 | 1,945 | 722 | 0.37 | 0.53 |
| | 1998 | 1,458.4 | 91.7 | 1,610 | 320.469 | 0.20 | 0.22 |
| | 1999 | 1,487.4 | 90.0 | 1,866 | 328.583 | 0.18 | 0.22 |
| | 2000 | 1,515.0 | 88.7 | 1,913 | 401.891 | 0.21 | 0.26 |
| | 2001 | 1,603.0 | 93.5 | 1,407 | 230.242 | 0.16 | 0.14 |
| | 2002 | 1,600.0 | 94.0 | 1,299 | 214.441 | 0.17 | 0.13 |
| | 2003 | 1,606.3 | 94.5 | 1,295 | 168.281 | 0.13 | 0.10 |
| | 2004 | 1,641.3 | 95.3 | 1,209 | 180.129 | 0.15 | 0.11 |
| | 2005 | 1,562.1 | 91.3 | 1,288 | 207.295 | 0.16 | 0.13 |
| | 2006 | 1,604.9 | 94.0 | 1,405 | 259.313 | 0.18 | 0.16 |
| | 2007 | 1,626.5 | 94.0 | 1,341 | 137.273 | 0.10 | 0.08 |
| | 2008 | 1,584.0 | 92.7 | 1,397 | 189.433 | 0.14 | 0.12 |
| | 2009 | 1,416.5 | 83.2 | 1,310 | 186.013 | 0.14 | 0.13 |
| | 2010 | 1,586.9 | 93.0 | 1,734 | 245.797 | 0.14 | 0.15 |
| **HOPE CREEK 1** | 1987 | 869.2 | 86.4 | 589 | 117 | 0.20 | 0.13 |
| Docket 50-354; | 1988 | 832.7 | 80.7 | 1,734 | 287 | 0.17 | 0.34 |
| NPF-57 | 1989 | 791.1 | 77.8 | 1,873 | 465 | 0.25 | 0.59 |
| 1st commercial operation 12/86 | 1990 | 966.4 | 91.6 | 1,394 | 196 | 0.14 | 0.20 |
| Type - BWR | 1991 | 882.5 | 84.2 | 1,700 | 373 | 0.22 | 0.42 |
| Capacity - 1,172 MWe | 1992 | 841.9 | 80.8 | 1,694 | 436 | 0.26 | 0.52 |
| | 1993 | 1,049.2 | 97.8 | 688 | 98 | 0.14 | 0.09 |
| | 1994 | 852.0 | 81.2 | 1,779 | 326 | 0.18 | 0.38 |
| | 1995 | 844.5 | 79.8 | 1,571 | 196 | 0.12 | 0.23 |
| | 1996 | 806.9 | 77.4 | 1,069 | 158 | 0.15 | 0.20 |
| | 1997 | 731.8 | 77.8 | 1,747 | 350 | 0.20 | 0.48 |
| | 1998 | 993.2 | 98.0 | 620 | 54.816 | 0.09 | 0.06 |
| | 1999 | 879.1 | 86.7 | 1,111 | 279.063 | 0.25 | 0.32 |
| | 2000 | 827.8 | 87.9 | 1,236 | 188.295 | 0.15 | 0.23 |
| | 2001 | 918.2 | 91.1 | 1,532 | 156.180 | 0.10 | 0.17 |
| | 2002 | 1,007.0 | 99.2 | 220 | 25.922 | 0.12 | 0.03 |
| | 2003 | 826.6 | 84.6 | 1,597 | 139.295 | 0.09 | 0.17 |
| | 2004 | 688.6 | 71.3 | 2,440 | 239.540 | 0.10 | 0.35 |
| | 2005 | 874.9 | 88.6 | 881 | 67.063 | 0.08 | 0.08 |
| | 2006 | 983.8 | 93.0 | 2,135 | 133.570 | 0.06 | 0.14 |
| | 2007 | 929.3 | 91.0 | 2,221 | 191.068 | 0.09 | 0.21 |
| | 2008 | 1,139.1 | 100.0 | 999 | 34.510 | 0.03 | 0.03 |
| | 2009 | 1,111.4 | 93.3 | 2,090 | 169.362 | 0.08 | 0.15 |
| | 2010 | 1,082.0 | 92.1 | 1,985 | 160.910 | 0.08 | 0.15 |
| **HUMBOLDT BAY**[6] | 1969 | 44.6 | | 125 | 164 | 1.31 | 3.68 |
| Docket 50-133; | 1970 | 49.3 | | 115 | 209 | 1.82 | 4.24 |
| DPR-7 | 1971 | 39.6 | | 140 | 292 | 2.09 | 7.37 |
| 1st commercial operation 8/63 | 1972 | 43.1 | | 127 | 253 | 1.99 | 5.87 |
| Type - BWR | 1973 | 50.1 | | 210 | 266 | 1.27 | 5.31 |
| Capacity - (63) MWe | 1974 | 43.4 | 83.8 | 296 | 318 | 1.07 | 7.33 |
| | 1975 | 45.3 | 83.9 | 265 | 339 | 1.28 | 7.48 |
| | 1976 | 23.5 | 46.4 | 523 | 683 | 1.31 | 29.06 |
| | 1977 | 0.0 | 0.0 | 1,063 | 1,905 | 1.79 | --- |
| | 1978 | 0.0 | 0.0 | 320 | 335 | 1.05 | --- |
| | 1979 | 0.0 | 0.0 | 135 | 31 | 0.23 | --- |
| | 1980 | 0.0 | 0.0 | 142 | 22 | 0.15 | --- |
| | 1981 | 0.0 | 0.0 | 75 | 9 | 0.12 | --- |
| | 1982 | 0.0 | 0.0 | 71 | 19 | 0.27 | --- |
| | 1983 | 0.0 | 0.0 | 84 | 17 | 0.20 | --- |

[6] Humboldt Bay had been shut down since 1976, and, in 1984, it was decided that it would not be placed in operation again. Therefore, it is no longer included in the count of operating reactors. Parentheses indicate plant capacity when plant was operational.

| Reporting Organization | Year | Megawatt Years (MW-yr) | Unit Availability Factor | Total Personnel with Measurable Doses | Collective Dose per Site (person-rem) | Average Measurable Dose (rem) | Collective Dose/ MW-yr |
|---|---|---|---|---|---|---|---|
| **HUMBOLDT BAY**[6] | 1984 | "Data not available" | | | | | |
| (continued) | 1985 | 0.0 | 0.0 | 178 | 51 | 0.29 | --- |
| | 1986 | 0.0 | 0.0 | 115 | 50 | 0.43 | --- |
| | 1987 | "Data not available" | | | | | |
| | 1988 | 0.0 | 0.0 | 10 | 1 | 0.10 | --- |
| | 1989 | 0.0 | 0.0 | 0 | 0 | 0.00 | --- |
| | 1990 | 0.0 | 0.0 | 0 | 0 | 0.00 | --- |
| | 1991 | 0.0 | 0.0 | 0 | 0 | 0.00 | --- |
| | 1992 | 0.0 | 0.0 | 8 | 0 | 0.00 | --- |
| | 1993 | 0.0 | 0.0 | 24 | 1 | 0.04 | --- |
| | 1994 | 0.0 | 0.0 | 21 | 1 | 0.05 | --- |
| | 1995 | 0.0 | 0.0 | 42 | 2 | 0.05 | --- |
| | 1996 | 0.0 | 0.0 | 66 | 5 | 0.08 | --- |
| | 1997 | 0.0 | 0.0 | 105 | 16 | 0.15 | --- |
| | 1998 | 0.0 | 0.0 | 38 | 0.929 | 0.02 | --- |
| | 1999 | 0.0 | 0.0 | 28 | 0.720 | 0.03 | --- |
| | 2000 | 0.0 | 0.0 | 20 | 0.911 | 0.05 | --- |
| | 2001 | 0.0 | 0.0 | 10 | 0.360 | 0.04 | --- |
| | 2002 | 0.0 | 0.0 | 18 | 1.504 | 0.08 | --- |
| | 2003 | 0.0 | 0.0 | 14 | 0.351 | 0.03 | --- |
| | 2004 | 0.0 | 0.0 | 11 | 0.454 | 0.04 | --- |
| | 2005 | 0.0 | 0.0 | 11 | 0.547 | 0.05 | --- |
| | 2006 | 0.0 | 0.0 | 40 | 4.086 | 0.10 | --- |
| | 2007 | 0.0 | 0.0 | 45 | 3.271 | 0.07 | --- |
| | 2008 | 0.0 | 0.0 | 56 | 2.051 | 0.04 | --- |
| | 2009 | 0.0 | 0.0 | 30 | 0.631 | 0.02 | --- |
| | 2010 | 0.0 | 0.0 | 136 | 7.691 | 0.06 | --- |
| **INDIAN POINT 1**[7]**, 2, 3**[8] | 1969 | 206.2 | | | 298 | | 1.45 |
| Docket 50-3, 50-247, 50-286; | 1970 | 43.3 | | | 1,639 | | 37.85 |
| DPR-5, DPR-26, DPR-64 | 1971 | 154.0 | | | 768 | | 4.99 |
| 1st commercial operation | 1972 | 142.3 | | | 967 | | 6.80 |
| 10/62, 8/74, 8/76 | 1973 | 0.0 | | 2,998 | 5,262 | 1.76 | --- |
| Type - PWRs | 1974 | 556.1 | 59.4 | 1,019 | 910 | 0.89 | 1.64 |
| Capacity - (265), 998, 1,030 MWe | 1975 | 584.4 | 74.8 | 891 | 705 | 0.79 | 1.21 |
| | 1976 | 273.9 | 34.8 | 1,590 | 1,950 | 1.23 | 7.12 |
| | 1977 | 1,278.3 | 75.3 | 1,391 | 1,070 | 0.77 | 0.84 |
| | 1978 | 1,172.3 | 67.8 | 1,909 | 2,006 | 1.05 | 1.71 |
| **INDIAN POINT 1**[7]**, 2** | 1979 | 574.0 | 71.4 | 1,349 | 1,279 | 0.95 | 2.23 |
| Docket 50-3, 50-247; | 1980 | 510.8 | 64.8 | 1,577 | 971 | 0.62 | 1.90 |
| DPR-5, DPR-26 | 1981 | 367.5 | 46.0 | 2,595 | 2,731 | 1.05 | 7.43 |
| 1st commercial operation | 1982 | 532.4 | 65.4 | 2,144 | 1,635 | 0.76 | 3.07 |
| 10/62, 8/74 | 1983 | 702.6 | 84.0 | 1,057 | 486 | 0.46 | 0.69 |
| Type - PWRs | 1984 | 416.7 | 51.9 | 2,919 | 2,644 | 0.91 | 6.35 |
| Capacity - (265), 998 MWe | 1985 | 791.4 | 95.7 | 708 | 192 | 0.27 | 0.24 |
| | 1986 | 457.5 | 56.2 | 1,926 | 1,250 | 0.65 | 2.73 |
| | 1987 | 611.4 | 73.4 | 1,980 | 1,217 | 0.61 | 1.99 |
| | 1988 | 719.3 | 86.9 | 890 | 235 | 0.26 | 0.33 |
| | 1989 | 532.5 | 64.6 | 2,093 | 1,436 | 0.69 | 2.70 |
| | 1990 | 618.0 | 66.6 | 1,061 | 608 | 0.57 | 0.98 |
| | 1991 | 461.2 | 55.7 | 1,810 | 1,468 | 0.81 | 3.18 |
| | 1992 | 930.9 | 99.1 | 489 | 97 | 0.20 | 0.10 |
| | 1993 | 702.1 | 75.7 | 1,514 | 675 | 0.45 | 0.96 |
| | 1994 | 903.8 | 100.0 | 381 | 48 | 0.13 | 0.05 |
| | 1995 | 582.4 | 70.8 | 1,690 | 548 | 0.32 | 0.94 |

[6] Humboldt Bay had been shut down since 1976, and, in 1984, it was decided that it would not be placed in operation again. Therefore, it is no longer included in the count of operating reactors. Parentheses indicate plant capacity when plant was operational.

[7] Indian Point 1 was defueled in 1975, and in 1984, it was decided that it would not be placed in operation again. Therefore, it is no longer included in the count of operating reactors. Parentheses indicate plant capacity when plant was operational.

[8] Indian Point 3 was purchased by a different utility in 1979 and, subsequently, reported its dose separately. Indian Point 1, 2, and 3 have been owned by the same utility since 2001 and report together.

| Reporting Organization | Year | Megawatt Years (MW-yr) | Unit Availability Factor | Total Personnel with Measurable Doses | Collective Dose per Site (person-rem) | Average Measurable Dose (rem) | Collective Dose/ MW-yr |
|---|---|---|---|---|---|---|---|
| **INDIAN POINT 1[7], 2** | 1996 | 927.8 | 94.8 | 388 | 54 | 0.14 | 0.06 |
| (continued) | 1997 | 360.6 | 45.1 | 1,340 | 367 | 0.27 | 1.02 |
| | 1998 | 282.8 | 31.5 | 1,154 | 289.600 | 0.25 | 1.03 |
| | 1999 | 831.8 | 88.2 | 350 | 40.931 | 0.12 | 0.05 |
| | 2000 | 115.4 | 13.0 | 2,003 | 567.224 | 0.28 | 4.92 |
| | 2001 | 887.2 | 97.2 | 399 | 22.067 | 0.06 | 0.02 |
| | 2002 | 860.0 | 91.3 | 1,361 | 248.487 | 0.18 | 0.29 |
| | 2003 | 953.0 | 98.9 | 241 | 11.778 | 0.05 | 0.01 |
| **INDIAN POINT 1[7]** | 2004 | 0.0 | 0.0 | 156 | 3 | 0.02 | --- |
| Docket 50-3; | 2005 | 0.0 | 0.0 | 151 | 6.692 | 0.04 | --- |
| DPR-05 | 2006 | 0.0 | 0.0 | 193 | 7.670 | 0.04 | --- |
| 1st commercial operation 10/62 | 2007 | 0.0 | 0.0 | 210 | 2.554 | 0.01 | --- |
| Type - PWR | 2008 | 0.0 | 0.0 | 234 | 4.322 | 0.02 | --- |
| Capacity - (265) MWe | 2009 | 0.0 | 0.0 | 140 | 0.404 | 0.00 | --- |
| | 2010 | 0.0 | 0.0 | 157 | 0.833 | 0.01 | --- |
| **INDIAN POINT 2** | 2004 | 855.3 | 91.0 | 1,136 | 195.630 | 0.17 | 0.23 |
| Docket 50-247; | 2005 | 1,007.2 | 100.0 | 470 | 11.418 | 0.02 | 0.01 |
| DPR-26 | 2006 | 911.5 | 91.0 | 1,327 | 286.908 | 0.22 | 0.32 |
| 1st commercial operation 8/74 | 2007 | 1,009.2 | 100.0 | 649 | 7.009 | 0.01 | 0.01 |
| Type - PWR | 2008 | 934.1 | 92.6 | 1,013 | 139.683 | 0.14 | 0.15 |
| Capacity - 998 MWe | 2009 | 1,005.0 | 99.4 | 569 | 10.091 | 0.02 | 0.01 |
| | 2010 | 832.8 | 84.1 | 1,446 | 197.279 | 0.14 | 0.24 |
| **INDIAN POINT 3[8]** | 1979 | 574.0 | 66.5 | 808 | 636 | 0.79 | 1.11 |
| Docket 50-286; | 1980 | 367.3 | 53.2 | 977 | 308 | 0.32 | 0.84 |
| DPR-64 | 1981 | 367.5 | 59.8 | 677 | 364 | 0.54 | 0.99 |
| 1st commercial operation 8/76 | 1982 | 171.5 | 22.5 | 1,477 | 1,226 | 0.83 | 7.15 |
| Type - PWR | 1983 | 7.8 | 2.6 | 941 | 607 | 0.65 | 77.82 |
| Capacity - 1,030 MWe | 1984 | 714.4 | 76.3 | 658 | 230 | 0.35 | 0.32 |
| | 1985 | 566.5 | 66.0 | 1,093 | 570 | 0.52 | 1.01 |
| | 1986 | 655.3 | 73.4 | 588 | 202 | 0.34 | 0.31 |
| | 1987 | 574.6 | 62.7 | 1,308 | 500 | 0.38 | 0.87 |
| | 1988 | 792.5 | 83.3 | 451 | 93 | 0.21 | 0.12 |
| | 1989 | 587.8 | 61.1 | 1,800 | 876 | 0.49 | 1.49 |
| | 1990 | 595.3 | 62.9 | 1,066 | 358 | 0.34 | 0.60 |
| | 1991 | 862.8 | 87.5 | 299 | 40 | 0.13 | 0.05 |
| | 1992 | 561.7 | 61.4 | 1,003 | 212 | 0.21 | 0.38 |
| | 1993 | 140.5 | 14.9 | 478 | 60 | 0.13 | 0.43 |
| | 1994 | 0.0 | 0.0 | 529 | 58 | 0.11 | --- |
| | 1995 | 174.8 | 21.4 | 638 | 67 | 0.11 | 0.38 |
| | 1996 | 695.3 | 74.8 | 289 | 22 | 0.08 | 0.03 |
| | 1997 | 495.1 | 54.9 | 1,608 | 234 | 0.15 | 0.47 |
| | 1998 | 874.0 | 95.3 | 213 | 14.774 | 0.07 | 0.02 |
| | 1999 | 829.8 | 88.3 | 893 | 116.920 | 0.13 | 0.14 |
| | 2000 | 960.0 | 99.3 | 143 | 8.693 | 0.06 | 0.00 |
| | 2001 | 903.9 | 93.1 | 1,014 | 118.115 | 0.12 | 0.13 |
| | 2002 | 960.0 | 98.5 | 156 | 6.797 | 0.04 | 0.01 |
| | 2003 | 866.2 | 89.8 | 902 | 96.059 | 0.11 | 0.11 |
| | 2004 | 995.8 | 100.0 | 234 | 4.232 | 0.02 | 0.00 |
| | 2005 | 915.0 | 91.7 | 893 | 73.862 | 0.08 | 0.08 |
| | 2006 | 1,024.5 | 100.0 | 307 | 2.793 | 0.01 | 0.00 |
| | 2007 | 890.1 | 88.0 | 1,322 | 102.960 | 0.08 | 0.12 |
| | 2008 | 1,043.1 | 100.0 | 443 | 3.045 | 0.01 | 0.00 |
| | 2009 | 879.2 | 88.1 | 1,284 | 68.999 | 0.05 | 0.08 |
| | 2010 | 1,026.4 | 99.5 | 516 | 3.103 | 0.01 | 0.00 |

[7] Indian Point 1 was defueled in 1975, and in 1984, it was decided that it would not be placed in operation again. Therefore, it is no longer included in the count of operating reactors. Parentheses indicate plant capacity when plant was operational.

[8] Indian Point 3 was purchased by a different utility in 1979 and, subsequently, reported its dose separately. Indian Point 1, 2, and 3 have been owned by the same utility since 2001 and report together.

| Reporting Organization | Year | Megawatt Years (MW-yr) | Unit Availability Factor | Total Personnel with Measurable Doses | Collective Dose per Site (person-rem) | Average Measurable Dose (rem) | Collective Dose/ MW-yr |
|---|---|---|---|---|---|---|---|
| **KEWAUNEE** | 1975 | 401.9 | 88.2 | 104 | 28 | 0.27 | 0.07 |
| Docket 50-305; | 1976 | 405.9 | 78.9 | 381 | 270 | 0.71 | 0.67 |
| DPR-43 | 1977 | 425.0 | 79.9 | 312 | 140 | 0.45 | 0.33 |
| 1st commercial operation 6/74 | 1978 | 466.6 | 89.5 | 335 | 154 | 0.46 | 0.33 |
| Type - PWR | 1979 | 412.0 | 79.0 | 343 | 127 | 0.37 | 0.31 |
| Capacity - 556 MWe | 1980 | 433.8 | 82.1 | 401 | 165 | 0.41 | 0.38 |
| | 1981 | 451.8 | 86.7 | 383 | 141 | 0.37 | 0.31 |
| | 1982 | 458.4 | 87.6 | 353 | 101 | 0.29 | 0.22 |
| | 1983 | 444.1 | 83.7 | 445 | 165 | 0.37 | 0.37 |
| | 1984 | 455.3 | 85.7 | 482 | 139 | 0.29 | 0.31 |
| | 1985 | 443.1 | 82.4 | 519 | 176 | 0.34 | 0.40 |
| | 1986 | 461.7 | 85.8 | 502 | 169 | 0.34 | 0.37 |
| | 1987 | 480.0 | 89.7 | 755 | 226 | 0.30 | 0.47 |
| | 1988 | 467.5 | 88.3 | 705 | 210 | 0.30 | 0.45 |
| | 1989 | 449.1 | 84.9 | 570 | 239 | 0.42 | 0.53 |
| | 1990 | 468.8 | 87.9 | 490 | 145 | 0.30 | 0.31 |
| | 1991 | 441.8 | 83.4 | 495 | 221 | 0.45 | 0.50 |
| | 1992 | 471.4 | 88.0 | 450 | 122 | 0.27 | 0.26 |
| | 1993 | 457.1 | 86.8 | 436 | 106 | 0.24 | 0.23 |
| | 1994 | 475.6 | 88.8 | 364 | 72 | 0.20 | 0.15 |
| | 1995 | 455.6 | 87.8 | 415 | 109 | 0.26 | 0.24 |
| | 1996 | 380.4 | 71.8 | 474 | 126 | 0.27 | 0.33 |
| | 1997 | 269.8 | 56.0 | 278 | 56 | 0.20 | 0.21 |
| | 1998 | 423.0 | 87.2 | 384 | 88.205 | 0.23 | 0.21 |
| | 1999 | 505.1 | 100.0 | 103 | 5.055 | 0.05 | 0.01 |
| | 2000 | 432.6 | 88.8 | 394 | 99.864 | 0.25 | 0.23 |
| | 2001 | 394.1 | 80.8 | 1,110 | 200.245 | 0.18 | 0.51 |
| | 2002 | 509.0 | 97.4 | 102 | 4.449 | 0.04 | 0.01 |
| | 2003 | 473.5 | 90.5 | 439 | 73.108 | 0.17 | 0.15 |
| | 2004 | 441.0 | 81.0 | 565 | 91.168 | 0.16 | 0.21 |
| | 2005 | 346.4 | 62.7 | 97 | 4.000 | 0.04 | 0.01 |
| | 2006 | 419.4 | 77.0 | 539 | 74.734 | 0.14 | 0.18 |
| | 2007 | 528.0 | 95.0 | 145 | 11.126 | 0.08 | 0.02 |
| | 2008 | 499.5 | 88.9 | 598 | 92.951 | 0.16 | 0.19 |
| | 2009 | 515.4 | 92.0 | 595 | 56.215 | 0.09 | 0.11 |
| | 2010 | 569.7 | 100.0 | 135 | 4.690 | 0.03 | 0.01 |
| **LACROSSE**[9] | 1970 | 15.3 | | | 111 | | 7.25 |
| Docket 50-409; | 1971 | 33.1 | | 218 | 158 | 0.72 | 4.77 |
| DPR-45 | 1972 | 29.2 | | 151 | 172 | 1.14 | 5.89 |
| 1st commercial operation 11/69 | 1973 | 24.4 | | 157 | 221 | 1.41 | 9.06 |
| Type - BWR | 1974 | 37.9 | 81.0 | 115 | 139 | 1.21 | 3.67 |
| Capacity - (48) MWe | 1975 | 32.0 | 69.6 | 165 | 234 | 1.42 | 7.31 |
| | 1976 | 21.2 | 47.6 | 118 | 110 | 0.93 | 5.19 |
| | 1977 | 11.3 | 33.7 | 141 | 225 | 1.60 | 19.91 |
| | 1978 | 21.6 | 62.0 | 182 | 164 | 0.90 | 7.59 |
| | 1979 | 24.0 | 71.8 | 153 | 186 | 1.22 | 7.75 |
| | 1980 | 26.4 | 68.5 | 124 | 218 | 1.76 | 8.26 |
| | 1981 | 29.6 | 76.0 | 187 | 123 | 0.66 | 4.16 |
| | 1982 | 17.2 | 44.6 | 148 | 205 | 1.39 | 11.92 |
| | 1983 | 24.8 | 59.7 | 160 | 313 | 1.96 | 12.62 |
| | 1984 | 38.5 | 80.5 | 288 | 252 | 0.88 | 6.55 |
| | 1985 | 39.2 | 86.7 | 373 | 173 | 0.46 | 4.41 |
| | 1986 | 19.6 | 46.1 | 260 | 290 | 1.12 | 14.80 |
| | 1987 | 0.0 | 0.0 | 127 | 68 | 0.54 | --- |
| | 1988 | 0.0 | 0.0 | 49 | 31 | 0.63 | --- |
| | 1989 | 0.0 | 0.0 | 60 | 15 | 0.25 | --- |
| | 1990 | 0.0 | 0.0 | 51 | 9 | 0.18 | --- |
| | 1991 | 0.0 | 0.0 | 42 | 8 | 0.19 | --- |
| | 1992 | 0.0 | 0.0 | 28 | 6 | 0.21 | --- |
| | 1993 | 0.0 | 0.0 | 48 | 8 | 0.17 | --- |

[9] LaCrosse ended commercial operation in 1987 and will not be put in commercial operation again. Therefore, it is no longer included in the count of operating reactors. Parentheses indicate plant capacity when plant was operational.

| Reporting Organization | Year | Megawatt Years (MW-yr) | Unit Availability Factor | Total Personnel with Measurable Doses | Collective Dose per Site (person-rem) | Average Measurable Dose (rem) | Collective Dose/ MW-yr |
|---|---|---|---|---|---|---|---|
| **LACROSSE**[9] | 1994 | 0.0 | 0.0 | 65 | 8 | 0.12 | --- |
| (continued) | 1995 | 0.0 | 0.0 | 31 | 3 | 0.10 | --- |
| | 1996 | 0.0 | 0.0 | 25 | 4 | 0.15 | --- |
| | 1997 | 0.0 | 0.0 | 23 | 2 | 0.09 | --- |
| | 1998 | 0.0 | 0.0 | 27 | 1.530 | 0.06 | --- |
| | 1999 | 0.0 | 0.0 | 66 | 3.725 | 0.06 | --- |
| | 2000 | 0.0 | 0.0 | 37 | 3.548 | 0.10 | --- |
| | 2001 | 0.0 | 0.0 | 45 | 2.782 | 0.06 | --- |
| | 2002 | 0.0 | 0.0 | 47 | 2.314 | 0.05 | --- |
| | 2003 | 0.0 | 0.0 | 65 | 1.836 | 0.03 | --- |
| | 2004 | 0.0 | 0.0 | 56 | 0.918 | 0.02 | --- |
| | 2005 | 0.0 | 0.0 | 51 | 8.139 | 0.16 | --- |
| | 2006 | 0.0 | 0.0 | 0 | 0.000 | --- | --- |
| | 2007 | 0.0 | 0.0 | 86 | 37.092 | 0.43 | --- |
| | 2008 | 0.0 | 0.0 | 40 | 1.759 | 0.04 | --- |
| | 2009 | 0.0 | 0.0 | 48 | 1.307 | 0.03 | --- |
| | 2010 | 0.0 | 0.0 | 78 | 2.971 | 0.04 | --- |
| **LASALLE 1, 2** | 1984 | 677.8 | 77.8 | 1,245 | 252 | 0.20 | 0.37 |
| Docket 50-373, 50-374; | 1985 | 987.9 | 53.0 | 1,635 | 685 | 0.42 | 0.69 |
| NPF-11, NPF-18 | 1986 | 929.5 | 50.6 | 1,614 | 898 | 0.56 | 0.97 |
| 1st commercial operation | 1987 | 1,030.0 | 59.3 | 1,744 | 1,396 | 0.80 | 1.36 |
| 1/84, 6/84 | 1988 | 1,317.6 | 71.6 | 2,737 | 2,471 | 0.90 | 1.88 |
| Type - BWRs | 1989 | 1,503.5 | 73.1 | 2,475 | 1,386 | 0.56 | 0.92 |
| Capacity - 1,111, 1,111 MWe | 1990 | 1,754.3 | 84.6 | 1,830 | 948 | 0.52 | 0.54 |
| | 1991 | 1,837.0 | 86.7 | 1,985 | 806 | 0.41 | 0.44 |
| | 1992 | 1,447.4 | 72.0 | 2,418 | 1,167 | 0.48 | 0.81 |
| | 1993 | 1,542.0 | 76.0 | 1,701 | 854 | 0.50 | 0.55 |
| | 1994 | 1,580.0 | 77.6 | 1,812 | 726 | 0.40 | 0.46 |
| | 1995 | 1,696.6 | 82.1 | 1,623 | 512 | 0.32 | 0.30 |
| | 1996 | 1,053.8 | 54.3 | 2,782 | 819 | 0.29 | 0.78 |
| | 1997 | 0.0 | 0.0 | 1,661 | 316 | 0.19 | --- |
| | 1998 | 380.9 | 19.3 | 2,099 | 422.249 | 0.20 | 1.11 |
| | 1999 | 1,671.9 | 81.8 | 2,689 | 576.354 | 0.21 | 0.34 |
| | 2000 | 2,138.6 | 97.1 | 1,831 | 260.320 | 0.14 | 0.12 |
| | 2001 | 2,223.8 | 98.9 | 535 | 82.721 | 0.15 | 0.04 |
| | 2002 | 2,040.0 | 92.1 | 2,012 | 449.587 | 0.22 | 0.22 |
| | 2003 | 2,100.2 | 94.8 | 2,253 | 464.427 | 0.21 | 0.22 |
| | 2004 | 2,162.1 | 96.0 | 2,366 | 359.470 | 0.15 | 0.17 |
| | 2005 | 2,130.4 | 95.0 | 2,097 | 334.558 | 0.16 | 0.16 |
| | 2006 | 2,181.3 | 97.0 | 2,006 | 248.454 | 0.12 | 0.11 |
| | 2007 | 2,166.7 | 98.0 | 1,953 | 228.373 | 0.12 | 0.11 |
| | 2008 | 2,145.8 | 96.4 | 2,402 | 217.567 | 0.09 | 0.10 |
| | 2009 | 2,141.0 | 95.7 | 1,986 | 296.659 | 0.15 | 0.14 |
| | 2010 | 2,184.1 | 96.5 | 2,386 | 384.434 | 0.16 | 0.18 |
| **LIMERICK 1, 2** | 1987 | 636.1 | 70.2 | 2,156 | 174 | 0.08 | 0.27 |
| Docket 50-352, 50-353; | 1988 | 794.9 | 96.5 | 950 | 52 | 0.05 | 0.07 |
| NPF-39, NPF-85 | 1989 | 628.4 | 66.0 | 1,818 | 266 | 0.15 | 0.42 |
| 1st commercial operation | 1990 | 1,527.7 | 78.2 | 1,422 | 175 | 0.12 | 0.11 |
| 2/86, 1/90 | 1991 | 1,810.9 | 86.8 | 1,151 | 106 | 0.09 | 0.06 |
| Type - BWRs | 1992 | 1,741.4 | 84.8 | 1,559 | 330 | 0.21 | 0.19 |
| Capacity - 1,092, 1,096 MWe | 1993 | 1,913.2 | 91.6 | 1,287 | 217 | 0.17 | 0.11 |
| | 1994 | 1,944.4 | 94.9 | 1,543 | 275 | 0.18 | 0.14 |
| | 1995 | 1,957.1 | 93.0 | 1,581 | 260 | 0.16 | 0.13 |
| | 1996 | 2,026.2 | 93.3 | 1,654 | 234 | 0.14 | 0.12 |
| | 1997 | 2,001.7 | 95.8 | 1,463 | 234 | 0.16 | 0.12 |
| | 1998 | 1,907.2 | 89.5 | 1,854 | 357.139 | 0.19 | 0.19 |
| | 1999 | 2,089.6 | 94.2 | 1,800 | 271.547 | 0.15 | 0.13 |

[9] LaCrosse ended commercial operation in 1987 and will not be put in commercial operation again. Therefore, it is no longer included in the count of operating reactors. Parentheses indicate plant capacity when plant was operational.

| Reporting Organization | Year | Megawatt Years (MW-yr) | Unit Availability Factor | Total Personnel with Measurable Doses | Collective Dose per Site (person-rem) | Average Measurable Dose (rem) | Collective Dose/ MW-yr |
|---|---|---|---|---|---|---|---|
| **LIMERICK 1, 2** | 2000 | 2,154.9 | 95.8 | 1,279 | 260.611 | 0.20 | 0.12 |
| (continued) | 2001 | 2,205.9 | 97.3 | 1,127 | 210.336 | 0.19 | 0.10 |
| | 2002 | 2,197.0 | 97.1 | 1,248 | 160.324 | 0.13 | 0.07 |
| | 2003 | 2,213.6 | 97.2 | 1,298 | 147.047 | 0.11 | 0.07 |
| | 2004 | 2,218.9 | 97.6 | 1,265 | 149.433 | 0.12 | 0.07 |
| | 2005 | 2,168.9 | 96.3 | 1,460 | 187.609 | 0.13 | 0.09 |
| | 2006 | 2,207.2 | 97.0 | 1,509 | 193.429 | 0.13 | 0.09 |
| | 2007 | 2,185.8 | 96.0 | 1,570 | 197.104 | 0.13 | 0.09 |
| | 2008 | 2,169.2 | 96.0 | 1,393 | 176.825 | 0.13 | 0.08 |
| | 2009 | 2,211.4 | 97.2 | 1,606 | 234.742 | 0.15 | 0.11 |
| | 2010 | 2,165.2 | 96.7 | 1,525 | 167.797 | 0.11 | 0.08 |
| **MAINE YANKEE[10]** | 1973 | 408.7 | | 782 | 117 | 0.15 | 0.29 |
| Docket 50-309; | 1974 | 432.6 | 68.7 | 619 | 420 | 0.68 | 0.97 |
| DPR-36 | 1975 | 542.9 | 79.9 | 440 | 319 | 0.72 | 0.59 |
| 1st commercial operation 12/72 | 1976 | 712.2 | 95.0 | 244 | 85 | 0.35 | 0.12 |
| Type - PWR | 1977 | 617.6 | 82.2 | 508 | 245 | 0.48 | 0.40 |
| Capacity - (860) MWe | 1978 | 642.7 | 84.1 | 638 | 420 | 0.66 | 0.65 |
| | 1979 | 537.0 | 68.4 | 393 | 154 | 0.39 | 0.29 |
| | 1980 | 527.0 | 72.2 | 735 | 462 | 0.63 | 0.88 |
| | 1981 | 624.2 | 78.2 | 868 | 424 | 0.49 | 0.68 |
| | 1982 | 542.5 | 69.1 | 1,295 | 619 | 0.48 | 1.14 |
| | 1983 | 677.1 | 83.6 | 592 | 165 | 0.28 | 0.24 |
| | 1984 | 605.7 | 74.4 | 1,262 | 884 | 0.70 | 1.46 |
| | 1985 | 635.4 | 79.2 | 1,009 | 700 | 0.69 | 1.10 |
| | 1986 | 737.6 | 87.8 | 495 | 100 | 0.20 | 0.14 |
| | 1987 | 478.1 | 65.3 | 1,100 | 722 | 0.66 | 1.51 |
| | 1988 | 591.9 | 79.1 | 1,058 | 725 | 0.69 | 1.22 |
| | 1989 | 819.2 | 93.7 | 375 | 99 | 0.26 | 0.12 |
| | 1990 | 573.0 | 71.0 | 1,359 | 682 | 0.50 | 1.19 |
| | 1991 | 738.1 | 86.6 | 426 | 105 | 0.25 | 0.14 |
| | 1992 | 631.7 | 79.1 | 1,189 | 461 | 0.39 | 0.73 |
| | 1993 | 674.8 | 79.8 | 1,016 | 377 | 0.37 | 0.56 |
| | 1994 | 782.8 | 90.9 | 297 | 84 | 0.28 | 0.11 |
| | 1995 | 23.6 | 3.7 | 1,167 | 653 | 0.56 | 27.67 |
| | 1996 | 602.9 | 78.1 | 408 | 56 | 0.14 | 0.09 |
| | 1997 | 0.0 | 0.0 | 991 | 153 | 0.15 | --- |
| | 1998 | 0.0 | 0.0 | 438 | 163.008 | 0.37 | --- |
| | 1999 | 0.0 | 0.0 | 365 | 135.057 | 0.37 | --- |
| | 2000 | 0.0 | 0.0 | 490 | 121.133 | 0.25 | --- |
| | 2001 | 0.0 | 0.0 | 412 | 68.121 | 0.17 | --- |
| | 2002 | 0.0 | 0.0 | 452 | 66.226 | 0.15 | --- |
| | 2003 | 0.0 | 0.0 | 342 | 43.775 | 0.13 | --- |
| | 2004 | 0.0 | 0.0 | 190 | 21.313 | 0.11 | --- |
| | 2005 | 0.0 | 0.0 | 2 | 0.048 | 0.02 | --- |
| | 2006 | 0.0 | 0.0 | 0 | 0.000 | --- | --- |
| | 2007 | 0.0 | 0.0 | 0 | 0.000 | --- | --- |
| | 2008 | 0.0 | 0.0 | 1 | 0.013 | 0.01 | --- |
| | 2009 | 0.0 | 0.0 | 3 | 0.137 | 0.05 | --- |
| | 2010 | 0.0 | 0.0 | 1 | 0.084 | 0.08 | --- |
| **MCGUIRE 1, 2** | 1982 | 524.9 | 80.4 | 1,560 | 169 | 0.11 | 0.32 |
| Docket 50-369, 50-370; | 1983 | 558.3 | 55.4 | 1,751 | 521 | 0.30 | 0.93 |
| NPF-9, NPF-17 | 1984 | 764.1 | 68.5 | 1,663 | 507 | 0.30 | 0.66 |
| 1st commercial operation | 1985 | 808.4 | 77.0 | 2,217 | 771 | 0.35 | 0.95 |
| 12/81, 3/84 | 1986 | 1,360.0 | 60.1 | 2,326 | 1,015 | 0.44 | 0.75 |
| Type - PWRs | 1987 | 1,774.7 | 79.2 | 2,865 | 1,043 | 0.36 | 0.59 |
| Capacity - 1,100, 1,100 MWe | 1988 | 1,830.7 | 80.2 | 2,808 | 1,104 | 0.39 | 0.60 |
| | 1989 | 1,810.2 | 80.8 | 1,994 | 620 | 0.31 | 0.34 |
| | 1990 | 1,340.3 | 61.3 | 2,289 | 727 | 0.32 | 0.54 |
| | 1991 | 1,945.1 | 85.0 | 1,723 | 361 | 0.21 | 0.19 |
| | 1992 | 1,696.8 | 74.4 | 1,619 | 418 | 0.26 | 0.25 |

[10] Maine Yankee was shut down in August 1997 and is no longer included in the count of operating reactors. Parentheses indicate plant capacity when plant was operational.

| Reporting Organization | Year | Megawatt Years (MW-yr) | Unit Availability Factor | Total Personnel with Measurable Doses | Collective Dose per Site (person-rem) | Average Measurable Dose (rem) | Collective Dose/ MW-yr |
|---|---|---|---|---|---|---|---|
| **MCGUIRE 1, 2** | 1993 | 1,470.4 | 66.2 | 1,685 | 463 | 0.27 | 0.31 |
| (continued) | 1994 | 1,848.0 | 80.2 | 1,637 | 397 | 0.24 | 0.21 |
| | 1995 | 2,132.3 | 92.9 | 1,259 | 138 | 0.11 | 0.06 |
| | 1996 | 1,881.8 | 82.8 | 1,622 | 238 | 0.15 | 0.13 |
| | 1997 | 1,558.2 | 73.0 | 2,193 | 492 | 0.22 | 0.32 |
| | 1998 | 2,139.8 | 95.1 | 1,045 | 142.245 | 0.14 | 0.07 |
| | 1999 | 1,961.7 | 88.9 | 1,274 | 256.524 | 0.20 | 0.13 |
| | 2000 | 2,100.1 | 94.2 | 940 | 132.513 | 0.14 | 0.06 |
| | 2001 | 2,113.3 | 93.9 | 963 | 136.581 | 0.14 | 0.06 |
| | 2002 | 2,051.0 | 91.7 | 1,167 | 180.618 | 0.16 | 0.09 |
| | 2003 | 2,156.2 | 96.0 | 841 | 71.323 | 0.08 | 0.03 |
| | 2004 | 2,075.7 | 91.8 | 1,116 | 196.193 | 0.18 | 0.09 |
| | 2005 | 1,993.9 | 89.2 | 1,401 | 173.972 | 0.12 | 0.09 |
| | 2006 | 2,100.2 | 93.0 | 1,218 | 108.285 | 0.09 | 0.05 |
| | 2007 | 2,011.4 | 89.0 | 1,375 | 156.035 | 0.11 | 0.08 |
| | 2008 | 1,943.3 | 86.2 | 1,613 | 165.767 | 0.10 | 0.09 |
| | 2009 | 2,170.6 | 95.3 | 1,165 | 79.773 | 0.07 | 0.04 |
| | 2010 | 2,151.9 | 94.8 | 1,225 | 81.321 | 0.07 | 0.04 |
| **MILLSTONE 1"** | 1972 | 377.6 | | 612 | 596 | 0.97 | 1.58 |
| Docket 50-245; | 1973 | 225.1 | | 1,184 | 663 | 0.56 | 2.95 |
| DPR-21 | 1974 | 430.3 | 79.1 | 2,477 | 1,430 | 0.58 | 3.32 |
| 1st commercial operation 3/71 | 1975 | 465.4 | 75.6 | 2,587 | 2,022 | 0.78 | 4.34 |
| Type - BWR | 1976 | 449.8 | 76.1 | 1,387 | 1,194 | 0.86 | 2.65 |
| Capacity - (641) MWe | 1977 | 575.7 | 89.6 | 1,075 | 394 | 0.37 | 0.68 |
| | 1978 | 556.6 | 87.6 | 1,391 | 1,416 | 1.02 | 2.54 |
| | 1979 | 505.0 | 77.3 | 2,001 | 1,795 | 0.90 | 3.55 |
| | 1980 | 405.8 | 69.0 | 3,024 | 2,157 | 0.71 | 5.32 |
| | 1981 | 304.3 | 51.6 | 2,506 | 1,496 | 0.60 | 4.92 |
| | 1982 | 490.2 | 79.9 | 1,370 | 929 | 0.68 | 1.90 |
| | 1983 | 640.1 | 95.6 | 309 | 244 | 0.79 | 0.38 |
| | 1984 | 516.1 | 78.8 | 1,992 | 836 | 0.42 | 1.62 |
| | 1985 | 548.5 | 83.6 | 732 | 608 | 0.83 | 1.11 |
| | 1986 | 626.8 | 95.4 | 389 | 150 | 0.39 | 0.24 |
| | 1987 | 523.4 | 79.6 | 1,588 | 684 | 0.43 | 1.31 |
| | 1988 | 658.8 | 98.6 | 327 | 144 | 0.44 | 0.22 |
| | 1989 | 554.6 | 84.2 | 852 | 462 | 0.54 | 0.83 |
| | 1990 | 608.3 | 91.6 | 365 | 131 | 0.36 | 0.22 |
| | 1991 | 213.1 | 35.4 | 1,154 | 409 | 0.35 | 1.92 |
| | 1992 | 431.8 | 68.1 | 348 | 99 | 0.28 | 0.23 |
| | 1993 | 627.9 | 96.8 | 305 | 81 | 0.27 | 0.13 |
| | 1994 | 394.0 | 63.6 | 1,321 | 391 | 0.30 | 0.99 |
| | 1995 | 520.6 | 80.0 | 910 | 620 | 0.68 | 1.19 |
| | 1996 | 0.0 | 0.0 | 747 | 431 | 0.58 | --- |
| | 1997 | -2.9 | 0.0 | 1,053 | 195 | 0.19 | --- |
| | 1998 | -2.7 | 0.0 | 347 | 12.741 | 0.04 | --- |
| | 1999 | 0.0 | 0.0 | 397 | 9.790 | 0.02 | --- |
| | 2000 | 0.0 | 0.0 | 478 | 59.955 | 0.13 | --- |
| | 2001 | 0.0 | 0.0 | 414 | 14.946 | 0.04 | --- |
| | 2002 | 0.0 | 0.0 | 185 | 4.151 | 0.02 | --- |
| | 2003 | 0.0 | 0.0 | 195 | 10.675 | 0.05 | --- |
| | 2004 | 0.0 | 0.0 | 147 | 11.152 | 0.08 | --- |
| | 2005 | 0.0 | 0.0 | 145 | 0.897 | 0.01 | --- |
| | 2006 | 0.0 | 0.0 | 4 | 0.607 | 0.15 | --- |
| | 2007 | 0.0 | 0.0 | 33 | 0.901 | 0.03 | --- |
| | 2008 | 0.0 | 0.0 | 0 | 0.222 | --- | --- |
| | 2009 | 0.0 | 0.0 | 0 | 0.114 | --- | --- |
| | 2010 | 0.0 | 0.0 | 0 | 0.142 | --- | --- |

[11]Millstone 1 was shut down on June 30, 1998, and is no longer included in the count of operating reactors. Parentheses indicate plant capacity when plant was operational.

| Reporting Organization | Year | Megawatt Years (MW-yr) | Unit Availability Factor | Total Personnel with Measurable Doses | Collective Dose per Site (person-rem) | Average Measurable Dose (rem) | Collective Dose/ MW-yr |
|---|---|---|---|---|---|---|---|
| **MILLSTONE 2, 3** | 1976 | 545.7 | 78.7 | 620 | 168 | 0.27 | 0.31 |
| Docket 50-336, 50-423; | 1977 | 518.7 | 65.7 | 667 | 242 | 0.36 | 0.47 |
| DPR-65; NPF-49 | 1978 | 536.6 | 67.3 | 1,420 | 1,444 | 1.02 | 2.69 |
| 1st commercial operation | 1979 | 520.0 | 62.8 | 525 | 471 | 0.90 | 0.91 |
| 12/75, 4/86 | 1980 | 579.3 | 69.2 | 893 | 637 | 0.71 | 1.10 |
| Type - PWRs | 1981 | 722.4 | 82.6 | 890 | 531 | 0.60 | 0.74 |
| Capacity - 878, 1,218 MWe | 1982 | 595.9 | 70.6 | 2,083 | 1,413 | 0.68 | 2.37 |
| | 1983 | 294.0 | 34.2 | 2,383 | 1,881 | 0.79 | 6.40 |
| | 1984 | 782.7 | 93.5 | 285 | 120 | 0.42 | 0.15 |
| | 1985 | 417.8 | 49.4 | 1,905 | 1,581 | 0.83 | 3.78 |
| | 1986 | 1,313.8 | 80.4 | 2,393 | 993 | 0.41 | 0.76 |
| | 1987 | 1,624.5 | 84.1 | 1,441 | 505 | 0.35 | 0.31 |
| | 1988 | 1,594.8 | 83.2 | 1,827 | 804 | 0.44 | 0.50 |
| | 1989 | 1,428.3 | 72.9 | 1,984 | 1,079 | 0.54 | 0.76 |
| | 1990 | 1,614.9 | 87.1 | 1,652 | 593 | 0.36 | 0.37 |
| | 1991 | 819.5 | 69.7 | 1,084 | 381 | 0.35 | 0.46 |
| | 1992 | 1,115.1 | 59.9 | 3,190 | 1,280 | 0.40 | 1.15 |
| | 1993 | 1,525.2 | 79.7 | 2,064 | 557 | 0.27 | 0.37 |
| | 1994 | 1,556.6 | 73.1 | 1,249 | 188 | 0.15 | 0.12 |
| | 1995 | 1,278.1 | 60.5 | 1,691 | 416 | 0.25 | 0.33 |
| | 1996 | 418.1 | 19.3 | 983 | 126 | 0.13 | 0.30 |
| | 1997 | 0.0 | 0.0 | 1,435 | 253 | 0.18 | --- |
| | 1998 | 374.9 | 20.9 | 1,179 | 112.543 | 0.10 | 0.30 |
| | 1999 | 1,446.3 | 73.3 | 1,688 | 252.138 | 0.15 | 0.17 |
| | 2000 | 1,865.8 | 92.4 | 1,385 | 142.664 | 0.10 | 0.08 |
| | 2001 | 1,759.3 | 92.0 | 1,327 | 174.238 | 0.13 | 0.10 |
| | 2002 | 1,703.0 | 87.5 | 1,548 | 292.197 | 0.19 | 0.17 |
| | 2003 | 1,834.6 | 91.0 | 1,274 | 322.923 | 0.25 | 0.18 |
| | 2004 | 1,887.5 | 95.0 | 803 | 136.459 | 0.17 | 0.07 |
| | 2005 | 1,777.1 | 88.8 | 1,329 | 202.490 | 0.15 | 0.11 |
| | 2006 | 1,898.5 | 93.0 | 1,160 | 174.164 | 0.15 | 0.09 |
| | 2007 | 1,875.1 | 94.0 | 1,150 | 163.780 | 0.14 | 0.09 |
| | 2008 | 1,761.1 | 87.7 | 1,467 | 272.693 | 0.18 | 0.16 |
| | 2009 | 1,906.1 | 89.6 | 983 | 159.203 | 0.16 | 0.08 |
| | 2010 | 1,916.8 | 93.1 | 718 | 81.589 | 0.11 | 0.04 |
| **MONTICELLO** | 1972 | 424.4 | | 99 | 61 | 0.62 | 0.14 |
| Docket 50-263; | 1973 | 389.5 | | 401 | 176 | 0.44 | 0.45 |
| DPR-22 | 1974 | 349.3 | 74.9 | 842 | 349 | 0.41 | 1.00 |
| 1st commercial operation 6/71 | 1975 | 344.8 | 72.2 | 1,353 | 1,353 | 1.00 | 3.92 |
| Type - BWR | 1976 | 476.4 | 91.5 | 325 | 263 | 0.81 | 0.55 |
| Capacity - 578 MWe | 1977 | 425.6 | 79.9 | 860 | 1,000 | 1.16 | 2.35 |
| | 1978 | 459.4 | 87.2 | 679 | 375 | 0.55 | 0.82 |
| | 1979 | 522.0 | 97.6 | 372 | 157 | 0.42 | 0.30 |
| | 1980 | 411.8 | 78.2 | 1,114 | 531 | 0.48 | 1.29 |
| | 1981 | 389.3 | 72.6 | 1,446 | 1,004 | 0.69 | 2.58 |
| | 1982 | 291.1 | 63.3 | 1,307 | 993 | 0.76 | 3.41 |
| | 1983 | 494.6 | 96.3 | 416 | 121 | 0.29 | 0.24 |
| | 1984 | 33.7 | 9.2 | 1,872 | 2,462 | 1.32 | 73.06 |
| | 1985 | 509.8 | 91.7 | 586 | 327 | 0.56 | 0.64 |
| | 1986 | 402.7 | 79.1 | 895 | 596 | 0.67 | 1.48 |
| | 1987 | 422.5 | 81.9 | 941 | 568 | 0.60 | 1.34 |
| | 1988 | 542.5 | 99.8 | 375 | 110 | 0.29 | 0.20 |
| | 1989 | 318.2 | 76.2 | 1,102 | 507 | 0.46 | 1.59 |
| | 1990 | 536.0 | 96.9 | 336 | 94 | 0.28 | 0.18 |
| | 1991 | 429.4 | 80.8 | 964 | 465 | 0.48 | 1.08 |
| | 1992 | 528.3 | 97.5 | 454 | 114 | 0.25 | 0.22 |
| | 1993 | 458.1 | 84.4 | 954 | 494 | 0.52 | 1.08 |
| | 1994 | 471.3 | 87.0 | 788 | 395 | 0.50 | 0.84 |
| | 1995 | 564.7 | 100.0 | 200 | 44 | 0.22 | 0.08 |
| | 1996 | 461.6 | 86.9 | 757 | 240 | 0.32 | 0.52 |
| | 1997 | 417.4 | 75.9 | 399 | 106 | 0.27 | 0.25 |
| | 1998 | 470.2 | 88.1 | 674 | 209.137 | 0.31 | 0.44 |

| Reporting Organization | Year | Megawatt Years (MW-yr) | Unit Availability Factor | Total Personnel with Measurable Doses | Collective Dose per Site (person-rem) | Average Measurable Dose (rem) | Collective Dose/ MW-yr |
|---|---|---|---|---|---|---|---|
| **MONTICELLO** | 1999 | 530.7 | 92.9 | 451 | 70.075 | 0.16 | 0.13 |
| (continued) | 2000 | 483.2 | 84.2 | 792 | 216.136 | 0.27 | 0.45 |
| | 2001 | 441.3 | 78.5 | 834 | 220.683 | 0.26 | 0.50 |
| | 2002 | 571.0 | 99.0 | 399 | 40.030 | 0.10 | 0.07 |
| | 2003 | 522.8 | 91.7 | 858 | 168.896 | 0.20 | 0.32 |
| | 2004 | 573.2 | 99.2 | 279 | 35.081 | 0.13 | 0.06 |
| | 2005 | 509.4 | 90.0 | 919 | 175.201 | 0.19 | 0.34 |
| | 2006 | 579.1 | 100.0 | 273 | 33.416 | 0.12 | 0.06 |
| | 2007 | 478.6 | 85.0 | 1,075 | 191.398 | 0.18 | 0.40 |
| | 2008 | 555.3 | 95.8 | 351 | 43.777 | 0.12 | 0.08 |
| | 2009 | 473.1 | 85.2 | 1,235 | 173.624 | 0.14 | 0.37 |
| | 2010 | 536.0 | 98.5 | 534 | 56.116 | 0.11 | 0.10 |
| **NINE MILE POINT 1, 2** | 1970 | 227.0 | | 821 | 44 | 0.05 | 0.19 |
| Docket 50-220, 50-410; | 1971 | 346.5 | | 1,006 | 195 | 0.19 | 0.56 |
| DPR-63; NPF-69 | 1972 | 381.8 | | 735 | 285 | 0.39 | 0.75 |
| 1st commercial operation | 1973 | 411.0 | | 550 | 567 | 1.03 | 1.38 |
| 12/69, 4/88 | 1974 | 385.9 | 70.5 | 740 | 824 | 1.11 | 2.14 |
| Type - BWRs | 1975 | 359.0 | 72.1 | 649 | 681 | 1.05 | 1.90 |
| Capacity - 565, 1,120 MWe | 1976 | 484.6 | 88.2 | 392 | 428 | 1.09 | 0.88 |
| | 1977 | 347.4 | 59.2 | 1,093 | 1,383 | 1.27 | 3.98 |
| | 1978 | 527.7 | 95.1 | 561 | 314 | 0.56 | 0.60 |
| | 1979 | 354.0 | 66.1 | 1,326 | 1,497 | 1.13 | 4.23 |
| | 1980 | 533.9 | 92.3 | 1,174 | 591 | 0.50 | 1.11 |
| | 1981 | 385.2 | 66.0 | 2,029 | 1,592 | 0.78 | 4.13 |
| | 1982 | 133.5 | 21.4 | 1,352 | 1,264 | 0.93 | 9.47 |
| | 1983 | 329.8 | 56.2 | 1,405 | 860 | 0.61 | 2.61 |
| | 1984 | 426.8 | 71.9 | 1,530 | 890 | 0.58 | 2.09 |
| | 1985 | 580.9 | 96.4 | 1,007 | 265 | 0.26 | 0.46 |
| | 1986 | 371.0 | 65.3 | 1,878 | 1,275 | 0.68 | 3.44 |
| | 1987 | 542.6 | 93.3 | 1,190 | 141 | 0.12 | 0.26 |
| | 1988 | 0.0 | 0.0 | 2,626 | 854 | 0.33 | --- |
| | 1989 | 527.5 | 29.7 | 2,737 | 564 | 0.21 | 1.07 |
| | 1990 | 656.2 | 46.6 | 2,405 | 699 | 0.29 | 1.07 |
| | 1991 | 1,250.8 | 79.7 | 1,543 | 292 | 0.19 | 0.23 |
| | 1992 | 965.9 | 61.8 | 1,800 | 563 | 0.31 | 0.58 |
| | 1993 | 1,380.2 | 84.6 | 2,352 | 633 | 0.27 | 0.46 |
| | 1994 | 1,589.6 | 95.9 | 800 | 149 | 0.19 | 0.09 |
| | 1995 | 1,382.2 | 82.5 | 2,304 | 759 | 0.33 | 0.55 |
| | 1996 | 1,598.6 | 91.6 | 1,596 | 290 | 0.18 | 0.18 |
| | 1997 | 1,321.5 | 74.8 | 1,425 | 429 | 0.30 | 0.32 |
| | 1998 | 1,387.3 | 87.0 | 1,744 | 378.484 | 0.22 | 0.27 |
| | 1999 | 1,409.5 | 81.3 | 1,709 | 446.699 | 0.26 | 0.32 |
| | 2000 | 1,443.9 | 88.1 | 1,783 | 282.838 | 0.16 | 0.20 |
| | 2001 | 1,506.9 | 88.9 | 1,371 | 343.197 | 0.25 | 0.23 |
| | 2002 | 1,517.0 | 90.4 | 2,449 | 516.663 | 0.21 | 0.34 |
| | 2003 | 1,585.6 | 91.4 | 1,501 | 374.775 | 0.25 | 0.24 |
| | 2004 | 1,551.9 | 92.0 | 1,362 | 448.509 | 0.33 | 0.29 |
| | 2005 | 1,656.5 | 94.5 | 1,366 | 401.719 | 0.29 | 0.24 |
| | 2006 | 1,647.1 | 96.0 | 1,130 | 229.551 | 0.20 | 0.14 |
| | 2007 | 1,598.3 | 93.0 | 1,826 | 329.307 | 0.18 | 0.21 |
| | 2008 | 1,642.1 | 95.8 | 1,391 | 301.824 | 0.22 | 0.18 |
| | 2009 | 1,706.2 | 97.1 | 1,456 | 237.552 | 0.16 | 0.14 |
| | 2010 | 1,627.1 | 95.2 | 1,703 | 375.424 | 0.22 | 0.23 |
| **NORTH ANNA 1, 2** | 1979 | 507.0 | 61.7 | 2,025 | 449 | 0.22 | 0.89 |
| Docket 50-338, 50-339; | 1980 | 681.8 | 86.5 | 2,086 | 218 | 0.10 | 0.32 |
| NPF-4, NPF-7 | 1981 | 1,241.9 | 71.5 | 2,416 | 680 | 0.28 | 0.55 |
| 1st commercial operation | 1982 | 777.7 | 45.8 | 2,872 | 1,915 | 0.67 | 2.46 |
| 6/78, 12/80 | 1983 | 1,338.4 | 76.1 | 2,228 | 665 | 0.30 | 0.50 |
| Type - PWRs | 1984 | 1,021.3 | 58.8 | 3,062 | 1,945 | 0.64 | 1.90 |
| Capacity - 903, 903 MWe | 1985 | 1,516.9 | 86.1 | 2,436 | 838 | 0.34 | 0.55 |
| | 1986 | 1,484.5 | 83.0 | 2,831 | 722 | 0.26 | 0.49 |
| | 1987 | 1,112.6 | 67.8 | 2,624 | 1,521 | 0.58 | 1.37 |

| Reporting Organization | Year | Megawatt Years (MW-yr) | Unit Availability Factor | Total Personnel with Measurable Doses | Collective Dose per Site (person-rem) | Average Measurable Dose (rem) | Collective Dose/ MW-yr |
|---|---|---|---|---|---|---|---|
| **NORTH ANNA 1, 2** | 1988 | 1,772.7 | 96.7 | 992 | 112 | 0.11 | 0.06 |
| (continued) | 1989 | 1,226.8 | 72.5 | 2,861 | 1,471 | 0.51 | 1.20 |
| | 1990 | 1,590.4 | 90.5 | 2,161 | 590 | 0.27 | 0.37 |
| | 1991 | 1,597.5 | 88.6 | 2,085 | 629 | 0.30 | 0.39 |
| | 1992 | 1,403.2 | 84.1 | 2,159 | 576 | 0.27 | 0.41 |
| | 1993 | 1,428.4 | 80.1 | 2,768 | 908 | 0.33 | 0.64 |
| | 1994 | 1,717.1 | 95.9 | 1,036 | 193 | 0.19 | 0.11 |
| | 1995 | 1,666.4 | 90.8 | 1,551 | 367 | 0.24 | 0.22 |
| | 1996 | 1,569.6 | 89.1 | 1,203 | 291 | 0.24 | 0.19 |
| | 1997 | 1,711.5 | 96.2 | 856 | 103 | 0.12 | 0.06 |
| | 1998 | 1,632.8 | 92.7 | 1,201 | 265.922 | 0.22 | 0.16 |
| | 1999 | 1,747.7 | 96.1 | 727 | 94.402 | 0.13 | 0.05 |
| | 2000 | 1,734.1 | 95.8 | 730 | 65.405 | 0.09 | 0.04 |
| | 2001 | 1,491.0 | 84.8 | 1,231 | 308.907 | 0.25 | 0.21 |
| | 2002 | 1,557.0 | 84.3 | 914 | 143.312 | 0.16 | 0.09 |
| | 2003 | 1,569.1 | 87.2 | 1,041 | 187.014 | 0.18 | 0.12 |
| | 2004 | 1,685.6 | 92.0 | 965 | 129.686 | 0.13 | 0.08 |
| | 2005 | 1,751.5 | 96.0 | 686 | 58.844 | 0.09 | 0.03 |
| | 2006 | 1,723.0 | 95.0 | 749 | 82.069 | 0.11 | 0.05 |
| | 2007 | 1,596.7 | 88.0 | 1,581 | 309.237 | 0.20 | 0.19 |
| | 2008 | 1,643.1 | 91.2 | 795 | 61.003 | 0.08 | 0.04 |
| | 2009 | 1,735.5 | 95.6 | 745 | 78.126 | 0.10 | 0.05 |
| | 2010 | 1,529.6 | 84.9 | 1,032 | 182.289 | 0.18 | 0.12 |
| **OCONEE 1, 2, 3** | 1974 | 650.6 | 60.1 | 844 | 517 | 0.61 | 0.79 |
| Docket 50-269, 50-270, 50-287; | 1975 | 1,838.3 | 75.5 | 829 | 497 | 0.60 | 0.27 |
| DPR-38, DPR-47, DPR-55 | 1976 | 1,561.4 | 63.0 | 1,215 | 1,026 | 0.84 | 0.66 |
| 1st commercial operation | 1977 | 1,566.4 | 65.9 | 1,595 | 1,329 | 0.83 | 0.85 |
| 7/73, 9/74, 12/74 | 1978 | 1,909.0 | 75.8 | 1,636 | 1,393 | 0.85 | 0.73 |
| Type - PWRs | 1979 | 1,708.0 | 67.7 | 2,100 | 1,001 | 0.48 | 0.59 |
| Capacity - 846, 846, 846 MWe | 1980 | 1,703.7 | 70.1 | 2,124 | 1,055 | 0.50 | 0.62 |
| | 1981 | 1,661.5 | 66.8 | 2,445 | 1,211 | 0.50 | 0.73 |
| | 1982 | 1,293.1 | 52.5 | 2,445 | 1,792 | 0.73 | 1.39 |
| | 1983 | 2,141.5 | 82.2 | 1,902 | 1,207 | 0.63 | 0.56 |
| | 1984 | 2,242.9 | 85.7 | 2,085 | 1,106 | 0.53 | 0.49 |
| | 1985 | 2,036.3 | 80.5 | 2,729 | 1,304 | 0.48 | 0.64 |
| | 1986 | 1,995.6 | 79.0 | 2,499 | 949 | 0.38 | 0.48 |
| | 1987 | 1,962.6 | 82.4 | 2,672 | 1,142 | 0.43 | 0.58 |
| | 1988 | 2,228.9 | 87.2 | 2,672 | 871 | 0.33 | 0.39 |
| | 1989 | 2,188.6 | 85.4 | 2,205 | 684 | 0.31 | 0.31 |
| | 1990 | 2,405.2 | 91.4 | 1,948 | 404 | 0.21 | 0.17 |
| | 1991 | 2,275.0 | 86.7 | 1,966 | 551 | 0.28 | 0.24 |
| | 1992 | 2,110.7 | 82.0 | 1,954 | 612 | 0.31 | 0.29 |
| | 1993 | 2,399.2 | 91.3 | 1,499 | 237 | 0.16 | 0.10 |
| | 1994 | 2,144.3 | 82.2 | 1,923 | 537 | 0.28 | 0.25 |
| | 1995 | 2,366.1 | 89.5 | 1,586 | 304 | 0.19 | 0.13 |
| | 1996 | 1,847.9 | 70.3 | 1,479 | 257 | 0.17 | 0.14 |
| | 1997 | 1,563.7 | 67.7 | 1,379 | 223 | 0.16 | 0.14 |
| | 1998 | 1,989.1 | 81.3 | 1,695 | 366.028 | 0.22 | 0.18 |
| | 1999 | 2,264.5 | 90.3 | 1,568 | 202.025 | 0.13 | 0.09 |
| | 2000 | 2,321.0 | 91.6 | 1,686 | 272.697 | 0.16 | 0.12 |
| | 2001 | 2,167.6 | 86.8 | 2,002 | 579.209 | 0.29 | 0.27 |
| | 2002 | 2,355.0 | 92.5 | 1,723 | 224.672 | 0.13 | 0.10 |
| | 2003 | 2,177.7 | 86.3 | 2,180 | 245.349 | 0.11 | 0.11 |
| | 2004 | 2,125.2 | 84.1 | 2,295 | 367.891 | 0.16 | 0.17 |
| | 2005 | 2,349.5 | 92.3 | 1,516 | 148.694 | 0.10 | 0.06 |
| | 2006 | 2,274.8 | 90.0 | 1,859 | 221.222 | 0.12 | 0.10 |
| | 2007 | 2,347.8 | 92.0 | 1,915 | 252.936 | 0.13 | 0.11 |
| | 2008 | 2298.5 | 90.9 | 1,924 | 186.335 | 0.10 | 0.08 |
| | 2009 | 2,385.7 | 92.6 | 1,830 | 180.868 | 0.10 | 0.08 |
| | 2010 | 2,391.1 | 93.3 | 1,953 | 193.088 | 0.10 | 0.08 |

| Reporting Organization | Year | Megawatt Years (MW-yr) | Unit Availability Factor | Total Personnel with Measurable Doses | Collective Dose per Site (person-rem) | Average Measurable Dose (rem) | Collective Dose/ MW-yr |
|---|---|---|---|---|---|---|---|
| **OYSTER CREEK** | 1970 | 413.6 | | 95 | 63 | 0.66 | 0.15 |
| Docket 50-219; | 1971 | 448.9 | | 249 | 240 | 0.96 | 0.53 |
| DPR-16 | 1972 | 515.0 | | 339 | 582 | 1.72 | 1.13 |
| 1st commercial operation 12/69 | 1973 | 424.6 | | 782 | 1,236 | 1.58 | 2.91 |
| Type - BWR | 1974 | 434.5 | 70.4 | 935 | 984 | 1.05 | 2.26 |
| Capacity - 619 MWe | 1975 | 373.6 | 73.3 | 1,210 | 1,140 | 0.94 | 3.05 |
| | 1976 | 456.5 | 79.3 | 1,582 | 1,078 | 0.68 | 2.36 |
| | 1977 | 385.7 | 70.1 | 1,673 | 1,614 | 0.96 | 4.18 |
| | 1978 | 431.8 | 74.3 | 1,411 | 1,279 | 0.91 | 2.96 |
| | 1979 | 541.0 | 85.9 | 842 | 467 | 0.55 | 0.86 |
| | 1980 | 232.9 | 41.4 | 1,966 | 1,733 | 0.88 | 7.44 |
| | 1981 | 314.8 | 59.8 | 1,689 | 917 | 0.54 | 2.91 |
| | 1982 | 242.7 | 62.5 | 1,270 | 865 | 0.68 | 3.56 |
| | 1983 | 27.9 | 11.5 | 2,303 | 2,257 | 0.98 | 80.90 |
| | 1984 | 37.1 | 9.6 | 2,369 | 2,054 | 0.87 | 55.36 |
| | 1985 | 446.1 | 89.4 | 2,342 | 748 | 0.32 | 1.68 |
| | 1986 | 157.3 | 31.5 | 3,740 | 2,436 | 0.65 | 15.49 |
| | 1987 | 371.0 | 64.2 | 1,932 | 522 | 0.27 | 1.41 |
| | 1988 | 419.6 | 65.9 | 2,875 | 1,504 | 0.52 | 3.58 |
| | 1989 | 287.5 | 57.3 | 2,395 | 910 | 0.38 | 3.17 |
| | 1990 | 511.8 | 89.1 | 1,941 | 310 | 0.16 | 0.61 |
| | 1991 | 351.6 | 60.5 | 3,089 | 1,185 | 0.38 | 3.37 |
| | 1992 | 536.3 | 85.9 | 2,771 | 657 | 0.24 | 1.23 |
| | 1993 | 551.9 | 87.8 | 2,560 | 416 | 0.16 | 0.75 |
| | 1994 | 431.7 | 70.8 | 2,382 | 844 | 0.35 | 1.96 |
| | 1995 | 615.4 | 97.4 | 761 | 90 | 0.12 | 0.15 |
| | 1996 | 515.0 | 82.6 | 1,833 | 449 | 0.24 | 0.87 |
| | 1997 | 579.1 | 94.3 | 509 | 50 | 0.10 | 0.09 |
| | 1998 | 490.8 | 82.4 | 1,408 | 308.323 | 0.22 | 0.63 |
| | 1999 | 615.1 | 100.0 | 466 | 41.664 | 0.09 | 0.07 |
| | 2000 | 444.9 | 83.3 | 2,044 | 614.379 | 0.30 | 1.38 |
| | 2001 | 595.0 | 97.6 | 442 | 45.817 | 0.10 | 0.08 |
| | 2002 | 573.0 | 94.0 | 1,468 | 265.810 | 0.18 | 0.46 |
| | 2003 | 598.4 | 97.2 | 416 | 43.363 | 0.10 | 0.07 |
| | 2004 | 551.8 | 91.6 | 1,346 | 226.880 | 0.17 | 0.41 |
| | 2005 | 611.9 | 99.5 | 316 | 27.813 | 0.09 | 0.05 |
| | 2006 | 530.2 | 90.0 | 1,443 | 189.950 | 0.13 | 0.36 |
| | 2007 | 579.7 | 97.0 | 464 | 46.590 | 0.10 | 0.08 |
| | 2008 | 531.0 | 91.0 | 1,511 | 211.932 | 0.14 | 0.40 |
| | 2009 | 568.3 | 96.4 | 382 | 37.272 | 0.10 | 0.07 |
| | 2010 | 525.7 | 89.9 | 1,655 | 206.284 | 0.12 | 0.39 |
| **PALISADES** | 1972 | 216.8 | | | 78 | | 0.36 |
| Docket 50-255; | 1973 | 286.8 | | 975 | 1,133 | 1.16 | 3.95 |
| DPR-20 | 1974 | 10.7 | 5.5 | 774 | 627 | 0.81 | 58.60 |
| 1st commercial operation 12/71 | 1975 | 302.0 | 64.5 | 495 | 306 | 0.62 | 1.01 |
| Type - PWR | 1976 | 346.9 | 55.2 | 742 | 696 | 0.94 | 2.01 |
| Capacity - 744 MWe | 1977 | 616.6 | 91.4 | 332 | 100 | 0.30 | 0.16 |
| | 1978 | 320.2 | 49.7 | 849 | 764 | 0.90 | 2.39 |
| | 1979 | 415.0 | 59.9 | 1,599 | 854 | 0.53 | 2.06 |
| | 1980 | 288.3 | 42.9 | 1,307 | 424 | 0.32 | 1.47 |
| | 1981 | 418.2 | 57.2 | 2,151 | 902 | 0.42 | 2.16 |
| | 1982 | 404.3 | 54.7 | 1,554 | 330 | 0.21 | 0.82 |
| | 1983 | 454.4 | 60.3 | 2,167 | 977 | 0.45 | 2.15 |
| | 1984 | 98.7 | 15.2 | 1,344 | 573 | 0.43 | 5.81 |
| | 1985 | 639.2 | 83.8 | 1,355 | 507 | 0.37 | 0.79 |
| | 1986 | 102.3 | 15.1 | 1,438 | 672 | 0.47 | 6.57 |
| | 1987 | 319.2 | 48.2 | 1,122 | 456 | 0.41 | 1.43 |
| | 1988 | 413.4 | 56.8 | 1,472 | 730 | 0.50 | 1.77 |
| | 1989 | 442.8 | 69.1 | 1,026 | 314 | 0.31 | 0.71 |
| | 1990 | 366.7 | 58.7 | 2,414 | 766 | 0.32 | 2.09 |
| | 1991 | 587.0 | 78.1 | 1,315 | 211 | 0.16 | 0.36 |
| | 1992 | 581.9 | 76.1 | 1,267 | 295 | 0.23 | 0.51 |

| Reporting Organization | Year | Megawatt Years (MW-yr) | Unit Availability Factor | Total Personnel with Measurable Doses | Collective Dose per Site (person-rem) | Average Measurable Dose (rem) | Collective Dose/ MW-yr |
|---|---|---|---|---|---|---|---|
| **PALISADES** | 1993 | 424.4 | 53.7 | 908 | 289 | 0.32 | 0.68 |
| (continued) | 1994 | 541.8 | 67.0 | 397 | 60 | 0.15 | 0.11 |
| | 1995 | 583.5 | 75.8 | 1,230 | 462 | 0.38 | 0.79 |
| | 1996 | 638.2 | 81.4 | 1,109 | 318 | 0.29 | 0.50 |
| | 1997 | 662.5 | 89.9 | 338 | 48 | 0.14 | 0.07 |
| | 1998 | 615.4 | 83.5 | 895 | 216.563 | 0.24 | 0.35 |
| | 1999 | 585.4 | 80.2 | 939 | 218.451 | 0.23 | 0.37 |
| | 2000 | 654.4 | 88.0 | 255 | 26.305 | 0.10 | 0.04 |
| | 2001 | 268.2 | 36.3 | 1,032 | 362.723 | 0.35 | 1.35 |
| | 2002 | 725.0 | 94.8 | 224 | 24.380 | 0.11 | 0.03 |
| | 2003 | 701.1 | 90.7 | 822 | 202.571 | 0.25 | 0.29 |
| | 2004 | 608.6 | 82.3 | 974 | 370.895 | 0.38 | 0.61 |
| | 2005 | 756.6 | 98.0 | 156 | 10.459 | 0.07 | 0.01 |
| | 2006 | 675.5 | 86.0 | 882 | 239.652 | 0.27 | 0.36 |
| | 2007 | 665.6 | 85.0 | 1,065 | 256.632 | 0.24 | 0.39 |
| | 2008 | 778.4 | 98.2 | 272 | 23.478 | 0.09 | 0.03 |
| | 2009 | 698.5 | 89.0 | 975 | 267.295 | 0.27 | 0.38 |
| | 2010 | 712.5 | 90.8 | 908 | 219.873 | 0.24 | 0.31 |
| **PALO VERDE 1, 2, 3** | 1987 | 1,638.1 | 66.1 | 1,792 | 669 | 0.37 | 0.41 |
| Docket 50-528, 50-529, 50-530; | 1988 | 1,700.9 | 65.5 | 2,173 | 688 | 0.32 | 0.40 |
| NPF-41, NPF-51, NPF-74 | 1989 | 965.3 | 26.5 | 2,615 | 720 | 0.28 | 0.75 |
| 1st commercial operation | 1990 | 2,500.9 | 67.5 | 2,236 | 499 | 0.22 | 0.20 |
| 1/86, 9/86, 1/88 | 1991 | 3,043.9 | 78.9 | 2,242 | 605 | 0.27 | 0.20 |
| Type - PWRs | 1992 | 3,102.3 | 82.0 | 1,981 | 541 | 0.27 | 0.17 |
| Capacity - 1,311, 1,314, | 1993 | 2,677.1 | 74.3 | 2,124 | 592 | 0.28 | 0.22 |
| 1,312 MWe | 1994 | 2,827.6 | 79.1 | 2,048 | 462 | 0.23 | 0.16 |
| | 1995 | 3,265.2 | 85.6 | 1,875 | 482 | 0.26 | 0.15 |
| | 1996 | 3,482.7 | 90.0 | 1,717 | 302 | 0.18 | 0.09 |
| | 1997 | 3,369.2 | 92.2 | 1,585 | 246 | 0.16 | 0.07 |
| | 1998 | 3,454.4 | 93.2 | 1,410 | 192.425 | 0.14 | 0.06 |
| | 1999 | 3,471.2 | 93.2 | 1,275 | 146.328 | 0.11 | 0.04 |
| | 2000 | 3,458.6 | 93.0 | 1,279 | 158.105 | 0.12 | 0.05 |
| | 2001 | 3,280.2 | 88.6 | 1,361 | 182.043 | 0.13 | 0.06 |
| | 2002 | 3,513.0 | 94.0 | 1,343 | 140.057 | 0.10 | 0.04 |
| | 2003 | 3,254.4 | 88.6 | 1,943 | 210.842 | 0.11 | 0.06 |
| | 2004 | 3,201.4 | 86.3 | 1,324 | 199.016 | 0.15 | 0.06 |
| | 2005 | 2,937.6 | 80.4 | 2,014 | 200.300 | 0.10 | 0.07 |
| | 2006 | 2,741.1 | 79.0 | 1,585 | 151.516 | 0.10 | 0.06 |
| | 2007 | 3,058.5 | 81.0 | 2,372 | 148.660 | 0.06 | 0.05 |
| | 2008 | 3,330.0 | 86.1 | 1,706 | 159.913 | 0.09 | 0.05 |
| | 2009 | 3,500.2 | 89.6 | 1,695 | 97.902 | 0.06 | 0.03 |
| | 2010 | 3,561.6 | 90.9 | 1,655 | 112.612 | 0.07 | 0.03 |
| **PEACH BOTTOM 2, 3** | 1975 | 1,234.3 | 80.9 | 971 | 228 | 0.23 | 0.18 |
| Docket 50-277, 50-278; | 1976 | 1,379.2 | 73.0 | 2,136 | 840 | 0.39 | 0.61 |
| DPR-44, DPR-56 | 1977 | 1,052.4 | 58.7 | 2,827 | 2,036 | 0.72 | 1.93 |
| 1st commercial operation | 1978 | 1,636.3 | 84.0 | 2,244 | 1,317 | 0.59 | 0.80 |
| 7/74, 12/74 | 1979 | 1,740.0 | 84.5 | 2,276 | 1,388 | 0.61 | 0.80 |
| Type - BWRs | 1980 | 1,374.2 | 66.3 | 2,774 | 2,302 | 0.83 | 1.68 |
| Capacity - 1,112, 1,112 MWe | 1981 | 1,161.8 | 58.0 | 2,857 | 2,506 | 0.88 | 2.16 |
| | 1982 | 1,583.3 | 76.9 | 2,734 | 1,977 | 0.72 | 1.25 |
| | 1983 | 824.7 | 41.0 | 3,107 | 2,963 | 0.95 | 3.59 |
| | 1984 | 1,165.8 | 57.5 | 3,313 | 2,450 | 0.74 | 2.10 |
| | 1985 | 682.7 | 37.5 | 4,209 | 3,354 | 0.80 | 4.91 |
| | 1986 | 1,395.0 | 71.7 | 2,454 | 1,080 | 0.44 | 0.77 |
| | 1987 | 365.7 | 20.3 | 4,363 | 2,195 | 0.50 | 6.00 |
| | 1988 | 0.0 | 0.0 | 4,204 | 2,327 | 0.55 | — |
| | 1989 | 491.0 | 35.0 | 2,301 | 728 | 0.32 | 1.48 |
| | 1990 | 1,684.0 | 85.7 | 1,585 | 377 | 0.24 | 0.22 |
| | 1991 | 1,210.9 | 62.3 | 2,702 | 934 | 0.35 | 0.77 |
| | 1992 | 1,516.6 | 78.7 | 1,911 | 502 | 0.26 | 0.33 |
| | 1993 | 1,654.0 | 81.9 | 1,757 | 552 | 0.31 | 0.33 |
| | 1994 | 1,927.4 | 93.8 | 2,133 | 579 | 0.27 | 0.30 |

| Reporting Organization | Year | Megawatt Years (MW-yr) | Unit Availability Factor | Total Personnel with Measurable Doses | Collective Dose per Site (person-rem) | Average Measurable Dose (rem) | Collective Dose/ MW-yr |
|---|---|---|---|---|---|---|---|
| **PEACH BOTTOM 2, 3** | 1995 | 1,955.9 | 95.1 | 1,940 | 398 | 0.21 | 0.20 |
| (continued) | 1996 | 2,012.4 | 96.9 | 1,657 | 282 | 0.17 | 0.14 |
| | 1997 | 1,956.3 | 95.0 | 1,872 | 490 | 0.26 | 0.25 |
| | 1998 | 1,881.2 | 93.2 | 1,903 | 366.040 | 0.19 | 0.19 |
| | 1999 | 2,057.2 | 96.0 | 1,630 | 319.307 | 0.20 | 0.16 |
| | 2000 | 2,058.3 | 96.7 | 1,729 | 330.928 | 0.19 | 0.16 |
| | 2001 | 2,037.1 | 95.8 | 1,445 | 344.283 | 0.24 | 0.17 |
| | 2002 | 2,105.0 | 96.7 | 1,915 | 333.056 | 0.17 | 0.16 |
| | 2003 | 2,072.4 | 94.9 | 1,641 | 355.969 | 0.22 | 0.17 |
| | 2004 | 2,148.8 | 96.4 | 1,422 | 264.727 | 0.19 | 0.12 |
| | 2005 | 2,102.0 | 95.6 | 1,801 | 306.201 | 0.17 | 0.15 |
| | 2006 | 2,169.1 | 97.0 | 1,513 | 247.676 | 0.16 | 0.11 |
| | 2007 | 2,163.8 | 97.0 | 1,906 | 384.795 | 0.20 | 0.18 |
| | 2008 | 2,115.3 | 95.1 | 1,816 | 212.741 | 0.12 | 0.10 |
| | 2009 | 2,130.4 | 95.5 | 2,032 | 310.517 | 0.15 | 0.15 |
| | 2010 | 2,145.3 | 96.2 | 1,716 | 219.372 | 0.13 | 0.10 |
| **PERRY** | 1988 | 869.3 | 79.0 | 782 | 105 | 0.13 | 0.12 |
| Docket 50-440; | 1989 | 642.2 | 57.0 | 1,883 | 767 | 0.41 | 1.19 |
| NPF-58 | 1990 | 792.7 | 67.1 | 1,537 | 638 | 0.42 | 0.80 |
| 1st commercial operation 11/87 | 1991 | 1,074.2 | 91.9 | 600 | 146 | 0.24 | 0.14 |
| Type - BWR | 1992 | 856.2 | 75.5 | 1,487 | 571 | 0.38 | 0.67 |
| Capacity - 1,240 MWe | 1993 | 479.2 | 48.2 | 1,235 | 278 | 0.23 | 0.58 |
| | 1994 | 550.8 | 50.2 | 2,098 | 691 | 0.33 | 1.25 |
| | 1995 | 1,090.9 | 95.6 | 587 | 64 | 0.11 | 0.06 |
| | 1996 | 895.6 | 77.2 | 1,622 | 307 | 0.19 | 0.34 |
| | 1997 | 930.6 | 84.7 | 1,524 | 272 | 0.18 | 0.29 |
| | 1998 | 1,163.1 | 99.3 | 385 | 41.945 | 0.11 | 0.04 |
| | 1999 | 1,041.7 | 89.9 | 1,758 | 326.014 | 0.19 | 0.31 |
| | 2000 | 1,148.2 | 97.1 | 501 | 55.827 | 0.11 | 0.05 |
| | 2001 | 885.9 | 79.6 | 1,392 | 258.268 | 0.19 | 0.29 |
| | 2002 | 1,136.0 | 95.0 | 436 | 70.258 | 0.16 | 0.06 |
| | 2003 | 973.7 | 83.8 | 1,880 | 607.384 | 0.32 | 0.62 |
| | 2004 | 1,164.3 | 95.9 | 496 | 73.481 | 0.15 | 0.06 |
| | 2005 | 872.9 | 73.8 | 1,734 | 416.608 | 0.24 | 0.48 |
| | 2006 | 1,195.8 | 99.0 | 488 | 65.152 | 0.13 | 0.05 |
| | 2007 | 919.7 | 79.0 | 1,650 | 505.121 | 0.31 | 0.55 |
| | 2008 | 1,215.9 | 97.9 | 528 | 52.058 | 0.10 | 0.04 |
| | 2009 | 869.2 | 73.3 | 1,818 | 614.959 | 0.34 | 0.71 |
| | 2010 | 1,213.3 | 98.5 | 278 | 32.186 | 0.12 | 0.03 |
| **PILGRIM 1** | 1973 | 484.0 | | 230 | 126 | 0.55 | 0.26 |
| Docket 50-293; | 1974 | 234.1 | 39.2 | 454 | 415 | 0.91 | 1.77 |
| DPR-35 | 1975 | 308.1 | 71.3 | 473 | 798 | 1.69 | 2.59 |
| 1st commercial operation 12/72 | 1976 | 287.8 | 60.7 | 1,317 | 2,648 | 2.01 | 9.20 |
| Type - BWR | 1977 | 316.6 | 61.4 | 1,875 | 3,142 | 1.68 | 9.92 |
| Capacity - 685 MWe | 1978 | 519.5 | 83.1 | 1,667 | 1,327 | 0.80 | 2.55 |
| | 1979 | 574.0 | 89.4 | 2,458 | 1,015 | 0.41 | 1.77 |
| | 1980 | 360.3 | 56.2 | 3,549 | 3,626 | 1.02 | 10.06 |
| | 1981 | 408.9 | 65.9 | 2,803 | 1,836 | 0.66 | 4.49 |
| | 1982 | 389.9 | 63.9 | 2,854 | 1,539 | 0.54 | 3.95 |
| | 1983 | 559.5 | 87.2 | 2,326 | 1,162 | 0.50 | 2.08 |
| | 1984 | 1.4 | 0.4 | 4,542 | 4,082 | 0.90 | 2,915.71 |
| | 1985 | 587.3 | 91.5 | 2,209 | 893 | 0.40 | 1.52 |
| | 1986 | 121.9 | 18.8 | 2,635 | 874 | 0.33 | 7.17 |
| | 1987 | 0.0 | 0.0 | 4,710 | 1,579 | 0.34 | --- |
| | 1988 | 0.0 | 0.0 | 2,073 | 392 | 0.19 | --- |
| | 1989 | 204.6 | 64.1 | 1,797 | 207 | 0.12 | 1.01 |
| | 1990 | 503.5 | 82.1 | 1,898 | 225 | 0.12 | 0.45 |
| | 1991 | 406.3 | 65.8 | 2,836 | 605 | 0.21 | 1.49 |
| | 1992 | 561.0 | 85.4 | 1,332 | 281 | 0.21 | 0.50 |
| | 1993 | 513.7 | 80.9 | 1,328 | 435 | 0.33 | 0.85 |
| | 1994 | 453.6 | 71.4 | 758 | 200 | 0.26 | 0.44 |

| Reporting Organization | Year | Megawatt Years (MW-yr) | Unit Availability Factor | Total Personnel with Measurable Doses | Collective Dose per Site (person-rem) | Average Measurable Dose (rem) | Collective Dose/ MW-yr |
|---|---|---|---|---|---|---|---|
| **PILGRIM 1** | 1995 | 531.7 | 80.7 | 1,294 | 482 | 0.37 | 0.91 |
| (continued) | 1996 | 631.3 | 95.4 | 517 | 116 | 0.22 | 0.18 |
| | 1997 | 492.1 | 80.7 | 1,655 | 588 | 0.36 | 1.19 |
| | 1998 | 650.5 | 100.0 | 530 | 71.446 | 0.13 | 0.11 |
| | 1999 | 510.7 | 84.4 | 1,222 | 344.270 | 0.28 | 0.67 |
| | 2000 | 627.5 | 98.3 | 422 | 50.797 | 0.12 | 0.08 |
| | 2001 | 585.6 | 91.0 | 1,113 | 179.585 | 0.16 | 0.31 |
| | 2002 | 657.0 | 100.0 | 463 | 38.280 | 0.08 | 0.06 |
| | 2003 | 566.6 | 87.5 | 1,437 | 250.192 | 0.17 | 0.44 |
| | 2004 | 676.1 | 99.5 | 427 | 41.109 | 0.10 | 0.06 |
| | 2005 | 623.2 | 93.7 | 1,212 | 206.089 | 0.17 | 0.33 |
| | 2006 | 665.4 | 100.0 | 654 | 43.531 | 0.07 | 0.07 |
| | 2007 | 584.5 | 90.0 | 1,407 | 240.526 | 0.17 | 0.41 |
| | 2008 | 668.1 | 99.0 | 377 | 22.568 | 0.06 | 0.03 |
| | 2009 | 616.0 | 91.7 | 1,301 | 264.215 | 0.20 | 0.43 |
| | 2010 | 675.5 | 100.0 | 303 | 25.739 | 0.08 | 0.04 |
| **POINT BEACH 1, 2** | 1971 | 393.4 | | | 164 | | 0.42 |
| Docket 50-266, 50-301; | 1972 | 378.3 | | | 580 | | 1.53 |
| DPR-24, DPR-27 | 1973 | 693.7 | | 501 | 588 | 1.17 | 0.85 |
| 1st commercial operation | 1974 | 760.2 | 81.3 | 400 | 295 | 0.74 | 0.39 |
| 12/70, 10/72 | 1975 | 801.2 | 82.9 | 339 | 459 | 1.35 | 0.57 |
| Type - PWRs | 1976 | 857.3 | 86.7 | 313 | 370 | 1.18 | 0.43 |
| Capacity - 516, 518 MWe | 1977 | 873.9 | 87.3 | 417 | 430 | 1.03 | 0.49 |
| | 1978 | 914.4 | 90.9 | 336 | 320 | 0.95 | 0.35 |
| | 1979 | 808.0 | 80.8 | 610 | 644 | 1.06 | 0.80 |
| | 1980 | 727.2 | 82.5 | 561 | 598 | 1.07 | 0.82 |
| | 1981 | 760.4 | 83.6 | 773 | 596 | 0.77 | 0.78 |
| | 1982 | 757.2 | 84.3 | 767 | 609 | 0.79 | 0.80 |
| | 1983 | 648.2 | 72.7 | 1,702 | 1,403 | 0.82 | 2.16 |
| | 1984 | 788.9 | 78.6 | 1,372 | 789 | 0.58 | 1.00 |
| | 1985 | 831.3 | 82.5 | 671 | 482 | 0.72 | 0.58 |
| | 1986 | 858.9 | 85.7 | 664 | 402 | 0.61 | 0.47 |
| | 1987 | 857.5 | 85.5 | 720 | 554 | 0.77 | 0.65 |
| | 1988 | 899.3 | 88.6 | 734 | 410 | 0.56 | 0.46 |
| | 1989 | 847.8 | 85.5 | 736 | 504 | 0.68 | 0.59 |
| | 1990 | 875.5 | 86.5 | 617 | 378 | 0.61 | 0.43 |
| | 1991 | 874.8 | 87.1 | 724 | 265 | 0.37 | 0.30 |
| | 1992 | 866.7 | 85.8 | 617 | 256 | 0.41 | 0.30 |
| | 1993 | 911.0 | 90.0 | 559 | 186 | 0.33 | 0.20 |
| | 1994 | 914.5 | 91.2 | 548 | 170 | 0.31 | 0.19 |
| | 1995 | 858.4 | 86.1 | 548 | 190 | 0.35 | 0.22 |
| | 1996 | 831.6 | 84.7 | 1,029 | 276 | 0.27 | 0.33 |
| | 1997 | 186.8 | 21.8 | 670 | 92 | 0.14 | 0.49 |
| | 1998 | 649.7 | 69.7 | 881 | 169.253 | 0.19 | 0.26 |
| | 1999 | 806.0 | 83.1 | 962 | 194.489 | 0.20 | 0.24 |
| | 2000 | 872.0 | 88.7 | 765 | 138.989 | 0.18 | 0.16 |
| | 2001 | 915.9 | 93.4 | 740 | 131.667 | 0.18 | 0.14 |
| | 2002 | 909.0 | 91.1 | 945 | 180.654 | 0.19 | 0.20 |
| | 2003 | 917.2 | 92.1 | 627 | 84.965 | 0.14 | 0.09 |
| | 2004 | 912.3 | 90.1 | 627 | 109.515 | 0.17 | 0.12 |
| | 2005 | 782.5 | 78.1 | 851 | 128.646 | 0.15 | 0.16 |
| | 2006 | 977.2 | 96.0 | 453 | 39.597 | 0.09 | 0.04 |
| | 2007 | 958.5 | 94.0 | 535 | 52.023 | 0.10 | 0.05 |
| | 2008 | 889.4 | 87.8 | 958 | 144.021 | 0.15 | 0.16 |
| | 2009 | 902.3 | 92.9 | 766 | 93.270 | 0.12 | 0.10 |
| | 2010 | 952.8 | 93.8 | 869 | 95.695 | 0.11 | 0.10 |

| Reporting Organization | Year | Megawatt Years (MW-yr) | Unit Availability Factor | Total Personnel with Measurable Doses | Collective Dose per Site (person-rem) | Average Measurable Dose (rem) | Collective Dose/ MW-yr |
|---|---|---|---|---|---|---|---|
| **PRAIRIE ISLAND 1, 2** | 1974 | 181.9 | 43.9 | 150 | 18 | 0.12 | 0.10 |
| Docket 50-282, 50-306; | 1975 | 836.0 | 83.3 | 477 | 123 | 0.26 | 0.15 |
| DPR-42, DPR-60 | 1976 | 725.2 | 76.6 | 818 | 447 | 0.55 | 0.62 |
| 1st commercial operation | 1977 | 922.9 | 87.2 | 718 | 300 | 0.42 | 0.33 |
| 12/73, 12/74 | 1978 | 941.1 | 92.2 | 546 | 221 | 0.40 | 0.23 |
| Type - PWRs | 1979 | 865.0 | 86.0 | 594 | 180 | 0.30 | 0.21 |
| Capacity - 522, 522 MWe | 1980 | 800.7 | 79.9 | 983 | 353 | 0.36 | 0.44 |
| | 1981 | 844.9 | 80.5 | 836 | 329 | 0.39 | 0.39 |
| | 1982 | 944.9 | 90.4 | 645 | 229 | 0.36 | 0.24 |
| | 1983 | 921.1 | 86.8 | 654 | 233 | 0.36 | 0.25 |
| | 1984 | 972.4 | 91.7 | 546 | 147 | 0.27 | 0.15 |
| | 1985 | 882.6 | 84.0 | 1,082 | 416 | 0.38 | 0.47 |
| | 1986 | 930.6 | 90.3 | 818 | 255 | 0.31 | 0.27 |
| | 1987 | 969.6 | 91.6 | 593 | 135 | 0.23 | 0.14 |
| | 1988 | 932.0 | 89.1 | 732 | 199 | 0.27 | 0.21 |
| | 1989 | 1,001.8 | 94.7 | 476 | 99 | 0.21 | 0.10 |
| | 1990 | 925.4 | 89.2 | 737 | 188 | 0.26 | 0.20 |
| | 1991 | 1,023.3 | 95.6 | 586 | 98 | 0.17 | 0.10 |
| | 1992 | 811.6 | 76.2 | 845 | 211 | 0.25 | 0.26 |
| | 1993 | 978.3 | 90.7 | 532 | 106 | 0.20 | 0.11 |
| | 1994 | 996.9 | 91.5 | 478 | 109 | 0.23 | 0.11 |
| | 1995 | 1,023.2 | 93.9 | 499 | 107 | 0.21 | 0.10 |
| | 1996 | 992.1 | 91.4 | 558 | 112 | 0.20 | 0.11 |
| | 1997 | 817.6 | 81.4 | 753 | 174 | 0.23 | 0.21 |
| | 1998 | 860.3 | 83.4 | 582 | 116.649 | 0.20 | 0.14 |
| | 1999 | 989.3 | 93.8 | 542 | 72.496 | 0.13 | 0.07 |
| | 2000 | 992.2 | 93.1 | 632 | 106.091 | 0.17 | 0.11 |
| | 2001 | 900.8 | 85.8 | 691 | 124.708 | 0.18 | 0.14 |
| | 2002 | 987.0 | 93.6 | 969 | 127.713 | 0.13 | 0.13 |
| | 2003 | 1,006.1 | 96.4 | 594 | 61.137 | 0.10 | 0.06 |
| | 2004 | 940.4 | 89.9 | 1,186 | 143.806 | 0.12 | 0.15 |
| | 2005 | 952.5 | 90.8 | 782 | 84.337 | 0.11 | 0.09 |
| | 2006 | 926.4 | 89.0 | 1,103 | 137.352 | 0.12 | 0.15 |
| | 2007 | 1,014.8 | 98.0 | 130 | 6.276 | 0.05 | 0.01 |
| | 2008 | 924.3 | 88.9 | 1,060 | 126.723 | 0.12 | 0.14 |
| | 2009 | 942.2 | 89.9 | 560 | 53.590 | 0.10 | 0.06 |
| | 2010 | 1,002.6 | 94.9 | 661 | 54.933 | 0.08 | 0.05 |
| **QUAD CITIES 1, 2** | 1974 | 958.1 | 72.3 | 678 | 482 | 0.71 | 0.50 |
| Docket 50-254, 50-265; | 1975 | 833.6 | 68.4 | 1,083 | 1,618 | 1.49 | 1.94 |
| DPR-29, DPR-30 | 1976 | 951.2 | 73.1 | 1,225 | 1,651 | 1.35 | 1.74 |
| 1st commercial operation | 1977 | 970.1 | 84.0 | 907 | 1,031 | 1.14 | 1.06 |
| 2/73, 3/73 | 1978 | 1,124.5 | 88.6 | 1,207 | 1,618 | 1.34 | 1.44 |
| Type - BWRs | 1979 | 1,075.0 | 84.6 | 1,688 | 2,158 | 1.28 | 2.01 |
| Capacity - 866, 871 MWe | 1980 | 866.9 | 64.4 | 3,089 | 4,838 | 1.57 | 5.58 |
| | 1981 | 1,156.9 | 81.1 | 2,246 | 3,146 | 1.40 | 2.72 |
| | 1982 | 1,018.7 | 76.0 | 2,314 | 3,757 | 1.62 | 3.69 |
| | 1983 | 1,088.5 | 79.2 | 1,802 | 2,491 | 1.38 | 2.29 |
| | 1984 | 994.6 | 65.7 | 1,678 | 1,579 | 0.94 | 1.59 |
| | 1985 | 1,268.0 | 82.7 | 1,184 | 990 | 0.84 | 0.78 |
| | 1986 | 1,093.2 | 71.0 | 1,451 | 950 | 0.65 | 0.87 |
| | 1987 | 1,126.6 | 75.3 | 1,429 | 720 | 0.50 | 0.64 |
| | 1988 | 1,173.7 | 84.1 | 1,486 | 827 | 0.56 | 0.70 |
| | 1989 | 1,196.3 | 85.9 | 1,721 | 900 | 0.52 | 0.75 |
| | 1990 | 1,148.9 | 77.8 | 2,186 | 1,028 | 0.47 | 0.89 |
| | 1991 | 1,044.5 | 73.2 | 1,722 | 509 | 0.30 | 0.49 |
| | 1992 | 960.8 | 68.0 | 2,413 | 1,157 | 0.48 | 1.20 |
| | 1993 | 974.9 | 67.0 | 2,150 | 849 | 0.39 | 0.87 |
| | 1994 | 681.5 | 48.7 | 2,163 | 1,128 | 0.52 | 1.66 |
| | 1995 | 1,002.5 | 70.4 | 2,041 | 736 | 0.36 | 0.73 |
| | 1996 | 876.6 | 60.1 | 2,248 | 1,025 | 0.46 | 1.17 |
| | 1997 | 935.3 | 66.5 | 2,474 | 654 | 0.26 | 0.70 |

| Reporting Organization | Year | Megawatt Years (MW-yr) | Unit Availability Factor | Total Personnel with Measurable Doses | Collective Dose per Site (person-rem) | Average Measurable Dose (rem) | Collective Dose/ MW-yr |
|---|---|---|---|---|---|---|---|
| QUAD CITIES 1, 2 | 1998 | 794.8 | 55.1 | 2,177 | 760.596 | 0.35 | 0.96 |
| (continued) | 1999 | 1,476.5 | 95.9 | 1,000 | 200.556 | 0.20 | 0.14 |
| | 2000 | 1,410.4 | 93.9 | 2,840 | 893.766 | 0.32 | 0.63 |
| | 2001 | 1,478.2 | 95.9 | 736 | 143.849 | 0.20 | 0.10 |
| | 2002 | 1,396.0 | 89.0 | 3,818 | 1,786.021 | 0.47 | 1.28 |
| | 2003 | 1,569.4 | 93.1 | 998 | 438.144 | 0.44 | 0.28 |
| | 2004 | 1,443.8 | 95.5 | 2,334 | 510.521 | 0.22 | 0.35 |
| | 2005 | 1,516.2 | 94.2 | 2,869 | 961.026 | 0.33 | 0.63 |
| | 2006 | 1,524.9 | 93.0 | 2,329 | 559.362 | 0.24 | 0.37 |
| | 2007 | 1,650.3 | 97.0 | 1,945 | 249.927 | 0.13 | 0.15 |
| | 2008 | 1,619.4 | 95.2 | 2,065 | 274.444 | 0.13 | 0.17 |
| | 2009 | 1,662.6 | 95.4 | 2,366 | 318.418 | 0.13 | 0.19 |
| | 2010 | 1,688.9 | 95.0 | 2,267 | 241.444 | 0.11 | 0.14 |
| RANCHO SECO[12] | 1976 | 268.1 | 30.4 | 297 | 58 | 0.20 | 0.22 |
| Docket 50-312; | 1977 | 706.4 | 77.1 | 515 | 391 | 0.76 | 0.55 |
| DPR-54 | 1978 | 607.7 | 80.5 | 508 | 323 | 0.64 | 0.53 |
| 1st commercial operation 4/75 | 1979 | 687.0 | 91.1 | 287 | 126 | 0.44 | 0.18 |
| Type - PWR | 1980 | 530.9 | 60.4 | 890 | 412 | 0.46 | 0.78 |
| Capacity - (873) MWe | 1981 | 321.2 | 40.2 | 772 | 402 | 0.52 | 1.25 |
| | 1982 | 409.5 | 53.3 | 766 | 337 | 0.44 | 0.82 |
| | 1983 | 347.9 | 46.8 | 1,338 | 787 | 0.59 | 2.26 |
| | 1984 | 460.0 | 58.3 | 802 | 222 | 0.28 | 0.48 |
| | 1985 | 238.7 | 30.8 | 1,764 | 756 | 0.43 | 3.17 |
| | 1986 | 0.0 | 0.0 | 1,513 | 402 | 0.27 | --- |
| | 1987 | 0.0 | 0.0 | 1,533 | 300 | 0.20 | --- |
| | 1988 | 355.8 | 63.1 | 693 | 78 | 0.11 | 0.22 |
| | 1989 | 179.9 | 54.7 | 603 | 81 | 0.13 | 0.45 |
| | 1990 | 0.0 | 0.0 | 111 | 13 | 0.12 | --- |
| | 1991 | 0.0 | 0.0 | 101 | 9 | 0.09 | --- |
| | 1992 | 0.0 | 0.0 | 70 | 7 | 0.10 | --- |
| | 1993 | 0.0 | 0.0 | 35 | 4 | 0.11 | --- |
| | 1994 | 0.0 | 0.0 | 18 | 1 | 0.06 | --- |
| | 1995 | 0.0 | 0.0 | 16 | 1 | 0.06 | --- |
| | 1996 | 0.0 | 0.0 | 16 | 1 | 0.04 | --- |
| | 1997 | 0.0 | 0.0 | 16 | 0 | 0.00 | --- |
| | 1998 | 0.0 | 0.0 | 61 | 2.661 | 0.04 | --- |
| | 1999 | 0.0 | 0.0 | 302 | 11.191 | 0.04 | --- |
| | 2000 | 0.0 | 0.0 | 219 | 25.795 | 0.12 | --- |
| | 2001 | 0.0 | 0.0 | 210 | 18.432 | 0.09 | --- |
| | 2002 | 0.0 | 0.0 | 193 | 27.346 | 0.14 | --- |
| | 2003 | 0.0 | 0.0 | 121 | 18.300 | 0.15 | --- |
| | 2004 | 0.0 | 0.0 | 122 | 14.890 | 0.12 | --- |
| | 2005 | 0.0 | 0.0 | 157 | 33.444 | 0.21 | --- |
| | 2006 | 0.0 | 0.0 | 143 | 31.793 | 0.22 | --- |
| | 2007 | 0.0 | 0.0 | 129 | 12.524 | 0.10 | --- |
| | 2008 | 0.0 | 0.0 | 84 | 2.434 | 0.03 | --- |
| RIVER BEND 1 | 1987 | 605.2 | 68.4 | 1,268 | 378 | 0.30 | 0.62 |
| Docket 50-458; | 1988 | 880.7 | 94.3 | 513 | 107 | 0.21 | 0.12 |
| NPF-47 | 1989 | 584.5 | 69.1 | 1,566 | 558 | 0.36 | 0.95 |
| 1st commercial operation 6/86 | 1990 | 682.2 | 78.0 | 1,616 | 489 | 0.30 | 0.72 |
| Type - BWR | 1991 | 814.7 | 87.2 | 780 | 144 | 0.18 | 0.18 |
| Capacity - 967 MWe | 1992 | 336.1 | 39.7 | 2,022 | 710 | 0.35 | 2.11 |
| | 1993 | 640.0 | 71.6 | 847 | 180 | 0.21 | 0.28 |
| | 1994 | 595.7 | 64.9 | 2,209 | 519 | 0.23 | 0.87 |
| | 1995 | 967.1 | 99.6 | 667 | 85 | 0.13 | 0.09 |
| | 1996 | 836.1 | 85.3 | 2,093 | 473 | 0.23 | 0.57 |
| | 1997 | 778.8 | 86.3 | 1,671 | 347 | 0.21 | 0.45 |

[12] Rancho Seco was shut down in June 1989 and is no longer in the count of operating reactors. Parentheses indicate plant capacity when plant was operational.

| Reporting Organization | Year | Megawatt Years (MW-yr) | Unit Availability Factor | Total Personnel with Measurable Doses | Collective Dose per Site (person-rem) | Average Measurable Dose (rem) | Collective Dose/ MW-yr |
|---|---|---|---|---|---|---|---|
| **RIVER BEND 1** | 1998 | 894.2 | 96.2 | 466 | 57.749 | 0.12 | 0.06 |
| (continued) | 1999 | 651.2 | 75.2 | 1,327 | 343.858 | 0.26 | 0.53 |
| | 2000 | 837.1 | 89.7 | 1,104 | 216.053 | 0.20 | 0.26 |
| | 2001 | 889.3 | 93.6 | 1,249 | 207.614 | 0.17 | 0.23 |
| | 2002 | 965.0 | 98.5 | 373 | 35.145 | 0.09 | 0.04 |
| | 2003 | 871.3 | 92.7 | 1,296 | 216.950 | 0.17 | 0.25 |
| | 2004 | 845.6 | 90.1 | 1,378 | 235.749 | 0.17 | 0.28 |
| | 2005 | 890.5 | 94.4 | 498 | 55.816 | 0.11 | 0.06 |
| | 2006 | 853.7 | 92.0 | 1,494 | 214.409 | 0.14 | 0.25 |
| | 2007 | 823.0 | 92.0 | 1,131 | 131.373 | 0.12 | 0.16 |
| | 2008 | 724.8 | 78.7 | 1,809 | 311.697 | 0.17 | 0.43 |
| | 2009 | 895.6 | 92.6 | 1,978 | 219.446 | 0.11 | 0.25 |
| | 2010 | 955.1 | 98.9 | 888 | 40.356 | 0.05 | 0.04 |
| **ROBINSON 2** | 1972 | 580.0 | | 245 | 215 | 0.88 | 0.37 |
| Docket 50-261; | 1973 | 455.1 | | 831 | 695 | 0.84 | 1.53 |
| DPR-23 | 1974 | 578.1 | 83.3 | 853 | 672 | 0.79 | 1.16 |
| 1st commercial operation 3/71 | 1975 | 501.8 | 72.7 | 849 | 1,142 | 1.35 | 2.28 |
| Type - PWR | 1976 | 585.5 | 84.7 | 597 | 715 | 1.20 | 1.22 |
| Capacity - 724 MWe | 1977 | 511.5 | 85.2 | 634 | 455 | 0.72 | 0.89 |
| | 1978 | 480.5 | 72.0 | 943 | 963 | 1.02 | 2.00 |
| | 1979 | 482.0 | 70.8 | 1,454 | 1,188 | 0.82 | 2.46 |
| | 1980 | 387.3 | 62.2 | 2,009 | 1,852 | 0.92 | 4.78 |
| | 1981 | 426.6 | 73.0 | 1,462 | 733 | 0.50 | 1.72 |
| | 1982 | 277.5 | 48.9 | 2,011 | 1,426 | 0.71 | 5.14 |
| | 1983 | 409.8 | 75.5 | 2,244 | 923 | 0.41 | 2.25 |
| | 1984 | 28.0 | 7.0 | 4,127 | 2,880 | 0.70 | 102.86 |
| | 1985 | 629.5 | 87.9 | 1,378 | 311 | 0.23 | 0.49 |
| | 1986 | 577.1 | 80.3 | 1,571 | 539 | 0.34 | 0.93 |
| | 1987 | 510.1 | 72.5 | 1,379 | 499 | 0.36 | 0.98 |
| | 1988 | 385.0 | 65.9 | 1,351 | 564 | 0.42 | 1.46 |
| | 1989 | 336.6 | 48.7 | 1,098 | 195 | 0.18 | 0.58 |
| | 1990 | 400.3 | 64.8 | 1,626 | 437 | 0.27 | 1.09 |
| | 1991 | 575.1 | 81.4 | 885 | 193 | 0.22 | 0.34 |
| | 1992 | 487.2 | 66.8 | 1,267 | 352 | 0.28 | 0.72 |
| | 1993 | 502.7 | 70.7 | 1,221 | 337 | 0.28 | 0.67 |
| | 1994 | 560.3 | 79.5 | 420 | 63 | 0.15 | 0.11 |
| | 1995 | 618.7 | 84.7 | 1,058 | 215 | 0.20 | 0.35 |
| | 1996 | 654.8 | 88.6 | 1,031 | 167 | 0.16 | 0.26 |
| | 1997 | 707.5 | 99.0 | 304 | 13 | 0.04 | 0.02 |
| | 1998 | 628.5 | 88.9 | 978 | 170.476 | 0.17 | 0.27 |
| | 1999 | 648.9 | 91.8 | 807 | 123.952 | 0.15 | 0.19 |
| | 2000 | 710.0 | 99.7 | 138 | 8.396 | 0.06 | 0.01 |
| | 2001 | 627.9 | 90.6 | 827 | 124.750 | 0.15 | 0.20 |
| | 2002 | 638.0 | 91.2 | 830 | 110.631 | 0.13 | 0.17 |
| | 2003 | 733.1 | 100.0 | 109 | 4.838 | 0.04 | 0.01 |
| | 2004 | 653.7 | 89.3 | 952 | 118.159 | 0.12 | 0.18 |
| | 2005 | 656.9 | 89.7 | 791 | 64.662 | 0.08 | 0.10 |
| | 2006 | 735.5 | 100.0 | 86 | 3.320 | 0.04 | 0.00 |
| | 2007 | 655.0 | 90.0 | 890 | 80.752 | 0.09 | 0.12 |
| | 2008 | 618.1 | 84.6 | 788 | 68.381 | 0.09 | 0.11 |
| | 2009 | 738.9 | 99.3 | 126 | 6.643 | 0.05 | 0.01 |
| | 2010 | 410.8 | 57.0 | 996 | 85.917 | 0.09 | 0.21 |
| **SALEM 1, 2** | 1978 | 546.4 | 55.6 | 574 | 122 | 0.21 | 0.22 |
| Docket 50-272, 50-311; | 1979 | 250.0 | 25.5 | 1,488 | 584 | 0.39 | 2.34 |
| DPR-70, DPR-75 | 1980 | 680.6 | 69.2 | 1,704 | 449 | 0.26 | 0.66 |
| 1st commercial operation | 1981 | 743.0 | 78.1 | 1,652 | 254 | 0.15 | 0.34 |
| 6/77, 10/81 | 1982 | 1,440.4 | 72.6 | 3,228 | 1,203 | 0.37 | 0.84 |
| Type - PWRs | 1983 | 742.0 | 30.5 | 2,383 | 581 | 0.24 | 0.78 |
| Capacity - 1,116, 1,134  MWe | 1984 | 650.1 | 31.8 | 1,395 | 681 | 0.49 | 1.05 |
| | 1985 | 1,657.7 | 75.8 | 1,112 | 204 | 0.18 | 0.12 |
| | 1986 | 1,484.3 | 70.4 | 3,554 | 599 | 0.17 | 0.40 |

| Reporting Organization | Year | Megawatt Years (MW-yr) | Unit Availability Factor | Total Personnel with Measurable Doses | Collective Dose per Site (person-rem) | Average Measurable Dose (rem) | Collective Dose/ MW-yr |
|---|---|---|---|---|---|---|---|
| **SALEM 1, 2** | 1987 | 1,478.2 | 73.3 | 2,543 | 600 | 0.24 | 0.41 |
| (continued) | 1988 | 1,591.6 | 73.6 | 1,609 | 503 | 0.31 | 0.32 |
| | 1989 | 1,675.4 | 79.5 | 2,944 | 338 | 0.11 | 0.20 |
| | 1990 | 1,362.6 | 65.1 | 3,636 | 272 | 0.07 | 0.20 |
| | 1991 | 1,726.4 | 79.3 | 4,201 | 458 | 0.11 | 0.27 |
| | 1992 | 1,200.9 | 61.1 | 4,376 | 431 | 0.10 | 0.36 |
| | 1993 | 1,366.3 | 65.4 | 3,559 | 408 | 0.11 | 0.30 |
| | 1994 | 1,367.4 | 73.8 | 950 | 188 | 0.20 | 0.14 |
| | 1995 | 558.1 | 29.3 | 1,195 | 218 | 0.18 | 0.39 |
| | 1996 | 0.0 | 0.0 | 1,671 | 300 | 0.18 | --- |
| | 1997 | 279.3 | 17.8 | 894 | 175 | 0.20 | 0.63 |
| | 1998 | 1,629.3 | 79.1 | 408 | 41.100 | 0.10 | 0.03 |
| | 1999 | 1,821.8 | 86.8 | 1,200 | 317.545 | 0.27 | 0.17 |
| | 2000 | 1,973.4 | 93.0 | 1,191 | 198.068 | 0.17 | 0.10 |
| | 2001 | 1,961.2 | 91.1 | 1,274 | 153.088 | 0.12 | 0.08 |
| | 2002 | 1,934.0 | 89.4 | 2,460 | 292.692 | 0.12 | 0.15 |
| | 2003 | 1,957.2 | 90.7 | 1,301 | 124.042 | 0.10 | 0.06 |
| | 2004 | 1,850.2 | 85.8 | 1,496 | 148.694 | 0.10 | 0.08 |
| | 2005 | 2,086.4 | 91.7 | 3,162 | 240.567 | 0.08 | 0.12 |
| | 2006 | 2,211.8 | 97.0 | 1,446 | 90.541 | 0.06 | 0.04 |
| | 2007 | 2,158.2 | 96.0 | 1,365 | 117.604 | 0.09 | 0.05 |
| | 2008 | 1,998.6 | 87.8 | 3,362 | 328.761 | 0.10 | 0.16 |
| | 2009 | 2,252.9 | 96.2 | 1,249 | 101.186 | 0.08 | 0.04 |
| | 2010 | 2,147.3 | 93.9 | 964 | 77.828 | 0.08 | 0.04 |
| **SAN ONOFRE 1[13], 2, 3** | 1969 | 314.1 | | 123 | 42 | 0.34 | 0.13 |
| Docket 50-206, 50-361, 50-362; | 1970 | 365.9 | | 251 | 155 | 0.62 | 0.42 |
| DPR-13; NPF-10, NPF-15 | 1971 | 362.1 | | 121 | 50 | 0.41 | 0.14 |
| 1st commercial operation | 1972 | 338.5 | | 326 | 256 | 0.79 | 0.76 |
| 1/68, 8/83, 4/84 | 1973 | 273.7 | | 570 | 353 | 0.62 | 1.29 |
| Type - PWRs | 1974 | 377.8 | 86.1 | 219 | 71 | 0.32 | 0.19 |
| Capacity - (436), 1,070, | 1975 | 389.0 | 87.4 | 424 | 292 | 0.69 | 0.75 |
| 1,080 MWe | 1976 | 297.9 | 70.2 | 1,330 | 880 | 0.66 | 2.95 |
| | 1977 | 281.2 | 63.7 | 985 | 847 | 0.86 | 3.01 |
| | 1978 | 323.2 | 80.2 | 764 | 401 | 0.52 | 1.24 |
| | 1979 | 401.0 | 90.2 | 521 | 139 | 0.27 | 0.35 |
| | 1980 | 97.3 | 22.3 | 3,063 | 2,386 | 0.78 | 24.52 |
| | 1981 | 95.9 | 26.7 | 2,902 | 3,223 | 1.11 | 33.61 |
| | 1982 | 61.6 | 15.7 | 3,055 | 832 | 0.27 | 13.51 |
| | 1983 | 0.0 | 0.0 | 1,701 | 155 | 0.09 | --- |
| | 1984 | 670.4 | 68.3 | 7,514 | 986 | 0.13 | 1.47 |
| | 1985 | 1,381.8 | 132.9 | 5,742 | 722 | 0.13 | 0.52 |
| | 1986 | 1,698.2 | 61.1 | 3,594 | 824 | 0.23 | 0.49 |
| | 1987 | 1,983.0 | 78.8 | 2,138 | 696 | 0.33 | 0.35 |
| | 1988 | 1,982.3 | 68.4 | 2,324 | 781 | 0.34 | 0.39 |
| | 1989 | 1,840.8 | 64.9 | 2,237 | 567 | 0.25 | 0.31 |
| | 1990 | 1,980.5 | 69.1 | 2,224 | 885 | 0.40 | 0.45 |
| | 1991 | 1,987.6 | 75.3 | 1,814 | 412 | 0.23 | 0.21 |
| | 1992 | 2,228.6 | 87.1 | 1,651 | 324 | 0.20 | 0.15 |
| | 1993 | 1,771.3 | 79.9 | 2,193 | 767 | 0.35 | 0.43 |
| | 1994 | 2,220.7 | 100.0 | 528 | 32 | 0.06 | 0.01 |
| | 1995 | 1,686.9 | 79.1 | 1,914 | 455 | 0.24 | 0.27 |
| | 1996 | 2,089.3 | 93.2 | 1,272 | 129 | 0.10 | 0.06 |
| | 1997 | 1,533.9 | 72.9 | 1,652 | 341 | 0.21 | 0.22 |
| | 1998 | 1,996.4 | 92.0 | 1,091 | 195.600 | 0.18 | 0.10 |
| **SAN ONOFRE 1[13]** | 1999 | 0.0 | 0.0 | 241 | 15.863 | 0.07 | --- |
| Docket 50-206; | 2000 | 0.0 | 0.0 | 416 | 71.214 | 0.17 | --- |
| DPR-13 | 2001 | 0.0 | 0.0 | 338 | 57.785 | 0.17 | --- |
| 1st commercial operation 1/68 | 2002 | 0.0 | 0.0 | 308 | 61.214 | 0.20 | --- |
| Type - PWR | 2003 | 0.0 | 0.0 | 226 | 35.596 | 0.16 | --- |
| Capacity - (436) MWe | 2004 | 0.0 | 0.0 | 169 | 14.899 | 0.09 | --- |

[13] San Onofre 1 was shut down in November 1992 and is no longer in the count of operating reactors. Parentheses indicate plant capacity when plant was operational.

| Reporting Organization | Year | Megawatt Years (MW-yr) | Unit Availability Factor | Total Personnel with Measurable Doses | Collective Dose per Site (person-rem) | Average Measurable Dose (rem) | Collective Dose/ MW-yr |
|---|---|---|---|---|---|---|---|
| **SAN ONOFRE 1[13]** | 2005 | 0.0 | 0.0 | 198 | 20.624 | 0.10 | --- |
| (continued) | 2006 | 0.0 | 0.0 | 183 | 22.490 | 0.12 | --- |
| | 2007 | 0.0 | 0.0 | 20 | 0.417 | 0.02 | --- |
| | 2008 | 0.0 | 0.0 | 2 | 0.043 | 0.02 | --- |
| **SAN ONOFRE 2, 3** | 1999 | 1,901.4 | 86.9 | 1,477 | 353.765 | 0.24 | 0.19 |
| Docket 50-361, 50-362; | 2000 | 2,067.2 | 94.7 | 1,073 | 115.499 | 0.11 | 0.06 |
| NPF-10, NPF-15 | 2001 | 1,727.2 | 78.9 | 1,083 | 131.384 | 0.12 | 0.08 |
| 1st commercial operation | 2002 | 2,056.0 | 93.4 | 1,140 | 136.443 | 0.12 | 0.07 |
| 8/83, 4/84 | 2003 | 2,084.3 | 94.0 | 1,275 | 163.804 | 0.13 | 0.08 |
| Type - PWRs | 2004 | 1,713.8 | 79.1 | 1,761 | 407.063 | 0.23 | 0.24 |
| Capacity - 1,070, 1,080 MWe | 2005 | 2,094.7 | 96.0 | 305 | 11.332 | 0.04 | 0.01 |
| | 2006 | 1,552.2 | 73.0 | 1,632 | 315.087 | 0.19 | 0.20 |
| | 2007 | 1,964.6 | 89.0 | 1,065 | 91.545 | 0.09 | 0.05 |
| | 2008 | 1,753.0 | 82.7 | 1,014 | 125.320 | 0.12 | 0.07 |
| | 2009 | 1,774.5 | 79.9 | 1,575 | 178.131 | 0.11 | 0.10 |
| | 2010 | 1,578.9 | 75.3 | 1,642 | 199.399 | 0.12 | 0.13 |
| **SEABROOK** | 1991 | 810.4 | 75.9 | 699 | 92 | 0.13 | 0.11 |
| Docket 50-443; | 1992 | 932.4 | 81.3 | 806 | 147 | 0.18 | 0.16 |
| NPF-86 | 1993 | 1,071.5 | 93.6 | 110 | 6 | 0.05 | 0.01 |
| 1st commercial operation 8/90 | 1994 | 736.4 | 63.5 | 852 | 113 | 0.13 | 0.15 |
| Type - PWR | 1995 | 995.5 | 87.5 | 800 | 102 | 0.13 | 0.10 |
| Capacity - 1,246 MWe | 1996 | 1,168.6 | 99.6 | 206 | 10 | 0.05 | 0.01 |
| | 1997 | 907.0 | 79.8 | 1,571 | 186 | 0.12 | 0.21 |
| | 1998 | 957.6 | 84.5 | 559 | 18.509 | 0.03 | 0.02 |
| | 1999 | 991.5 | 87.5 | 1,339 | 105.723 | 0.08 | 0.11 |
| | 2000 | 901.8 | 79.3 | 1,158 | 70.091 | 0.06 | 0.08 |
| | 2001 | 989.6 | 89.1 | 423 | 8.672 | 0.02 | 0.01 |
| | 2002 | 1,058.0 | 92.8 | 1,095 | 66.583 | 0.06 | 0.06 |
| | 2003 | 1,055.9 | 93.6 | 981 | 70.953 | 0.07 | 0.07 |
| | 2004 | 1,158.6 | 100.0 | 291 | 5.858 | 0.02 | 0.01 |
| | 2005 | 1,076.4 | 91.5 | 1,034 | 52.216 | 0.05 | 0.05 |
| | 2006 | 1,072.8 | 89.0 | 1,246 | 76.583 | 0.06 | 0.07 |
| | 2007 | 1,228.7 | 100.0 | 349 | 4.332 | 0.01 | 0.00 |
| | 2008 | 1,064.4 | 86.9 | 1,297 | 74.992 | 0.06 | 0.07 |
| | 2009 | 1,006.4 | 86.5 | 1,233 | 87.372 | 0.07 | 0.09 |
| | 2010 | 1,245.4 | 100.0 | 335 | 4.488 | 0.01 | 0.00 |
| **SEQUOYAH 1, 2** | 1982 | 583.5 | 52.8 | 1,968 | 570 | 0.29 | 0.98 |
| Docket 50-327, 50-328; | 1983 | 1,663.7 | 75.1 | 1,769 | 491 | 0.28 | 0.30 |
| DPR-77, DPR-79 | 1984 | 1,481.9 | 69.0 | 2,373 | 1,119 | 0.47 | 0.76 |
| 1st commercial operation | 1985 | 1,151.3 | 51.3 | 1,853 | 1,072 | 0.58 | 0.93 |
| 7/81, 6/82 | 1986 | 0.0 | 0.0 | 1,738 | 527 | 0.30 | --- |
| Type - PWR | 1987 | 0.0 | 0.0 | 2,080 | 420 | 0.20 | --- |
| Capacity - 1,152, 1,126 MWe | 1988 | 490.8 | 31.8 | 2,441 | 678 | 0.28 | 1.38 |
| | 1989 | 1,851.7 | 85.7 | 2,007 | 657 | 0.33 | 0.35 |
| | 1990 | 1,662.6 | 77.2 | 2,935 | 1,687 | 0.57 | 1.01 |
| | 1991 | 1,965.4 | 88.0 | 1,933 | 700 | 0.36 | 0.36 |
| | 1992 | 1,849.0 | 85.4 | 1,714 | 465 | 0.27 | 0.25 |
| | 1993 | 405.7 | 21.8 | 1,631 | 373 | 0.23 | 0.92 |
| | 1994 | 1,418.7 | 66.3 | 1,702 | 295 | 0.17 | 0.21 |
| | 1995 | 1,864.2 | 86.1 | 1,650 | 368 | 0.22 | 0.20 |
| | 1996 | 2,003.9 | 87.9 | 1,444 | 269 | 0.19 | 0.13 |
| | 1997 | 1,946.1 | 89.0 | 1,962 | 420 | 0.21 | 0.22 |
| | 1998 | 2,135.3 | 95.3 | 1,530 | 265.980 | 0.17 | 0.12 |
| | 1999 | 2,165.1 | 97.0 | 1,346 | 164.569 | 0.12 | 0.08 |
| | 2000 | 1,910.0 | 86.8 | 2,039 | 357.220 | 0.18 | 0.19 |
| | 2001 | 2,158.3 | 95.7 | 1,292 | 145.066 | 0.11 | 0.07 |
| | 2002 | 2,106.0 | 94.1 | 1,257 | 108.252 | 0.09 | 0.05 |

[13] San Onofre 1 was shut down in November 1992 and is no longer in the count of operating reactors. Parentheses indicate plant capacity when plant was operational.

| Reporting Organization | Year | Megawatt Years (MW-yr) | Unit Availability Factor | Total Personnel with Measurable Doses | Collective Dose per Site (person-rem) | Average Measurable Dose (rem) | Collective Dose/ MW-yr |
|---|---|---|---|---|---|---|---|
| **SEQUOYAH 1, 2** | 2003 | 1,776.4 | 80.0 | 2,484 | 430.889 | 0.17 | 0.24 |
| (continued) | 2004 | 2,135.2 | 93.9 | 1,161 | 85.941 | 0.07 | 0.04 |
| | 2005 | 2,162.9 | 94.9 | 1,125 | 95.133 | 0.08 | 0.04 |
| | 2006 | 2,054.9 | 91.0 | 1,752 | 242.016 | 0.14 | 0.12 |
| | 2007 | 2,129.1 | 94.0 | 1,197 | 123.540 | 0.10 | 0.06 |
| | 2008 | 2,153.6 | 94.3 | 960 | 83.730 | 0.09 | 0.04 |
| | 2009 | 2,026.8 | 90.1 | 1,415 | 166.776 | 0.12 | 0.08 |
| | 2010 | 2,054.9 | 92.2 | 828 | 56.956 | 0.07 | 0.03 |
| **SOUTH TEXAS 1, 2** | 1989 | 769.3 | 65.6 | 989 | 161 | 0.16 | 0.21 |
| Docket 50-498, 50-499; | 1990 | 1,504.1 | 65.9 | 1,136 | 206 | 0.18 | 0.14 |
| NPF-76, NPF-80 | 1991 | 1,741.5 | 72.4 | 1,144 | 257 | 0.22 | 0.15 |
| 1st commercial operation | 1992 | 2,096.0 | 83.8 | 923 | 147 | 0.16 | 0.07 |
| 8/88, 6/89 | 1993 | 163.1 | 8.3 | 1,138 | 251 | 0.22 | 1.54 |
| Type - PWRs | 1994 | 1,700.2 | 70.6 | 661 | 47 | 0.07 | 0.03 |
| Capacity - 1,251, 1,251 MWe | 1995 | 2,294.2 | 89.9 | 1,485 | 291 | 0.20 | 0.13 |
| | 1996 | 2,465.9 | 95.0 | 1,145 | 137 | 0.12 | 0.06 |
| | 1997 | 2,265.5 | 93.6 | 1,583 | 273 | 0.17 | 0.12 |
| | 1998 | 2,379.4 | 96.9 | 1,171 | 183.977 | 0.16 | 0.08 |
| | 1999 | 2,219.7 | 91.6 | 1,328 | 259.770 | 0.20 | 0.12 |
| | 2000 | 2,180.0 | 89.7 | 1,372 | 231.634 | 0.17 | 0.11 |
| | 2001 | 2,262.7 | 92.2 | 1,325 | 237.645 | 0.18 | 0.11 |
| | 2002 | 2,173.0 | 87.5 | 1,510 | 329.091 | 0.22 | 0.15 |
| | 2003 | 1,796.3 | 72.1 | 909 | 143.495 | 0.16 | 0.08 |
| | 2004 | 2,437.1 | 96.0 | 842 | 119.834 | 0.14 | 0.05 |
| | 2005 | 2,258.5 | 90.0 | 1,268 | 247.655 | 0.20 | 0.11 |
| | 2006 | 2,439.6 | 95.0 | 1,078 | 150.323 | 0.14 | 0.06 |
| | 2007 | 2,527.3 | 96.0 | 881 | 91.613 | 0.10 | 0.04 |
| | 2008 | 2,452.1 | 92.3 | 1,181 | 187.295 | 0.16 | 0.08 |
| | 2009 | 2,444.5 | 91.9 | 1,138 | 79.687 | 0.07 | 0.03 |
| | 2010 | 2,418.7 | 91.5 | 867 | 79.159 | 0.09 | 0.03 |
| **ST. LUCIE 1, 2** | 1977 | 649.1 | 84.7 | 445 | 152 | 0.34 | 0.23 |
| Docket 50-335, 50-389; | 1978 | 606.4 | 76.5 | 797 | 337 | 0.42 | 0.56 |
| DPR-67; NPF-16 | 1979 | 592.0 | 74.0 | 907 | 438 | 0.48 | 0.74 |
| 1st commercial operation | 1980 | 627.9 | 77.5 | 1,074 | 532 | 0.50 | 0.85 |
| 12/76, 8/83 | 1981 | 599.1 | 72.7 | 1,473 | 929 | 0.63 | 1.55 |
| Type - PWRs | 1982 | 816.8 | 94.0 | 1,045 | 272 | 0.26 | 0.33 |
| Capacity - 839, 839 MWe | 1983 | 290.3 | 15.4 | 2,211 | 1,204 | 0.54 | 4.15 |
| | 1984 | 1,183.0 | 69.6 | 2,090 | 1,263 | 0.60 | 1.07 |
| | 1985 | 1,445.8 | 82.5 | 1,971 | 1,344 | 0.68 | 0.93 |
| | 1986 | 1,588.6 | 89.1 | 1,279 | 491 | 0.38 | 0.31 |
| | 1987 | 1,407.9 | 81.9 | 2,012 | 951 | 0.47 | 0.68 |
| | 1988 | 1,639.7 | 93.0 | 1,448 | 611 | 0.42 | 0.37 |
| | 1989 | 1,493.1 | 85.1 | 1,414 | 495 | 0.35 | 0.33 |
| | 1990 | 1,188.4 | 70.0 | 1,876 | 777 | 0.41 | 0.65 |
| | 1991 | 1,592.8 | 90.8 | 1,282 | 479 | 0.37 | 0.30 |
| | 1992 | 1,511.9 | 87.3 | 1,251 | 264 | 0.21 | 0.17 |
| | 1993 | 1,227.6 | 77.7 | 1,462 | 492 | 0.34 | 0.40 |
| | 1994 | 1,424.8 | 85.0 | 1,896 | 505 | 0.27 | 0.35 |
| | 1995 | 1,306.6 | 76.0 | 1,498 | 413 | 0.28 | 0.32 |
| | 1996 | 1,473.4 | 86.5 | 1,433 | 385 | 0.27 | 0.26 |
| | 1997 | 1,394.6 | 83.6 | 2,314 | 646 | 0.28 | 0.46 |
| | 1998 | 1,572.5 | 94.2 | 1,170 | 134.459 | 0.11 | 0.09 |
| | 1999 | 1,569.1 | 93.8 | 1,107 | 176.878 | 0.16 | 0.11 |
| | 2000 | 1,630.0 | 96.0 | 990 | 98.691 | 0.10 | 0.06 |
| | 2001 | 1,527.5 | 91.6 | 1,375 | 228.071 | 0.17 | 0.15 |
| | 2002 | 1,633.0 | 96.6 | 992 | 155.946 | 0.16 | 0.10 |
| | 2003 | 1,524.7 | 91.5 | 937 | 141.734 | 0.15 | 0.09 |
| | 2004 | 1,492.0 | 89.3 | 1,157 | 159.436 | 0.14 | 0.11 |
| | 2005 | 1,408.4 | 85.1 | 2,262 | 406.171 | 0.18 | 0.29 |
| | 2006 | 1,542.4 | 93.0 | 1,226 | 119.963 | 0.10 | 0.08 |

| Reporting Organization | Year | Megawatt Years (MW-yr) | Unit Availability Factor | Total Personnel with Measurable Doses | Collective Dose per Site (person-rem) | Average Measurable Dose (rem) | Collective Dose/ MW-yr |
|---|---|---|---|---|---|---|---|
| **ST. LUCIE 1, 2** | 2007 | 1,302.1 | 78.0 | 2,447 | 409.958 | 0.17 | 0.32 |
| (continued) | 2008 | 1,566.5 | 92.7 | 1,127 | 112.234 | 0.10 | 0.07 |
| | 2009 | 1,490.6 | 88.8 | 1,139 | 132.861 | 0.12 | 0.09 |
| | 2010 | 1,440.2 | 88.4 | 1,357 | 197.359 | 0.15 | 0.14 |
| **SUMMER 1** | 1984 | 504.6 | 61.1 | 1,120 | 295 | 0.26 | 0.58 |
| Docket 50-395; | 1985 | 627.7 | 71.6 | 1,201 | 379 | 0.32 | 0.60 |
| NPF-12 | 1986 | 853.7 | 95.3 | 392 | 23 | 0.06 | 0.03 |
| 1st commercial operation 1/84 | 1987 | 618.7 | 71.0 | 1,075 | 560 | 0.52 | 0.91 |
| Type - PWR | 1988 | 605.3 | 69.1 | 1,127 | 511 | 0.45 | 0.84 |
| Capacity - 966 MWe | 1989 | 652.4 | 83.1 | 374 | 52 | 0.14 | 0.08 |
| | 1990 | 730.0 | 83.9 | 1,090 | 376 | 0.34 | 0.52 |
| | 1991 | 642.5 | 82.9 | 984 | 291 | 0.30 | 0.45 |
| | 1992 | 892.6 | 97.4 | 249 | 27 | 0.11 | 0.03 |
| | 1993 | 728.3 | 84.0 | 1,121 | 297 | 0.26 | 0.41 |
| | 1994 | 536.7 | 69.5 | 1,549 | 374 | 0.24 | 0.70 |
| | 1995 | 899.8 | 97.2 | 257 | 13 | 0.05 | 0.01 |
| | 1996 | 850.4 | 90.3 | 701 | 97 | 0.14 | 0.11 |
| | 1997 | 829.7 | 89.8 | 820 | 163 | 0.20 | 0.20 |
| | 1998 | 934.8 | 98.8 | 285 | 13.513 | 0.05 | 0.01 |
| | 1999 | 842.0 | 89.4 | 827 | 120.172 | 0.15 | 0.14 |
| | 2000 | 723.9 | 76.6 | 933 | 166.561 | 0.18 | 0.23 |
| | 2001 | 769.3 | 83.3 | 486 | 69.398 | 0.14 | 0.09 |
| | 2002 | 840.0 | 87.9 | 685 | 59.644 | 0.09 | 0.07 |
| | 2003 | 837.0 | 87.4 | 745 | 70.828 | 0.10 | 0.08 |
| | 2004 | 938.4 | 96.8 | 200 | 10.085 | 0.05 | 0.01 |
| | 2005 | 850.3 | 88.9 | 734 | 72.454 | 0.10 | 0.09 |
| | 2006 | 858.6 | 90.0 | 676 | 61.333 | 0.09 | 0.07 |
| | 2007 | 967.9 | 100.0 | 75 | 2.691 | 0.04 | 0.00 |
| | 2008 | 817.2 | 84.8 | 623 | 49.091 | 0.08 | 0.06 |
| | 2009 | 784.5 | 82.6 | 767 | 56.050 | 0.07 | 0.07 |
| | 2010 | 968.8 | 99.4 | 104 | 2.129 | 0.02 | 0.00 |
| **SURRY 1, 2** | 1973 | 420.6 | | 936 | 152 | 0.16 | 0.36 |
| Docket 50-280, 50-281; | 1974 | 717.4 | 49.8 | 1,715 | 884 | 0.52 | 1.23 |
| DPR-32, DPR-37 | 1975 | 1,079.0 | 70.8 | 1,948 | 1,649 | 0.85 | 1.53 |
| 1st commercial operation | 1976 | 930.7 | 60.4 | 2,753 | 3,165 | 1.15 | 3.40 |
| 12/72, 5/73 | 1977 | 1,139.0 | 72.2 | 1,860 | 2,307 | 1.24 | 2.03 |
| Type - PWRs | 1978 | 1,210.6 | 77.2 | 2,203 | 1,837 | 0.83 | 1.52 |
| Capacity - 799, 799 MWe | 1979 | 343.0 | 42.3 | 5,065 | 3,584 | 0.71 | 10.45 |
| | 1980 | 568.2 | 40.3 | 5,317 | 3,836 | 0.72 | 6.75 |
| | 1981 | 907.6 | 59.3 | 3,753 | 4,244 | 1.13 | 4.68 |
| | 1982 | 1,323.3 | 88.5 | 1,878 | 1,490 | 0.79 | 1.13 |
| | 1983 | 916.2 | 61.3 | 2,754 | 3,220 | 1.17 | 3.51 |
| | 1984 | 1,026.7 | 71.0 | 3,198 | 2,247 | 0.70 | 2.19 |
| | 1985 | 1,166.4 | 78.2 | 3,206 | 1,815 | 0.57 | 1.56 |
| | 1986 | 1,080.5 | 69.0 | 3,763 | 2,356 | 0.63 | 2.18 |
| | 1987 | 1,132.7 | 72.7 | 2,675 | 712 | 0.27 | 0.63 |
| | 1988 | 750.4 | 50.0 | 3,184 | 1,542 | 0.48 | 2.05 |
| | 1989 | 489.3 | 33.0 | 3,100 | 836 | 0.27 | 1.71 |
| | 1990 | 1,276.4 | 83.9 | 1,947 | 575 | 0.30 | 0.45 |
| | 1991 | 1,271.9 | 84.5 | 1,547 | 510 | 0.33 | 0.40 |
| | 1992 | 1,396.3 | 88.9 | 1,660 | 539 | 0.32 | 0.39 |
| | 1993 | 1,283.1 | 84.6 | 1,402 | 383 | 0.27 | 0.30 |
| | 1994 | 1,320.9 | 85.2 | 1,530 | 378 | 0.25 | 0.29 |
| | 1995 | 1,333.0 | 84.2 | 1,883 | 406 | 0.22 | 0.30 |
| | 1996 | 1,562.9 | 93.1 | 983 | 209 | 0.21 | 0.13 |
| | 1997 | 1,380.3 | 87.1 | 1,335 | 320 | 0.24 | 0.23 |
| | 1998 | 1,476.2 | 91.6 | 1,165 | 188.831 | 0.16 | 0.13 |
| | 1999 | 1,483.0 | 93.5 | 995 | 137.891 | 0.14 | 0.09 |
| | 2000 | 1,490.0 | 92.7 | 1,197 | 193.169 | 0.16 | 0.13 |
| | 2001 | 1,441.5 | 89.5 | 1,243 | 328.650 | 0.26 | 0.23 |
| | 2002 | 1,557.0 | 96.0 | 799 | 87.778 | 0.11 | 0.06 |

| Reporting Organization | Year | Megawatt Years (MW-yr) | Unit Availability Factor | Total Personnel with Measurable Doses | Collective Dose per Site (person-rem) | Average Measurable Dose (rem) | Collective Dose/ MW-yr |
|---|---|---|---|---|---|---|---|
| **SURRY 1, 2** | 2003 | 1,255.9 | 79.7 | 1,628 | 325.729 | 0.20 | 0.26 |
| (continued) | 2004 | 1,537.9 | 94.6 | 1,028 | 119.654 | 0.12 | 0.08 |
| | 2005 | 1,506.7 | 94.2 | 877 | 87.717 | 0.10 | 0.06 |
| | 2006 | 1,427.0 | 90.0 | 1,227 | 234.978 | 0.19 | 0.17 |
| | 2007 | 1,516.2 | 94.0 | 1,111 | 207.130 | 0.19 | 0.14 |
| | 2008 | 1,536.6 | 95.7 | 1,069 | 150.269 | 0.14 | 0.10 |
| | 2009 | 1,485.1 | 93.1 | 1,241 | 193.703 | 0.16 | 0.13 |
| | 2010 | 1,503.7 | 93.7 | 958 | 111.129 | 0.12 | 0.07 |
| **SUSQUEHANNA 1, 2** | 1984 | 719.9 | 72.6 | 2,827 | 308 | 0.11 | 0.43 |
| Docket 50-387, 50-388; | 1985 | 1,452.2 | 76.4 | 3,669 | 1,106 | 0.30 | 0.76 |
| NPF-14; NPF-22 | 1986 | 1,344.8 | 67.0 | 2,996 | 828 | 0.28 | 0.62 |
| 1st commercial operation | 1987 | 1,749.5 | 85.3 | 2,548 | 621 | 0.24 | 0.35 |
| 6/83, 2/85 | 1988 | 1,691.0 | 83.5 | 1,904 | 516 | 0.27 | 0.31 |
| Type - BWRs | 1989 | 1,572.5 | 77.1 | 2,063 | 704 | 0.34 | 0.45 |
| Capacity - 1,257, 1,190 MWe | 1990 | 1,746.9 | 85.4 | 1,691 | 440 | 0.26 | 0.25 |
| | 1991 | 1,878.0 | 89.8 | 1,844 | 507 | 0.27 | 0.27 |
| | 1992 | 1,604.2 | 79.7 | 1,885 | 724 | 0.38 | 0.45 |
| | 1993 | 1,602.1 | 77.3 | 1,488 | 335 | 0.23 | 0.21 |
| | 1994 | 1,814.4 | 85.4 | 1,580 | 442 | 0.28 | 0.24 |
| | 1995 | 1,850.8 | 85.3 | 1,773 | 476 | 0.27 | 0.26 |
| | 1996 | 1,998.7 | 90.7 | 1,430 | 289 | 0.20 | 0.14 |
| | 1997 | 1,918.9 | 89.6 | 1,646 | 433 | 0.26 | 0.23 |
| | 1998 | 1,879.6 | 88.3 | 1,575 | 360.778 | 0.23 | 0.19 |
| | 1999 | 1,896.0 | 89.6 | 1,787 | 431.397 | 0.24 | 0.23 |
| | 2000 | 1,994.6 | 92.6 | 1,812 | 331.163 | 0.18 | 0.17 |
| | 2001 | 2,027.6 | 94.2 | 1,807 | 288.413 | 0.16 | 0.14 |
| | 2002 | 1,973.0 | 91.6 | 1,890 | 259.968 | 0.14 | 0.13 |
| | 2003 | 2,050.8 | 93.4 | 1,934 | 250.096 | 0.13 | 0.12 |
| | 2004 | 2,058.8 | 92.7 | 2,144 | 272.202 | 0.13 | 0.13 |
| | 2005 | 2,086.6 | 93.5 | 1,898 | 181.360 | 0.10 | 0.09 |
| | 2006 | 2,040.4 | 91.0 | 1,873 | 184.901 | 0.10 | 0.09 |
| | 2007 | 2,089.2 | 93.0 | 2,303 | 263.021 | 0.11 | 0.13 |
| | 2008 | 2,174.1 | 94.2 | 1,895 | 192.892 | 0.10 | 0.09 |
| | 2009 | 2,231.1 | 94.7 | 1,956 | 266.597 | 0.14 | 0.12 |
| | 2010 | 2,121.6 | 90.4 | 1,950 | 176.161 | 0.09 | 0.08 |
| **THREE MILE ISLAND 1[14], 2[15]** | 1975 | 675.9 | 82.2 | 131 | 73 | 0.56 | 0.11 |
| Docket 50-289, 50-320; | 1976 | 530.0 | 65.4 | 819 | 286 | 0.35 | 0.54 |
| DPR-50, DPR-73 | 1977 | 664.5 | 80.9 | 1,122 | 360 | 0.32 | 0.54 |
| 1st commercial operation | 1978 | 690.0 | 85.1 | 1,929 | 504 | 0.26 | 0.73 |
| 9/74, 12/78 | 1979 | 266.0 | 21.9 | 3,975 | 1,392 | 0.35 | 5.23 |
| Type - PWRs | 1980 | 0.0 | 0.0 | 2,328 | 394 | 0.17 | --- |
| Capacity - 802, (880) MWe | 1981 | 0.0 | 0.0 | 2,103 | 376 | 0.18 | --- |
| | 1982 | 0.0 | 0.0 | 2,123 | 1,004 | 0.47 | --- |
| | 1983 | 0.0 | 0.0 | 1,592 | 1,159 | 0.73 | --- |
| | 1984 | 0.0 | 0.0 | 1,079 | 688 | 0.64 | --- |
| | 1985 | 103.6 | 10.6 | 1,890 | 857 | 0.45 | 8.27 |
| **THREE MILE ISLAND 1[14]** | 1986 | 585.2 | 70.9 | 1,360 | 213 | 0.16 | 0.36 |
| Docket 50-289; | 1987 | 610.7 | 73.6 | 1,259 | 149 | 0.12 | 0.24 |
| DPR-50 | 1988 | 661.0 | 77.8 | 1,012 | 210 | 0.21 | 0.32 |
| 1st commercial operation 9/74 | 1989 | 871.3 | 100.0 | 670 | 54 | 0.08 | 0.06 |
| Type - PWR | 1990 | 645.5 | 84.6 | 1,319 | 264 | 0.20 | 0.41 |
| Capacity - 802 MWe | 1991 | 688.7 | 86.4 | 1,542 | 198 | 0.13 | 0.29 |
| | 1992 | 836.8 | 100.0 | 558 | 34 | 0.06 | 0.04 |
| | 1993 | 722.0 | 88.5 | 1,835 | 206 | 0.11 | 0.29 |
| | 1994 | 798.7 | 95.5 | 434 | 40 | 0.09 | 0.05 |
| | 1995 | 772.9 | 90.8 | 1,220 | 213 | 0.17 | 0.28 |

[14] Three Mile Island 1 resumed commercial power generation in October 1985 after being under regulatory restraint since 1979.

[15] Three Mile Island 2 has been shut down since the 1979 accident but was still included in the count of reactors through 1988 since dose was still being accumulated to defuel and decontaminate the unit during this time period. Parentheses indicate plant capacity when plant was operational. Since 2001, the dose breakdowns for Three Mile Island 2 have been reported with those for Unit 1.

| Reporting Organization | Year | Megawatt Years (MW-yr) | Unit Availability Factor | Total Personnel with Measurable Doses | Collective Dose per Site (person-rem) | Average Measurable Dose (rem) | Collective Dose/ MW-yr |
|---|---|---|---|---|---|---|---|
| THREE MILE ISLAND 1[14] | 1996 | 857.4 | 100.0 | 267 | 16 | 0.06 | 0.02 |
| (continued) | 1997 | 675.7 | 84.3 | 1,049 | 204 | 0.19 | 0.30 |
| | 1998 | 805.8 | 100.0 | 280 | 16.722 | 0.06 | 0.02 |
| | 1999 | 722.4 | 89.7 | 1,171 | 154.936 | 0.13 | 0.21 |
| | 2000 | 813.4 | 100.0 | 183 | 8.689 | 0.05 | 0.01 |
| | 2001 | 616.7 | 84.2 | 1,196 | 196.699 | 0.16 | 0.32 |
| | 2002 | 833.0 | 100.0 | 172 | 6.533 | 0.04 | 0.01 |
| | 2003 | 706.4 | 87.1 | 1,230 | 155.101 | 0.13 | 0.22 |
| | 2004 | 828.0 | 100.0 | 105 | 3.573 | 0.03 | 0.00 |
| | 2005 | 769.1 | 93.2 | 955 | 65.576 | 0.07 | 0.09 |
| | 2006 | 825.0 | 99.0 | 125 | 5.155 | 0.04 | 0.01 |
| | 2007 | 758.6 | 92.0 | 1,266 | 114.203 | 0.09 | 0.15 |
| | 2008 | 838.5 | 100.0 | 64 | 2.219 | 0.03 | 0.00 |
| | 2009 | 672.6 | 81.7 | 2,019 | 241.780 | 0.12 | 0.36 |
| | 2010 | 757.3 | 93.1 | 790 | 38.994 | 0.05 | 0.05 |
| THREE MILE ISLAND 2[15] | 1986 | 0.0 | 0.0 | 1,497 | 915 | 0.61 | --- |
| Docket 50-320; | 1987 | 0.0 | 0.0 | 1,378 | 977 | 0.71 | --- |
| DPR-73 | 1988 | 0.0 | 0.0 | 1,247 | 917 | 0.74 | --- |
| 1st commercial operation 12/78 | 1989 | 0.0 | 0.0 | 1,014 | 639 | 0.63 | --- |
| Type - PWR | 1990 | 0.0 | 0.0 | 484 | 136 | 0.28 | --- |
| Capacity - (880) MWe | 1991 | 0.0 | 0.0 | 153 | 37 | 0.24 | --- |
| | 1992 | 0.0 | 0.0 | 315 | 157 | 0.50 | --- |
| | 1993 | 0.0 | 0.0 | 167 | 33 | 0.20 | --- |
| | 1994 | 0.0 | 0.0 | 259 | 7 | 0.03 | --- |
| | 1995 | 0.0 | 0.0 | 191 | 2 | 0.01 | --- |
| | 1996 | 0.0 | 0.0 | 122 | 2 | 0.02 | --- |
| | 1997 | 0.0 | 0.0 | 232 | 1 | 0.00 | --- |
| | 1998 | 0.0 | 0.0 | 105 | 0.697 | 0.01 | --- |
| | 1999 | 0.0 | 0.0 | 203 | 0.512 | 0.00 | --- |
| | 2000 | 0.0 | 0.0 | 70 | 0.401 | 0.01 | --- |
| | 2001 | 0.0 | 0.0 | 0 | 0.228 | --- | --- |
| | 2002 | 0.0 | 0.0 | 0 | | --- | --- |
| | 2003 | 0.0 | 0.0 | 0 | 0.260 | --- | --- |
| | 2004 | 0.0 | 0.0 | 0 | 0.216 | --- | --- |
| | 2005 | 0.0 | 0.0 | 0 | | --- | --- |
| | 2006 | 0.0 | 0.0 | 0 | 0.372 | --- | --- |
| | 2007 | 0.0 | 0.0 | 0 | 0.082 | --- | --- |
| | 2008 | 0.0 | 0.0 | 0 | 0.138 | --- | --- |
| | 2009 | 0.0 | 0.0 | 0 | 0.113 | --- | --- |
| | 2010 | 0.0 | 0.0 | 0 | 0.359 | --- | --- |
| TROJAN[16] | 1977 | 792.0 | 92.6 | 591 | 174 | 0.29 | 0.22 |
| Docket 50-344; | 1978 | 205.5 | 20.6 | 711 | 319 | 0.45 | 1.55 |
| NPF-1 | 1979 | 631.0 | 58.1 | 736 | 258 | 0.35 | 0.41 |
| 1st commercial operation 5/76 | 1980 | 727.5 | 72.5 | 1,159 | 421 | 0.36 | 0.58 |
| Type - PWR | 1981 | 775.6 | 74.1 | 1,311 | 609 | 0.46 | 0.79 |
| Capacity - (1,080) MWe | 1982 | 579.5 | 60.8 | 977 | 419 | 0.43 | 0.72 |
| | 1983 | 494.2 | 62.4 | 969 | 307 | 0.32 | 0.62 |
| | 1984 | 567.0 | 54.4 | 1,042 | 433 | 0.42 | 0.76 |
| | 1985 | 829.1 | 76.7 | 852 | 363 | 0.43 | 0.44 |
| | 1986 | 852.4 | 79.7 | 1,321 | 381 | 0.29 | 0.45 |
| | 1987 | 525.5 | 54.0 | 1,209 | 363 | 0.30 | 0.69 |
| | 1988 | 758.6 | 67.5 | 1,408 | 401 | 0.28 | 0.53 |
| | 1989 | 666.8 | 61.9 | 1,360 | 421 | 0.31 | 0.63 |

[14] Three Mile Island 1 resumed commercial power generation in October 1985 after being under regulatory restraint since 1979.

[15] Three Mile Island 2 has been shut down since the 1979 accident but was still included in the count of reactors through 1988 since dose was still being accumulated to defuel and decontaminate the unit during this time period. Parentheses indicate plant capacity when plant was operational. Since 2001, the dose breakdowns for Three Mile Island 2 have been reported with those for Unit 1.

[16] Trojan ended commercial operation as of January 1993 and will not be put in commercial operation again. It is no longer in the count of operating reactors. Parentheses indicate plant capacity when plant was operational. As of 2005, Trojan no longer reports under its reactor license but does report under its ISFSI license (see Appendix A).

| Reporting Organization | Year | Megawatt Years (MW-yr) | Unit Availability Factor | Total Personnel with Measurable Doses | Collective Dose per Site (person-rem) | Average Measurable Dose (rem) | Collective Dose/ MW-yr |
|---|---|---|---|---|---|---|---|
| **TROJAN**[16] | 1990 | 732.4 | 66.3 | 1,169 | 258 | 0.22 | 0.35 |
| (continued) | 1991 | 181.6 | 16.1 | 1,496 | 567 | 0.38 | 3.12 |
| | 1992 | 553.9 | 68.4 | 567 | 84 | 0.15 | 0.15 |
| | 1993 | 0.0 | 68.4 | 54 | 21 | 0.39 | --- |
| | 1994 | 0.0 | 0.0 | 51 | 9 | 0.18 | --- |
| | 1995 | 0.0 | 0.0 | 141 | 44 | 0.31 | --- |
| | 1996 | 0.0 | 0.0 | 112 | 41 | 0.37 | --- |
| | 1997 | 0.0 | 0.0 | 227 | 41 | 0.18 | --- |
| | 1998 | 0.0 | 0.0 | 283 | 46.417 | 0.16 | --- |
| | 1999 | 0.0 | 0.0 | 274 | 51.504 | 0.19 | --- |
| | 2000 | 0.0 | 0.0 | 127 | 17.631 | 0.14 | --- |
| | 2001 | 0.0 | 0.0 | 14 | 1.091 | 0.08 | --- |
| | 2002 | 0.0 | 0.0 | 13 | 0.536 | 0.04 | --- |
| | 2003 | 0.0 | 0.0 | 105 | 23.996 | 0.23 | --- |
| | 2004 | 0.0 | 0.0 | 5 | 0.079 | 0.02 | --- |
| **TURKEY POINT 3, 4** | 1973 | 401.9 | | 444 | 78 | 0.18 | 0.19 |
| Docket 50-250, 50-251; | 1974 | 953.6 | | 794 | 454 | 0.57 | 0.48 |
| DPR-31, DPR-41 | 1975 | 1,003.7 | 74.9 | 1,176 | 876 | 0.74 | 0.87 |
| 1st commercial operation | 1976 | 974.2 | 71.2 | 1,647 | 1,184 | 0.72 | 1.22 |
| 12/72, 9/73 | 1977 | 979.5 | 72.1 | 1,319 | 1,036 | 0.79 | 1.06 |
| Type - PWRs | 1978 | 1,000.2 | 78.8 | 1,336 | 1,032 | 0.77 | 1.03 |
| Capacity - 693, 693 MWe | 1979 | 811.0 | 62.4 | 2,002 | 1,680 | 0.84 | 2.07 |
| | 1980 | 990.6 | 73.6 | 1,803 | 1,651 | 0.92 | 1.67 |
| | 1981 | 654.0 | 46.8 | 2,932 | 2,251 | 0.77 | 3.44 |
| | 1982 | 915.7 | 65.2 | 2,956 | 2,119 | 0.72 | 2.31 |
| | 1983 | 878.4 | 62.8 | 2,930 | 2,681 | 0.92 | 3.05 |
| | 1984 | 946.7 | 68.5 | 2,010 | 1,255 | 0.62 | 1.33 |
| | 1985 | 1,034.9 | 74.7 | 1,905 | 1,253 | 0.66 | 1.21 |
| | 1986 | 754.1 | 54.9 | 1,808 | 946 | 0.52 | 1.25 |
| | 1987 | 431.3 | 36.6 | 1,980 | 1,371 | 0.69 | 3.18 |
| | 1988 | 809.8 | 59.5 | 1,841 | 738 | 0.40 | 0.91 |
| | 1989 | 689.9 | 56.8 | 1,625 | 433 | 0.27 | 0.63 |
| | 1990 | 933.1 | 69.0 | 2,099 | 730 | 0.35 | 0.78 |
| | 1991 | 258.2 | 21.0 | 2,087 | 939 | 0.45 | 3.64 |
| | 1992 | 968.9 | 75.5 | 1,374 | 325 | 0.24 | 0.34 |
| | 1993 | 1,244.8 | 91.0 | 1,271 | 275 | 0.22 | 0.22 |
| | 1994 | 1,172.9 | 87.2 | 1,489 | 476 | 0.32 | 0.41 |
| | 1995 | 1,320.3 | 94.6 | 1,142 | 215 | 0.19 | 0.16 |
| | 1996 | 1,307.8 | 94.0 | 1,157 | 187 | 0.16 | 0.14 |
| | 1997 | 1,220.9 | 88.6 | 1,581 | 414 | 0.26 | 0.34 |
| | 1998 | 1,323.0 | 94.5 | 1,045 | 156.415 | 0.15 | 0.12 |
| | 1999 | 1,352.5 | 96.5 | 919 | 127.567 | 0.14 | 0.09 |
| | 2000 | 1,283.7 | 92.2 | 1,292 | 219.852 | 0.17 | 0.17 |
| | 2001 | 1,324.1 | 95.0 | 827 | 101.575 | 0.12 | 0.08 |
| | 2002 | 1,374.0 | 97.9 | 793 | 73.764 | 0.09 | 0.05 |
| | 2003 | 1,253.2 | 91.6 | 1,442 | 247.053 | 0.17 | 0.20 |
| | 2004 | 1,231.0 | 89.9 | 1,089 | 117.404 | 0.11 | 0.10 |
| | 2005 | 1,143.0 | 84.9 | 1,136 | 109.996 | 0.10 | 0.10 |
| | 2006 | 1,251.8 | 90.0 | 1,321 | 149.208 | 0.11 | 0.12 |
| | 2007 | 1,281.5 | 91.0 | 1,085 | 107.601 | 0.10 | 0.08 |
| | 2008 | 1,294.9 | 92.0 | 1,067 | 97.357 | 0.09 | 0.08 |
| | 2009 | 1,219.7 | 87.6 | 1,359 | 166.217 | 0.12 | 0.14 |
| | 2010 | 1,290.9 | 91.9 | 1,025 | 86.749 | 0.08 | 0.07 |

[16] Trojan ended commercial operation as of January 1993 and will not be put in commercial operation again. It is no longer in the count of operating reactors. Parentheses indicate plant capacity when plant was operational. As of 2005, Trojan no longer reports under its reactor license but does report under its ISFSI license (see Appendix A).

| Reporting Organization | Year | Megawatt Years (MW-yr) | Unit Availability Factor | Total Personnel with Measurable Doses | Collective Dose per Site (person-rem) | Average Measurable Dose (rem) | Collective Dose/ MW-yr |
|---|---|---|---|---|---|---|---|
| **VERMONT YANKEE** | 1973 | 222.1 | | 244 | 85 | 0.35 | 0.38 |
| Docket 50-271; | 1974 | 303.5 | | 357 | 216 | 0.61 | 0.71 |
| DPR-28 | 1975 | 429.0 | 87.8 | 282 | 153 | 0.54 | 0.36 |
| 1st commercial operation 11/72 | 1976 | 389.6 | 77.1 | 815 | 411 | 0.50 | 1.05 |
| Type - BWR | 1977 | 423.5 | 85.1 | 641 | 258 | 0.40 | 0.61 |
| Capacity - 605 MWe | 1978 | 387.5 | 75.9 | 934 | 339 | 0.36 | 0.87 |
| | 1979 | 414.0 | 82.1 | 1,220 | 1,170 | 0.96 | 2.83 |
| | 1980 | 357.8 | 71.5 | 1,443 | 1,338 | 0.93 | 3.74 |
| | 1981 | 429.1 | 84.6 | 1,264 | 731 | 0.58 | 1.70 |
| | 1982 | 501.0 | 96.0 | 481 | 205 | 0.43 | 0.41 |
| | 1983 | 346.1 | 69.3 | 1,316 | 1,527 | 1.16 | 4.41 |
| | 1984 | 398.1 | 79.0 | 954 | 626 | 0.66 | 1.57 |
| | 1985 | 361.4 | 71.8 | 1,392 | 1,051 | 0.76 | 2.91 |
| | 1986 | 248.1 | 48.9 | 1,389 | 1,188 | 0.86 | 4.79 |
| | 1987 | 423.6 | 84.2 | 827 | 303 | 0.37 | 0.72 |
| | 1988 | 492.1 | 95.7 | 379 | 124 | 0.33 | 0.25 |
| | 1989 | 432.8 | 84.7 | 832 | 288 | 0.35 | 0.67 |
| | 1990 | 433.1 | 85.9 | 849 | 307 | 0.36 | 0.71 |
| | 1991 | 492.3 | 94.3 | 310 | 118 | 0.38 | 0.24 |
| | 1992 | 446.8 | 88.1 | 921 | 381 | 0.41 | 0.85 |
| | 1993 | 402.3 | 80.1 | 833 | 217 | 0.26 | 0.54 |
| | 1994 | 515.8 | 98.7 | 220 | 38 | 0.17 | 0.07 |
| | 1995 | 462.1 | 87.0 | 737 | 182 | 0.25 | 0.39 |
| | 1996 | 452.7 | 85.2 | 951 | 231 | 0.24 | 0.51 |
| | 1997 | 487.1 | 96.0 | 260 | 57 | 0.22 | 0.12 |
| | 1998 | 383.4 | 77.9 | 944 | 199.399 | 0.21 | 0.52 |
| | 1999 | 463.4 | 91.0 | 854 | 175.795 | 0.21 | 0.38 |
| | 2000 | 517.8 | 99.6 | 198 | 37.846 | 0.19 | 0.07 |
| | 2001 | 474.9 | 93.5 | 863 | 143.010 | 0.17 | 0.30 |
| | 2002 | 451.0 | 91.7 | 946 | 150.446 | 0.16 | 0.33 |
| | 2003 | 505.9 | 98.8 | 359 | 54.348 | 0.15 | 0.11 |
| | 2004 | 439.2 | 87.2 | 1,379 | 211.529 | 0.15 | 0.48 |
| | 2005 | 467.5 | 94.2 | 1,105 | 198.003 | 0.18 | 0.42 |
| | 2006 | 582.9 | 100.0 | 380 | 49.537 | 0.13 | 0.08 |
| | 2007 | 537.0 | 93.0 | 1,191 | 171.200 | 0.14 | 0.32 |
| | 2008 | 557.3 | 94.1 | 1,402 | 213.680 | 0.15 | 0.38 |
| | 2009 | 611.9 | 100.0 | 392 | 61.105 | 0.16 | 0.10 |
| | 2010 | 548.6 | 91.2 | 1,071 | 206.321 | 0.19 | 0.38 |
| **VOGTLE 1, 2** | 1988 | 820.4 | 77.7 | 1,108 | 138 | 0.12 | 0.17 |
| Docket 50-424; 50-425; | 1989 | 1,045.8 | 96.0 | 427 | 32 | 0.07 | 0.03 |
| NPF-68, NPF-81 | 1990 | 1,710.9 | 82.7 | 1,602 | 466 | 0.29 | 0.27 |
| 1st commercial operation | 1991 | 1,966.5 | 89.2 | 1,357 | 362 | 0.27 | 0.18 |
| 6/87, 5/89 | 1992 | 2,047.9 | 90.0 | 1,262 | 426 | 0.34 | 0.21 |
| Type - PWRs | 1993 | 2,060.4 | 88.3 | 1,338 | 367 | 0.27 | 0.18 |
| Capacity - 1,150, 1,152 MWe | 1994 | 2,170.1 | 91.3 | 1,048 | 217 | 0.21 | 0.10 |
| | 1995 | 2,285.4 | 95.2 | 953 | 199 | 0.21 | 0.09 |
| | 1996 | 2,056.8 | 86.5 | 1,395 | 452 | 0.32 | 0.22 |
| | 1997 | 2,121.1 | 91.4 | 994 | 158 | 0.16 | 0.07 |
| | 1998 | 2,123.9 | 92.3 | 994 | 162.210 | 0.16 | 0.08 |
| | 1999 | 2,106.0 | 91.5 | 1,359 | 228.942 | 0.17 | 0.11 |
| | 2000 | 2,223.9 | 95.6 | 899 | 121.312 | 0.14 | 0.05 |
| | 2001 | 2,231.5 | 96.2 | 870 | 129.270 | 0.15 | 0.06 |
| | 2002 | 1,942.0 | 85.3 | 1,152 | 243.957 | 0.21 | 0.13 |
| | 2003 | 2,179.9 | 94.8 | 806 | 84.344 | 0.10 | 0.04 |
| | 2004 | 2,200.7 | 95.7 | 765 | 80.763 | 0.11 | 0.04 |
| | 2005 | 2,027.9 | 88.6 | 1,099 | 151.096 | 0.14 | 0.07 |
| | 2006 | 2,048.8 | 89.0 | 892 | 115.509 | 0.13 | 0.06 |
| | 2007 | 2,089.9 | 92.0 | 951 | 120.515 | 0.13 | 0.06 |
| | 2008 | 2,023.9 | 89.3 | 1,185 | 137.620 | 0.12 | 0.07 |
| | 2009 | 2,201.6 | 95.7 | 931 | 79.681 | 0.09 | 0.04 |
| | 2010 | 2,238.6 | 95.8 | 924 | 89.182 | 0.10 | 0.04 |

| Reporting Organization | Year | Megawatt Years (MW-yr) | Unit Availability Factor | Total Personnel with Measurable Doses | Collective Dose per Site (person-rem) | Average Measurable Dose (rem) | Collective Dose/ MW-yr |
|---|---|---|---|---|---|---|---|
| **WATERFORD 3** | 1986 | 875.7 | 79.1 | 1,244 | 223 | 0.18 | 0.25 |
| Docket 50-382; | 1987 | 891.8 | 82.5 | 959 | 156 | 0.16 | 0.17 |
| NPF-38 | 1988 | 784.3 | 75.4 | 1,246 | 259 | 0.21 | 0.33 |
| 1st commercial operation 9/85 | 1989 | 909.8 | 82.6 | 1,306 | 265 | 0.20 | 0.29 |
| Type - PWR | 1990 | 1,027.9 | 92.8 | 432 | 47 | 0.11 | 0.05 |
| Capacity - 1,152 MWe | 1991 | 870.6 | 79.8 | 1,301 | 364 | 0.28 | 0.42 |
| | 1992 | 909.6 | 83.2 | 1,213 | 226 | 0.19 | 0.25 |
| | 1993 | 1,088.3 | 99.4 | 195 | 15 | 0.08 | 0.01 |
| | 1994 | 949.1 | 87.0 | 1,167 | 191 | 0.16 | 0.20 |
| | 1995 | 927.4 | 83.4 | 1,092 | 153 | 0.14 | 0.16 |
| | 1996 | 1,064.8 | 94.2 | 342 | 27 | 0.08 | 0.03 |
| | 1997 | 767.2 | 71.2 | 1,186 | 148 | 0.13 | 0.19 |
| | 1998 | 984.1 | 91.9 | 282 | 24.032 | 0.09 | 0.02 |
| | 1999 | 849.5 | 79.6 | 833 | 123.198 | 0.15 | 0.14 |
| | 2000 | 965.1 | 88.8 | 825 | 131.701 | 0.16 | 0.14 |
| | 2001 | 1,086.0 | 99.6 | 91 | 4.677 | 0.05 | 0.00 |
| | 2002 | 1,007.0 | 93.2 | 811 | 109.439 | 0.13 | 0.11 |
| | 2003 | 968.0 | 90.9 | 710 | 95.332 | 0.13 | 0.10 |
| | 2004 | 1,099.1 | 100.0 | 60 | 2.517 | 0.04 | 0.00 |
| | 2005 | 900.9 | 80.2 | 902 | 136.318 | 0.15 | 0.15 |
| | 2006 | 1,059.3 | 92.0 | 1,190 | 109.682 | 0.09 | 0.10 |
| | 2007 | 1,130.2 | 96.0 | 469 | 20.125 | 0.04 | 0.02 |
| | 2008 | 1,030.7 | 88.0 | 1,268 | 134.221 | 0.11 | 0.13 |
| | 2009 | 1,023.4 | 88.0 | 1,479 | 255.088 | 0.17 | 0.25 |
| | 2010 | 1,173.1 | 100.0 | 216 | 4.913 | 0.02 | 0.00 |
| **WATTS BAR 1** | 1997 | 867.6 | 83.8 | 1,103 | 113 | 0.10 | 0.13 |
| Docket 50-390; | 1998 | 1,105.1 | 99.1 | 96 | 3.106 | 0.03 | 0.00 |
| NPF-90 | 1999 | 943.1 | 87.2 | 975 | 98.946 | 0.10 | 0.10 |
| 1st commercial operation 5/96 | 2000 | 1,033.3 | 92.8 | 1,053 | 122.453 | 0.12 | 0.12 |
| Type - PWR | 2001 | 1,095.9 | 96.5 | 197 | 5.912 | 0.03 | 0.01 |
| Capacity - 1,121 MWe | 2002 | 1,034.0 | 92.1 | 909 | 93.598 | 0.10 | 0.09 |
| | 2003 | 973.3 | 86.7 | 1,392 | 165.741 | 0.12 | 0.17 |
| | 2004 | 1,122.1 | 99.1 | 220 | 5.893 | 0.03 | 0.01 |
| | 2005 | 1,003.7 | 90.0 | 1,244 | 143.506 | 0.12 | 0.14 |
| | 2006 | 764.5 | 70.0 | 2,070 | 322.682 | 0.16 | 0.42 |
| | 2007 | 1,150.6 | 100.0 | 128 | 4.414 | 0.03 | 0.00 |
| | 2008 | 923.5 | 83.2 | 887 | 70.648 | 0.08 | 0.08 |
| | 2009 | 1,051.1 | 92.1 | 853 | 63.846 | 0.07 | 0.06 |
| | 2010 | 1,111.7 | 98.3 | 129 | 6.193 | 0.05 | 0.01 |
| **WOLF CREEK 1** | 1986 | 832.8 | 73.3 | 682 | 143 | 0.21 | 0.17 |
| Docket 50-482; | 1987 | 778.8 | 71.1 | 675 | 138 | 0.20 | 0.18 |
| NPF-42 | 1988 | 794.7 | 70.7 | 1,010 | 297 | 0.29 | 0.37 |
| 1st commercial operation 9/85 | 1989 | 1,108.4 | 99.5 | 186 | 18 | 0.10 | 0.02 |
| Type - PWR | 1990 | 940.2 | 81.0 | 798 | 195 | 0.24 | 0.21 |
| Capacity - 1,160 MWe | 1991 | 707.6 | 71.9 | 1,010 | 331 | 0.33 | 0.47 |
| | 1992 | 1,010.8 | 86.7 | 446 | 78 | 0.17 | 0.08 |
| | 1993 | 940.5 | 80.6 | 975 | 183 | 0.19 | 0.19 |
| | 1994 | 1,017.2 | 86.8 | 1,082 | 235 | 0.22 | 0.23 |
| | 1995 | 1,198.0 | 98.7 | 242 | 14 | 0.06 | 0.01 |
| | 1996 | 980.6 | 81.2 | 986 | 171 | 0.17 | 0.17 |
| | 1997 | 964.3 | 83.8 | 989 | 265 | 0.27 | 0.27 |
| | 1998 | 1,187.3 | 100.0 | 184 | 10.382 | 0.06 | 0.01 |
| | 1999 | 1,045.3 | 90.1 | 812 | 147.704 | 0.18 | 0.14 |
| | 2000 | 1,032.7 | 89.5 | 861 | 143.417 | 0.17 | 0.14 |
| | 2001 | 1,177.9 | 100.0 | 105 | 5.176 | 0.05 | 0.00 |
| | 2002 | 1,029.0 | 88.7 | 816 | 99.987 | 0.12 | 0.10 |
| | 2003 | 1,013.5 | 87.2 | 820 | 88.941 | 0.11 | 0.09 |
| | 2004 | 1,153.5 | 98.8 | 93 | 3.388 | 0.04 | 0.00 |
| | 2005 | 1,004.2 | 86.7 | 856 | 106.870 | 0.12 | 0.11 |
| | 2006 | 1,067.4 | 91.0 | 789 | 96.788 | 0.12 | 0.09 |

| Reporting Organization | Year | Megawatt Years (MW-yr) | Unit Availability Factor | Total Personnel with Measurable Doses | Collective Dose per Site (person-rem) | Average Measurable Dose (rem) | Collective Dose/ MW-yr |
|---|---|---|---|---|---|---|---|
| **WOLF CREEK 1** | 2007 | 1,183.7 | 100.0 | 91 | 4.307 | 0.05 | 0.00 |
| (continued) | 2008 | 968.3 | 83.1 | 911 | 94.997 | 0.10 | 0.10 |
| | 2009 | 1,001.0 | 86.9 | 1,504 | 73.637 | 0.05 | 0.07 |
| | 2010 | 1,090.8 | 94.2 | 463 | 10.516 | 0.02 | 0.01 |
| **YANKEE ROWE**[17] | 1969 | 138.3 | | 193 | 215 | 1.11 | 1.55 |
| Docket 50-29; | 1970 | 146.1 | | 355 | 255 | 0.72 | 1.75 |
| DPR-3 | 1971 | 173.5 | | 155 | 90 | 0.58 | 0.52 |
| 1st commercial operation 7/61 | 1972 | 78.7 | | 282 | 255 | 0.90 | 3.24 |
| Type - PWR | 1973 | 127.1 | | 133 | 99 | 0.74 | 0.78 |
| Capacity - (175) MWe | 1974 | 111.3 | | 243 | 205 | 0.84 | 1.84 |
| | 1975 | 145.1 | 82.4 | 249 | 116 | 0.47 | 0.80 |
| | 1976 | 152.2 | 89.8 | 152 | 59 | 0.39 | 0.39 |
| | 1977 | 124.6 | 73.9 | 725 | 356 | 0.49 | 2.86 |
| | 1978 | 145.0 | 81.0 | 565 | 282 | 0.50 | 1.94 |
| | 1979 | 149.0 | 81.6 | 441 | 127 | 0.29 | 0.85 |
| | 1980 | 35.6 | 22.0 | 502 | 213 | 0.42 | 5.98 |
| | 1981 | 109.0 | 74.4 | 515 | 302 | 0.59 | 2.77 |
| | 1982 | 108.6 | 73.4 | 814 | 474 | 0.58 | 4.36 |
| | 1983 | 163.5 | 91.4 | 395 | 68 | 0.17 | 0.42 |
| | 1984 | 124.8 | 71.4 | 654 | 348 | 0.53 | 2.79 |
| | 1985 | 144.3 | 85.3 | 653 | 211 | 0.32 | 1.46 |
| | 1986 | 169.7 | 95.0 | 384 | 45 | 0.12 | 0.27 |
| | 1987 | 138.7 | 82.7 | 593 | 217 | 0.37 | 1.56 |
| | 1988 | 136.4 | 85.2 | 738 | 227 | 0.31 | 1.66 |
| | 1989 | 159.4 | 92.9 | 496 | 62 | 0.13 | 0.39 |
| | 1990 | 101.1 | 61.5 | 702 | 246 | 0.35 | 2.43 |
| | 1991 | 121.2 | 72.3 | 162 | 40 | 0.25 | 0.33 |
| | 1992 | 0.0 | 0.0 | 324 | 94 | 0.29 | --- |
| | 1993 | 0.0 | 0.0 | 313 | 163 | 0.52 | --- |
| | 1994 | 0.0 | 0.0 | 222 | 156 | 0.70 | --- |
| | 1995 | 0.0 | 0.0 | 191 | 78 | 0.41 | --- |
| | 1996 | 0.0 | 0.0 | 239 | 95 | 0.40 | --- |
| | 1997 | 0.0 | 0.0 | 323 | 65 | 0.20 | --- |
| | 1998 | 0.0 | 0.0 | 125 | 4.603 | 0.04 | --- |
| | 1999 | 0.0 | 0.0 | 83 | 2.291 | 0.02 | --- |
| | 2000 | 0.0 | 0.0 | 38 | 2.406 | 0.06 | --- |
| | 2001 | 0.0 | 0.0 | 48 | 3.969 | 0.08 | --- |
| | 2002 | 0.0 | 0.0 | 128 | 20.024 | 0.16 | --- |
| | 2003 | 0.0 | 0.0 | 136 | 30.934 | 0.23 | --- |
| | 2004 | 0.0 | 0.0 | 70 | 6.502 | 0.09 | --- |
| | 2005 | 0.0 | 0.0 | 63 | 1.456 | 0.02 | --- |
| | 2006 | 0.0 | 0.0 | 45 | 0.975 | 0.02 | --- |
| | 2007 | 0.0 | 0.0 | 0 | 0.000 | --- | --- |
| | 2008 | 0.0 | 0.0 | 1 | 0.019 | 0.02 | --- |
| | 2009 | 0.0 | 0.0 | 5 | 0.114 | 0.02 | --- |
| | 2010 | 0.0 | 0.0 | 3 | 0.083 | 0.03 | --- |
| **ZION 1**[18]**, 2** | 1974 | 425.3 | 71.1 | 306 | 56 | 0.18 | 0.13 |
| Docket 50-295; 50-304; | 1975 | 1,181.5 | 74.9 | 436 | 127 | 0.29 | 0.11 |
| DPR-39, DPR-48 | 1976 | 1,134.9 | 61.9 | 774 | 571 | 0.74 | 0.50 |
| 1st commercial operation | 1977 | 1,358.6 | 75.0 | 784 | 1,003 | 1.28 | 0.74 |
| 12/73, 9/74 | 1978 | 1,613.5 | 80.2 | 1,104 | 1,017 | 0.92 | 0.63 |
| Type - PWRs | 1979 | 1,238.0 | 67.6 | 1,472 | 1,274 | 0.87 | 1.03 |
| Capacity - (1,040), (1,040) MWe | 1980 | 1,411.2 | 74.1 | 1,363 | 920 | 0.67 | 0.65 |
| | 1981 | 1,366.9 | 72.3 | 1,754 | 1,720 | 0.98 | 1.26 |
| | 1982 | 1,186.4 | 64.3 | 1,575 | 2,103 | 1.34 | 1.77 |

[17] Yankee Rowe ended commercial operation as of October 1991 and will not be put in commercial operation again. It is no longer in the count of operating reactors. Parentheses indicate plant capacity when plant was operational.

[18] Zion 1, 2 were shut down in December 1997 and are no longer included in the count of operating reactors. Parentheses indicate plant capacity when plant was operational.

| Reporting Organization | Year | Megawatt Years (MW-yr) | Unit Availability Factor | Total Personnel with Measurable Doses | Collective Dose per Site (person-rem) | Average Measurable Dose (rem) | Collective Dose/ MW-yr |
|---|---|---|---|---|---|---|---|
| **ZION 1[18], 2** | 1983 | 1,222.3 | 69.4 | 1,285 | 1,311 | 1.02 | 1.07 |
| (continued) | 1984 | 1,389.9 | 69.6 | 1,110 | 786 | 0.71 | 0.57 |
| | 1985 | 1,187.9 | 62.9 | 1,498 | 1,166 | 0.78 | 0.98 |
| | 1986 | 1,462.0 | 73.2 | 967 | 474 | 0.49 | 0.32 |
| | 1987 | 1,337.0 | 71.0 | 1,046 | 653 | 0.62 | 0.49 |
| | 1988 | 1,549.1 | 78.3 | 1,926 | 1,260 | 0.65 | 0.81 |
| | 1989 | 1,514.1 | 77.6 | 1,282 | 624 | 0.49 | 0.41 |
| | 1990 | 860.4 | 46.9 | 1,385 | 696 | 0.50 | 0.81 |
| | 1991 | 1,125.7 | 58.2 | 902 | 173 | 0.19 | 0.15 |
| | 1992 | 1,128.8 | 59.0 | 1,732 | 1,043 | 0.60 | 0.92 |
| | 1993 | 1,458.2 | 70.9 | 1,772 | 643 | 0.36 | 0.44 |
| | 1994 | 1,224.9 | 59.9 | 1,176 | 306 | 0.26 | 0.25 |
| | 1995 | 1,471.6 | 72.4 | 1,807 | 797 | 0.44 | 0.54 |
| | 1996 | 1,538.4 | 75.8 | 1,567 | 437 | 0.28 | 0.28 |
| | 1997 | 123.2 | 7.1 | 924 | 119 | 0.13 | 0.97 |
| | 1998 | 0.0 | 0.0 | 246 | 12.417 | 0.05 | --- |
| | 1999 | 0.0 | 0.0 | 67 | 4.194 | 0.06 | --- |
| | 2000 | 0.0 | 0.0 | 26 | 3.015 | 0.12 | --- |
| | 2001 | 0.0 | 0.0 | 6 | 0.274 | 0.05 | --- |
| | 2002 | 0.0 | 0.0 | 12 | 0.276 | 0.02 | --- |
| | 2003 | 0.0 | 0.0 | 2 | 0.049 | 0.02 | --- |
| | 2004 | 0.0 | 0.0 | 6 | 0.167 | 0.03 | --- |
| | 2005 | 0.0 | 0.0 | 5 | 0.109 | 0.02 | --- |
| | 2006 | 0.0 | 0.0 | 7 | 0.109 | 0.02 | --- |
| | 2007 | 0.0 | 0.0 | 8 | 0.224 | 0.03 | --- |
| | 2008 | 0.0 | 0.0 | 7 | 0.147 | 0.02 | --- |
| | 2009 | 0.0 | 0.0 | 0 | 0.000 | --- | --- |
| | 2010 | 0.0 | 0.0 | 17 | 0.562 | 0.03 | --- |

[18] Zion 1, 2 were shut down in December 1997 and are no longer included in the count of operating reactors. Parentheses indicate plant capacity when plant was operational.

Appendix D*

# DOSE PERFORMANCE TRENDS BY REACTOR SITE

## 1973–2010

* Appendix D only contains data on plants still operating in 2010.

## GRAPHICAL REPRESENTATION OF
## DOSE TRENDS IN APPENDIX D

Each page of Appendix D presents a graph of selected dose performance trends from 1973 through 2010. The graphs illustrate the history of the collective dose per reactor for the site, the rolling three-year average collective dose per reactor, and the electricity generated at the site. These data are plotted, beginning with each plant's first full year of commercial operation and continuing through 2010. Data for years when a plant was not in commercial operation have been included when available. However, any data reported prior to 1973 are not included. The three-year average collective dose per reactor data is included because the data provide an overall indication of each plant's general trend in collective dose.

The three-year average collective dose per reactor is also one of the metrics used by NRC in the Reactor Oversight Program to evaluate a licensee's ALARA program. This average is determined by summing the collective dose for the current year and the previous two years and then dividing this sum by the number of reactors reporting during those years. Depicting dose trends by using a three-year average reduces the sporadic effects on annual doses of refueling operations (usually an 18- to 24-month cycle) and occasional high-dose maintenance activities and provides a more representative depiction of collective dose trends over the life of a plant. The annual average collective dose per reactor for all reactors of the same type is also shown on the graph.

## ARKANSAS 1, 2
### Dose Performance Trends

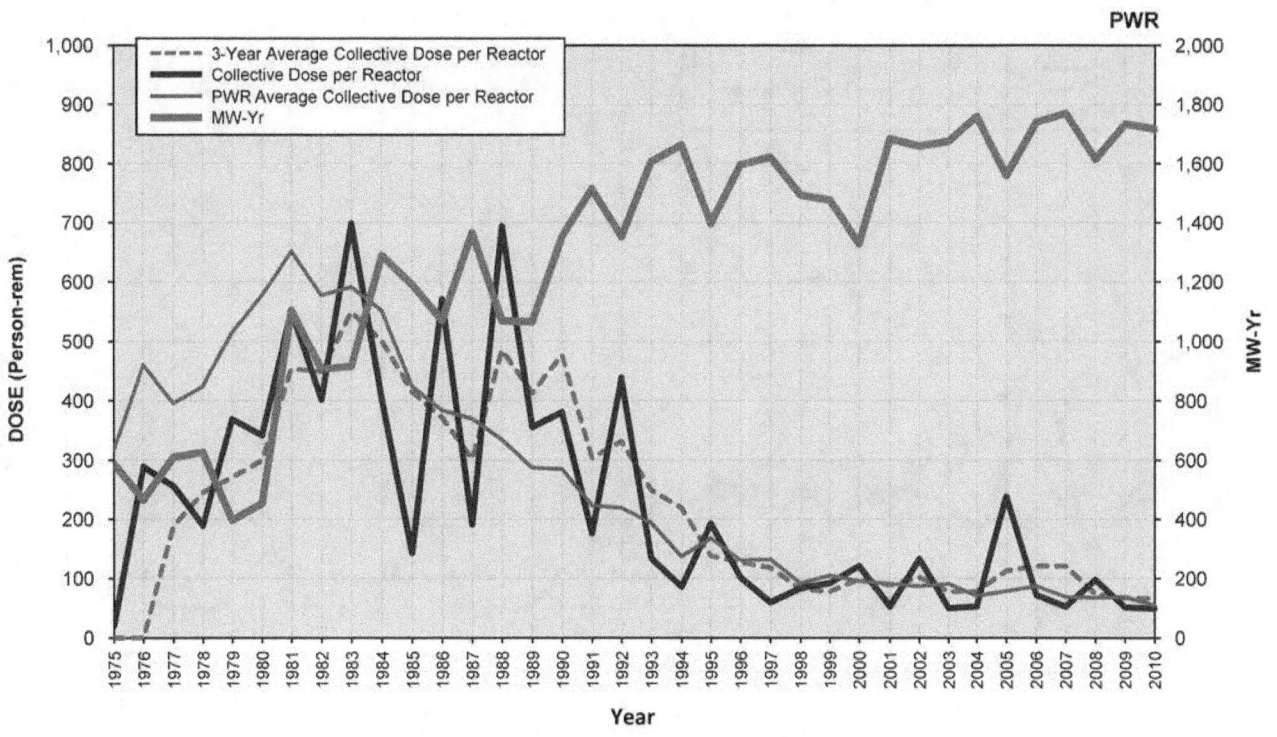

## BEAVER VALLEY 1, 2
### Dose Performance Trends

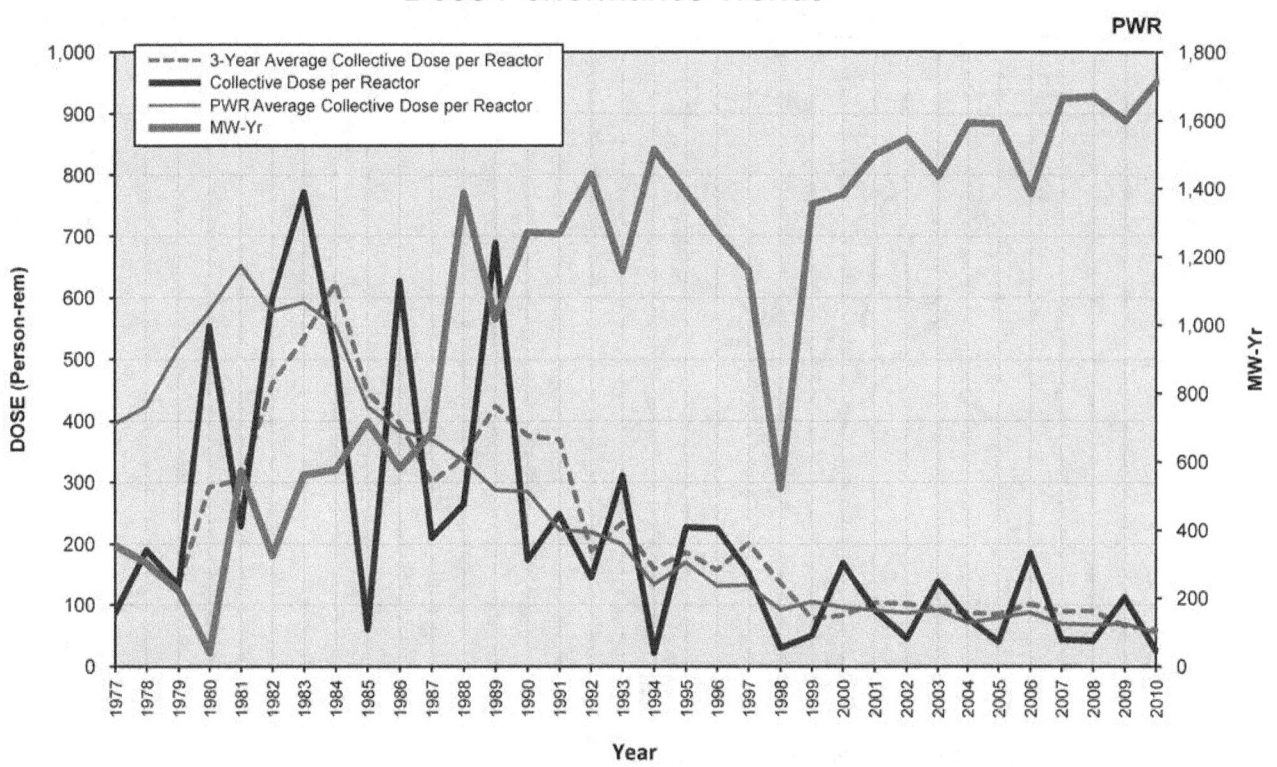

# BRAIDWOOD 1, 2
## Dose Performance Trends

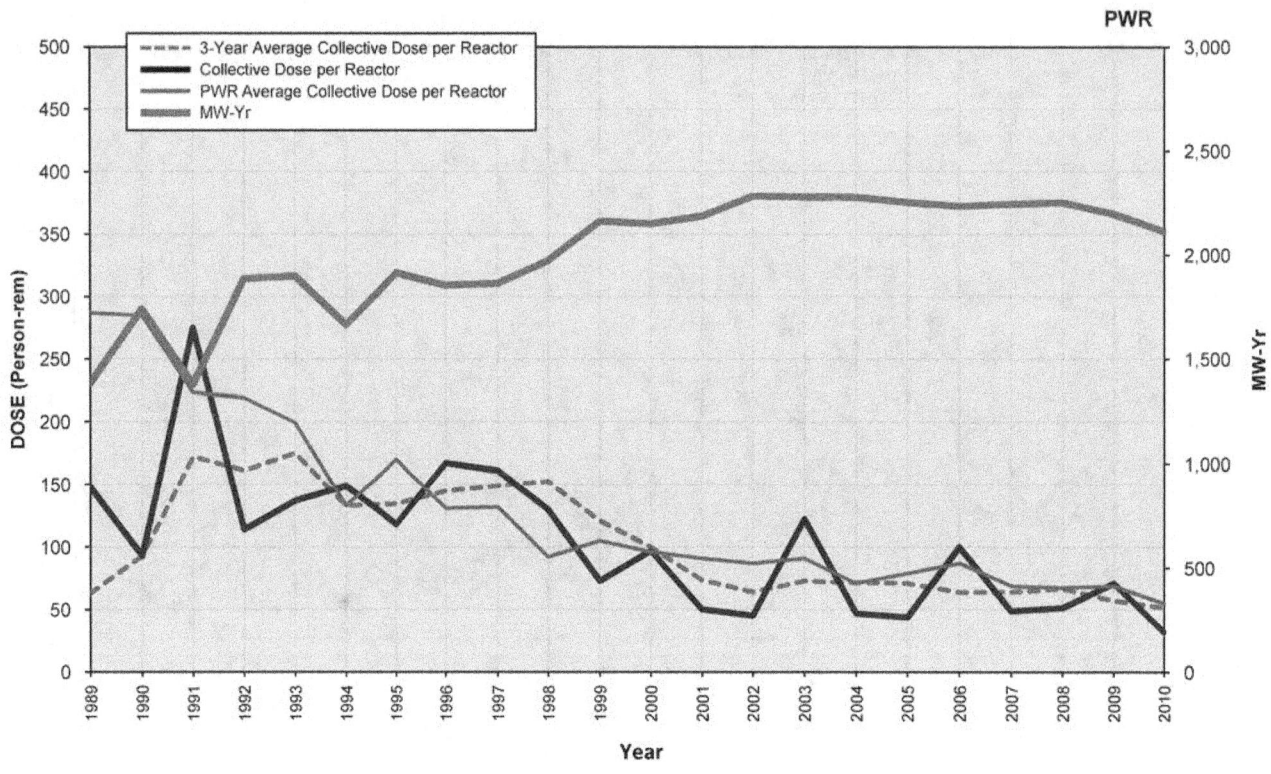

# BROWNS FERRY 1, 2, 3
## Dose Performance Trends

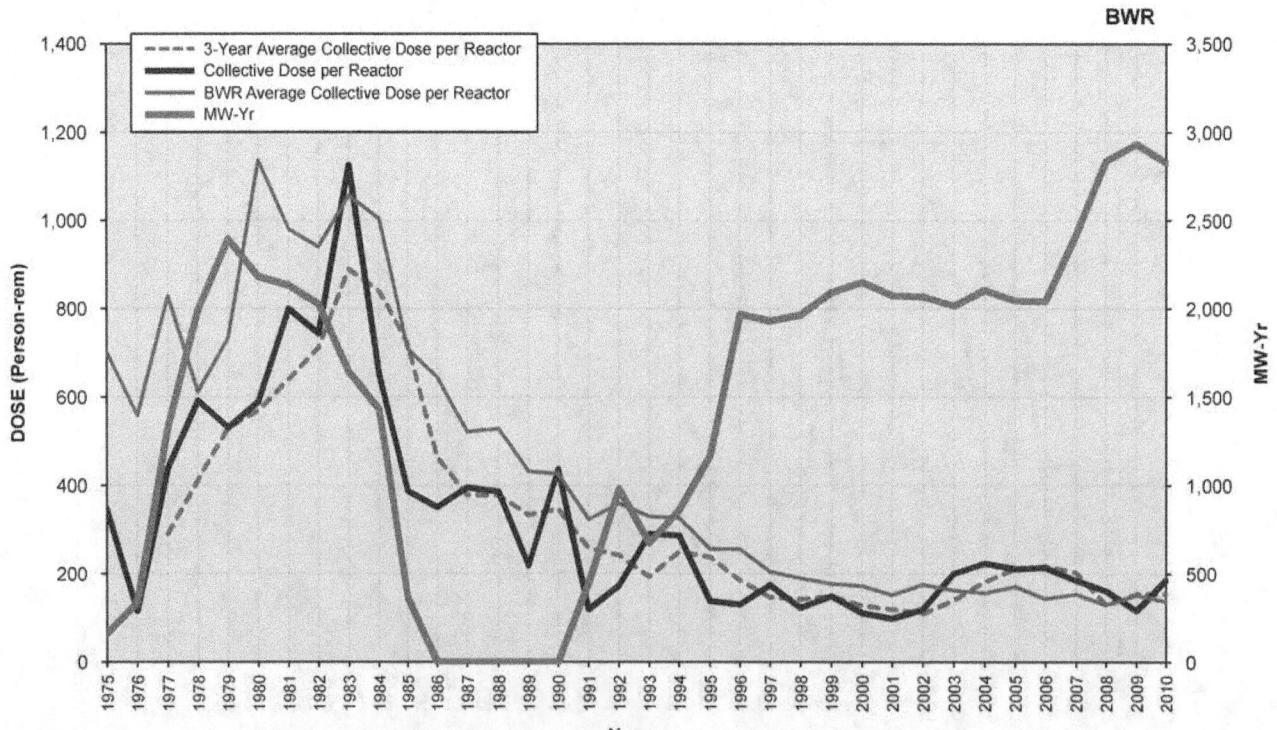

NOTE: Browns Ferry Unit 1 resumed power generation in 2007.

# BRUNSWICK 1, 2
## Dose Performance Trends

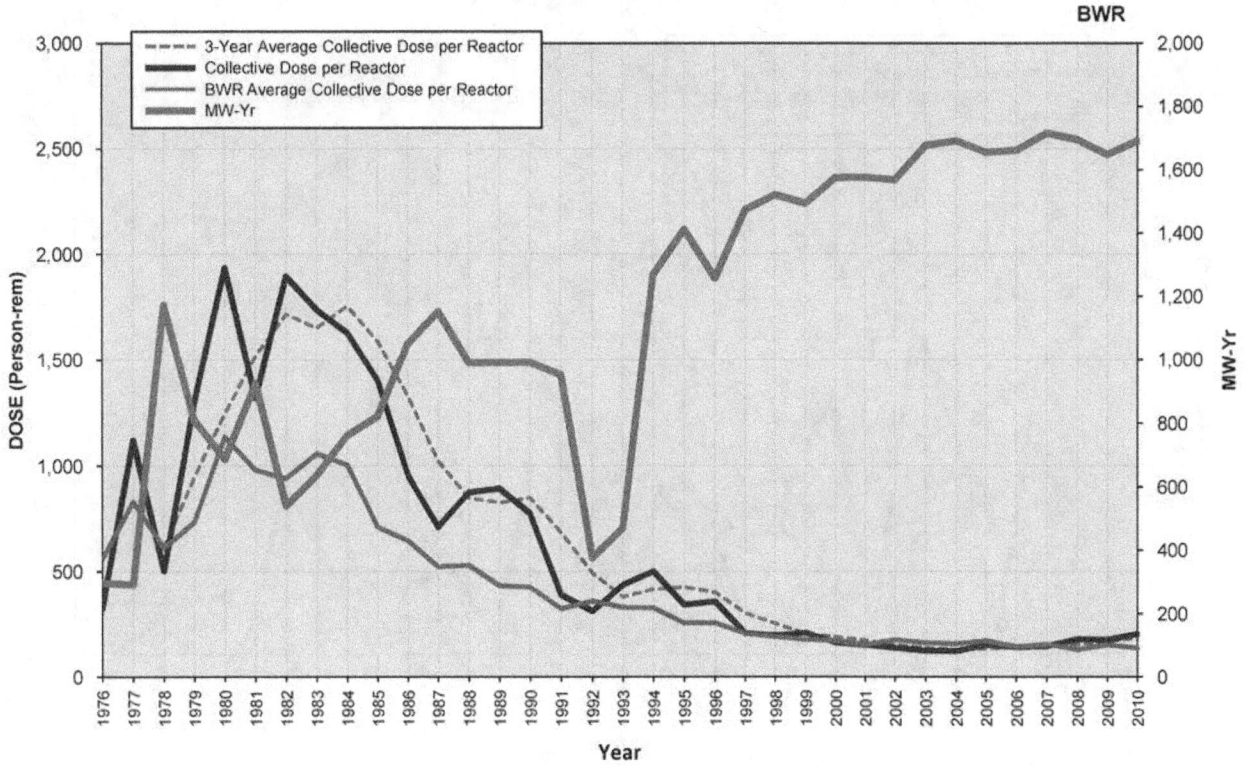

# BYRON 1, 2
## Dose Performance Trends

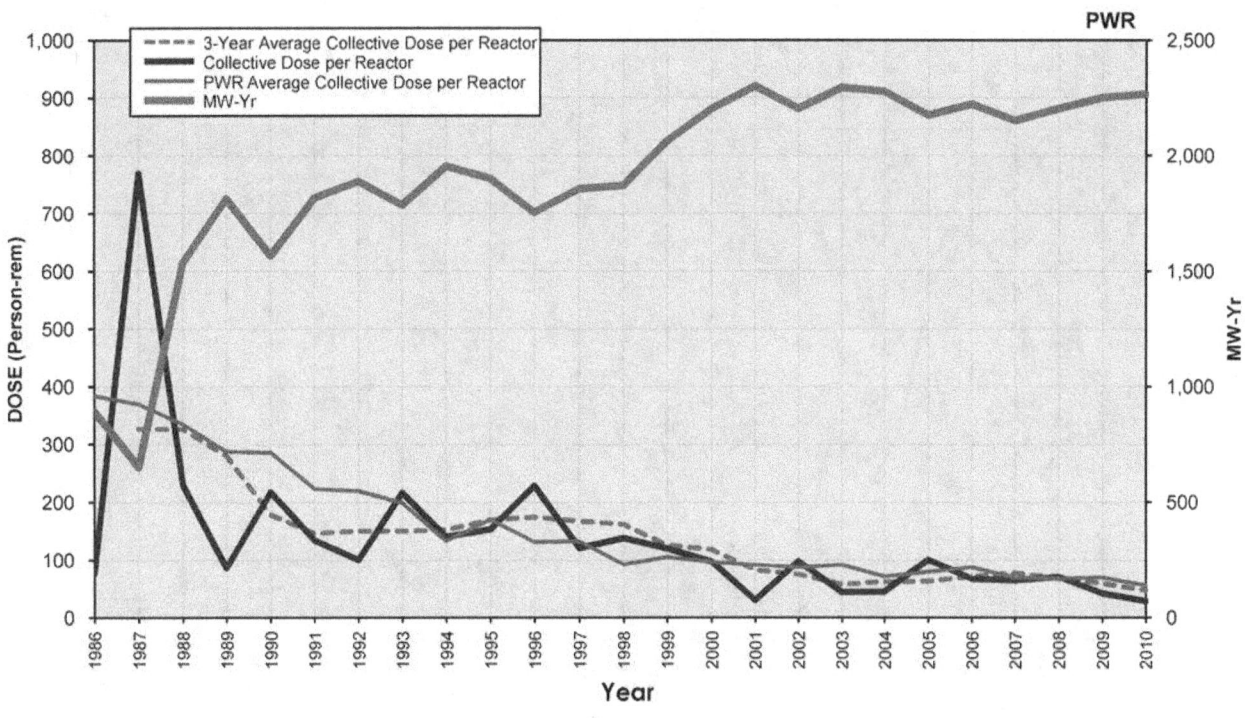

## CALLAWAY 1
### Dose Performance Trends

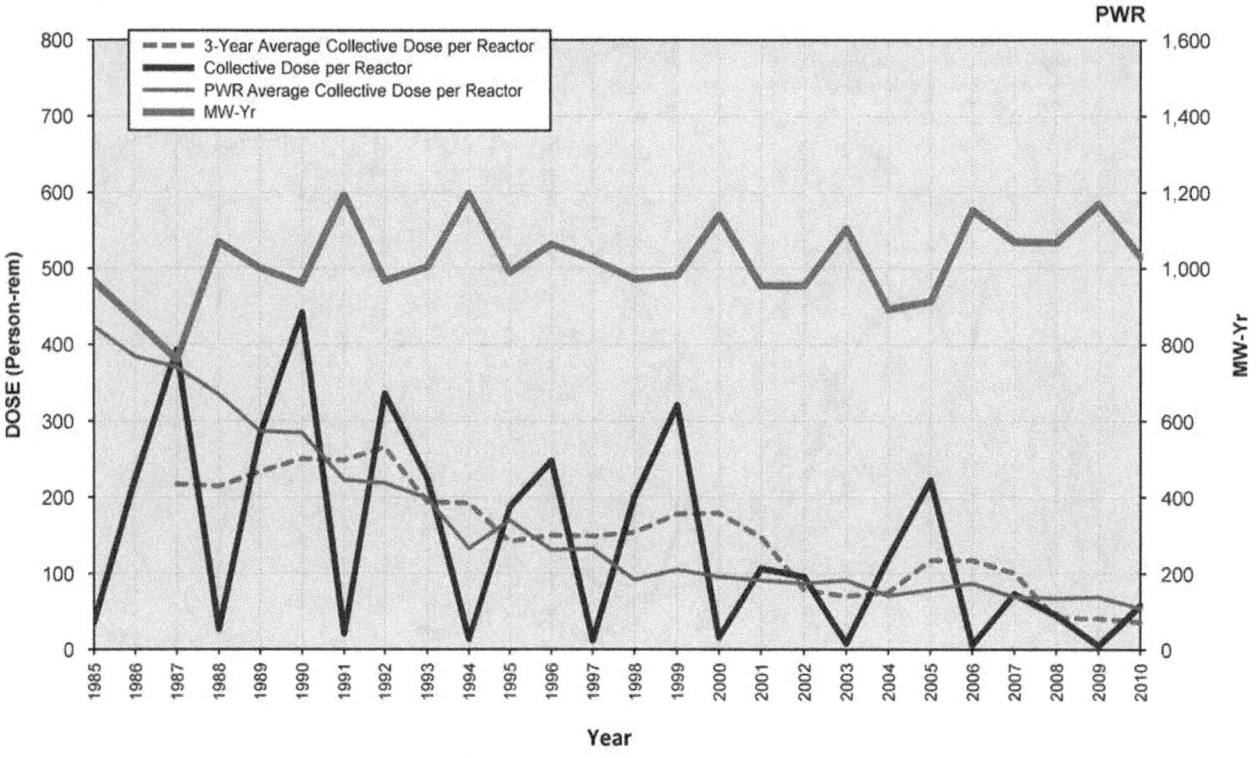

## CALVERT CLIFFS 1, 2
### Dose Performance Trends

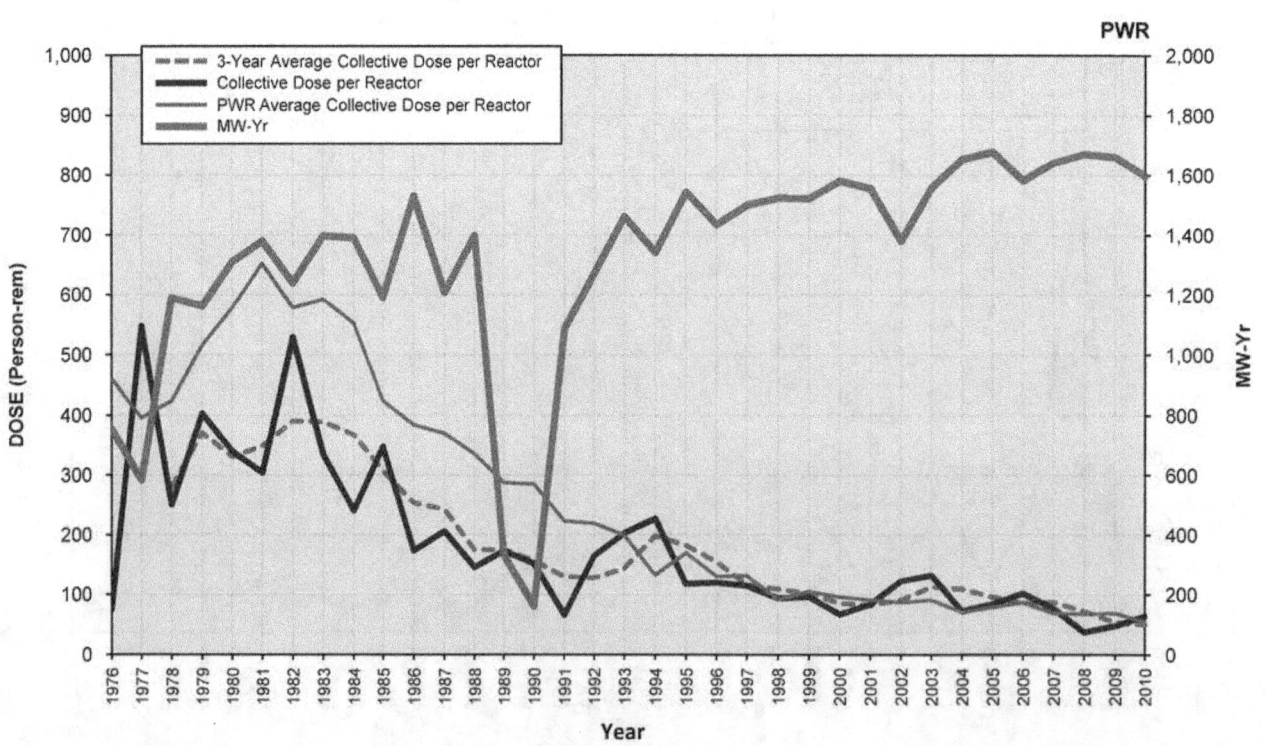

## CATAWBA 1, 2
### Dose Performance Trends

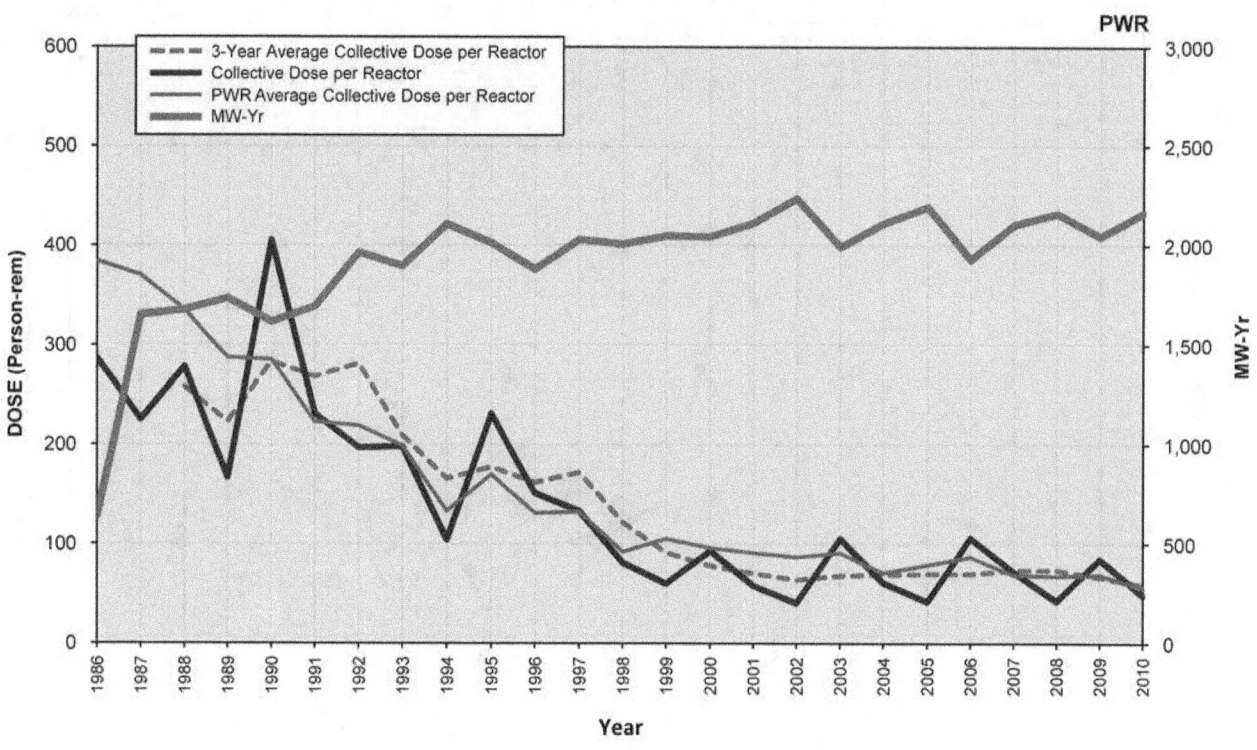

## CLINTON
### Dose Performance Trends

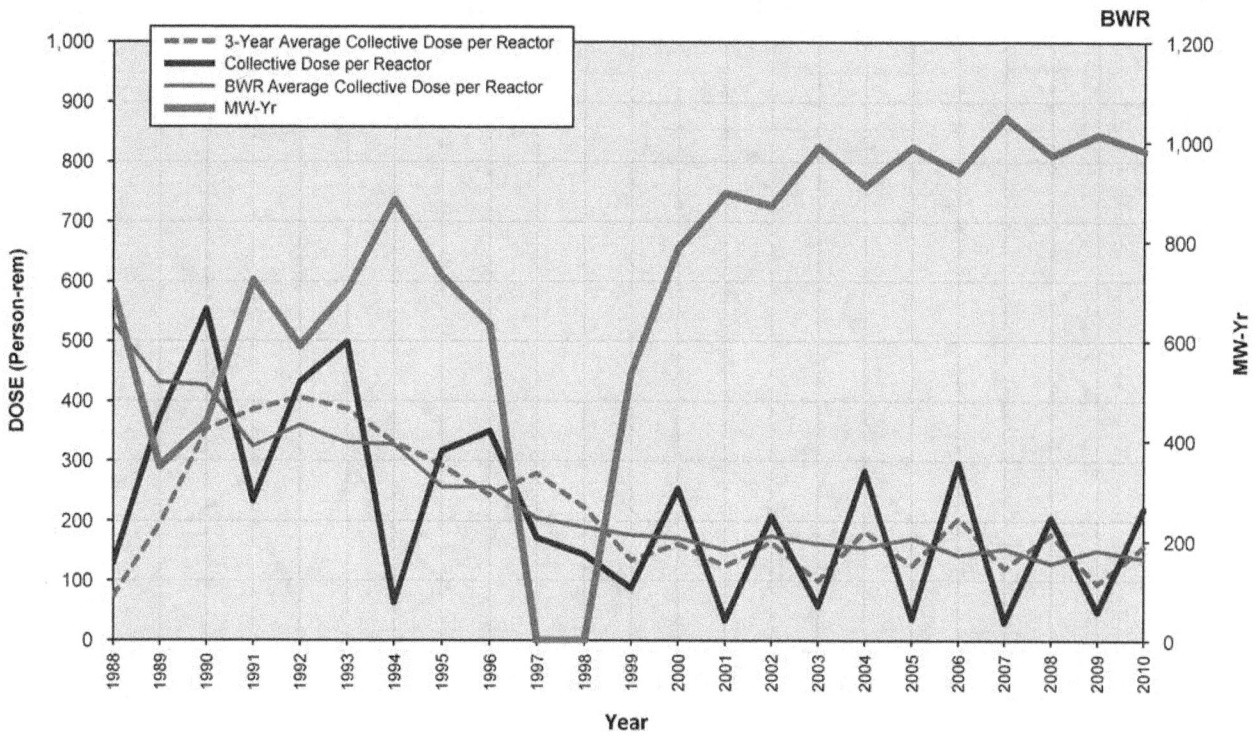

## COLUMBIA GENERATING
### Dose Performance Trends

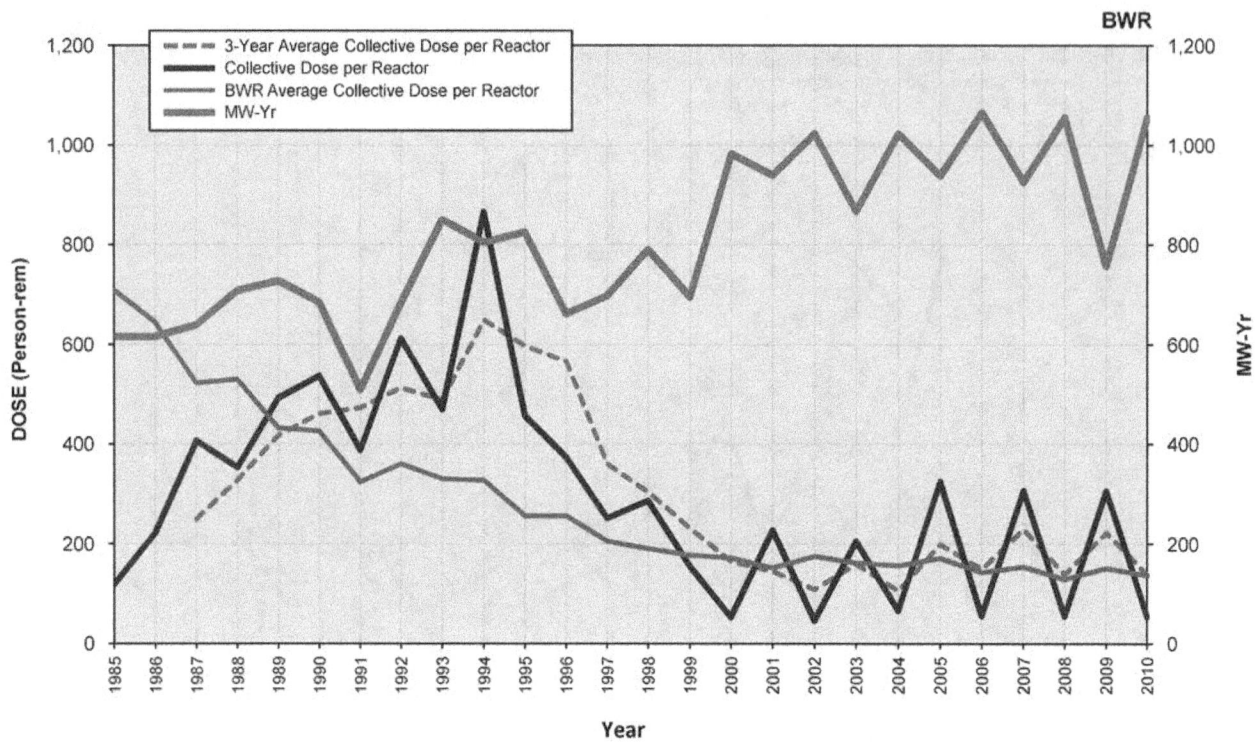

## COMANCHE PEAK 1, 2
### Dose Performance Trends

## COOK 1, 2
### Dose Performance Trends

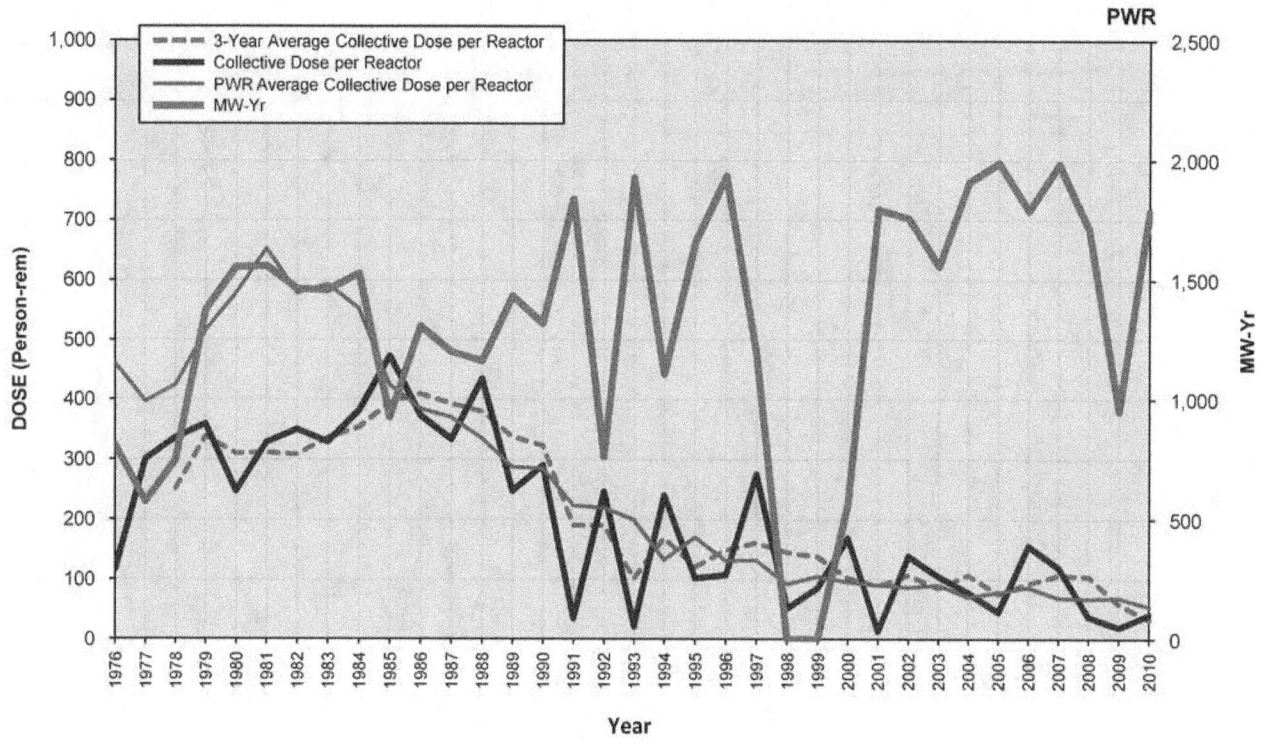

## COOPER STATION
### Dose Performance Trends

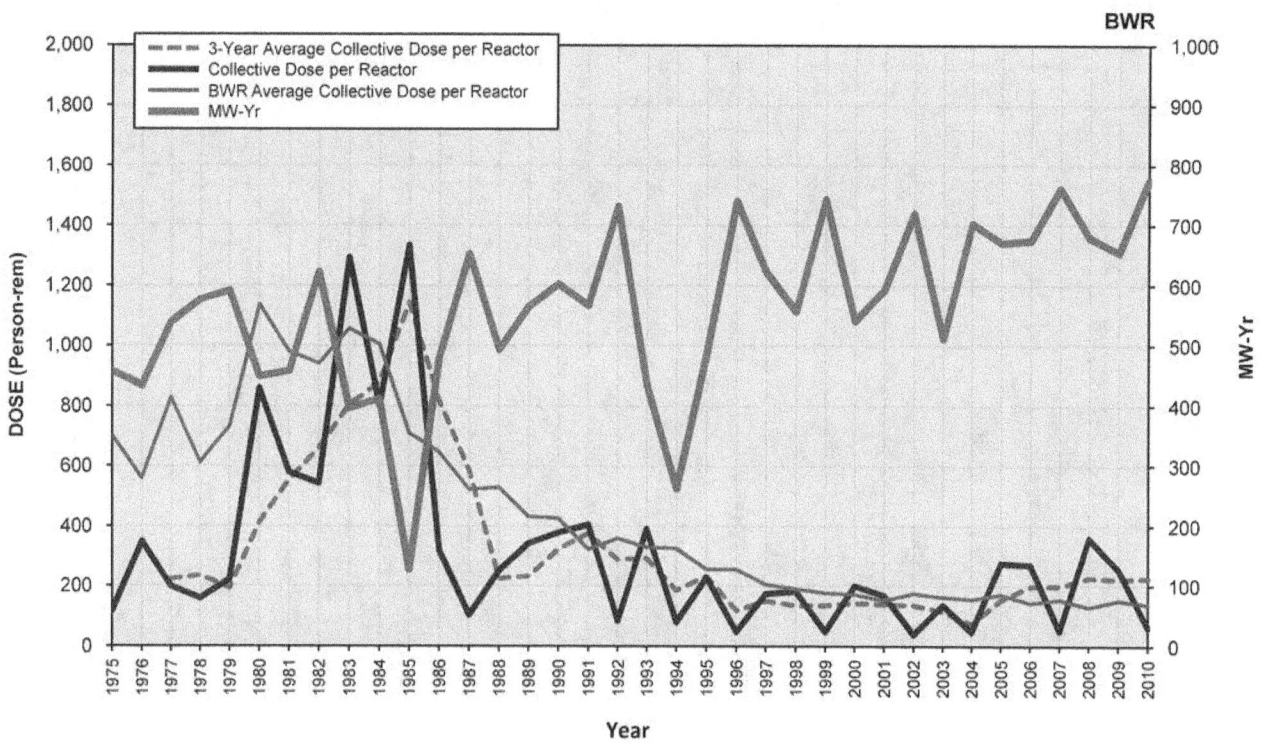

## CRYSTAL RIVER 3
### Dose Performance Trends

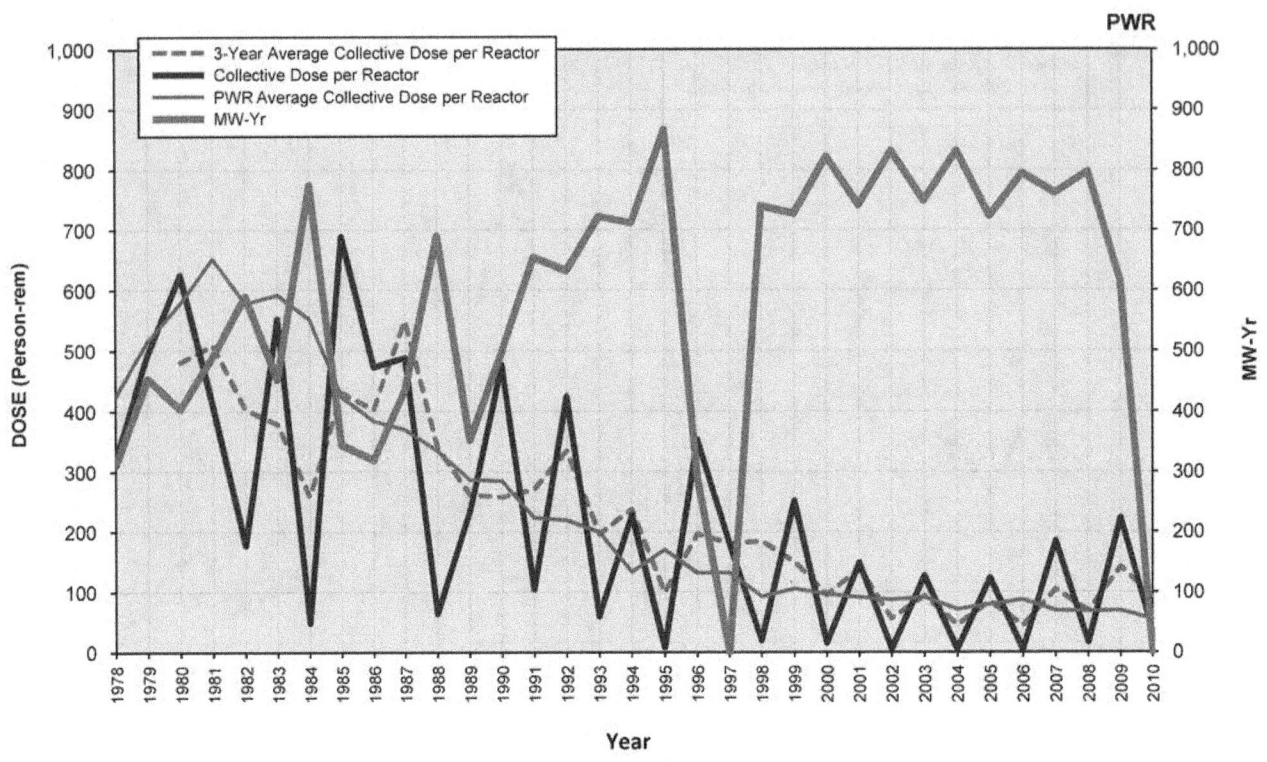

## DAVIS-BESSE 1
### Dose Performance Trends

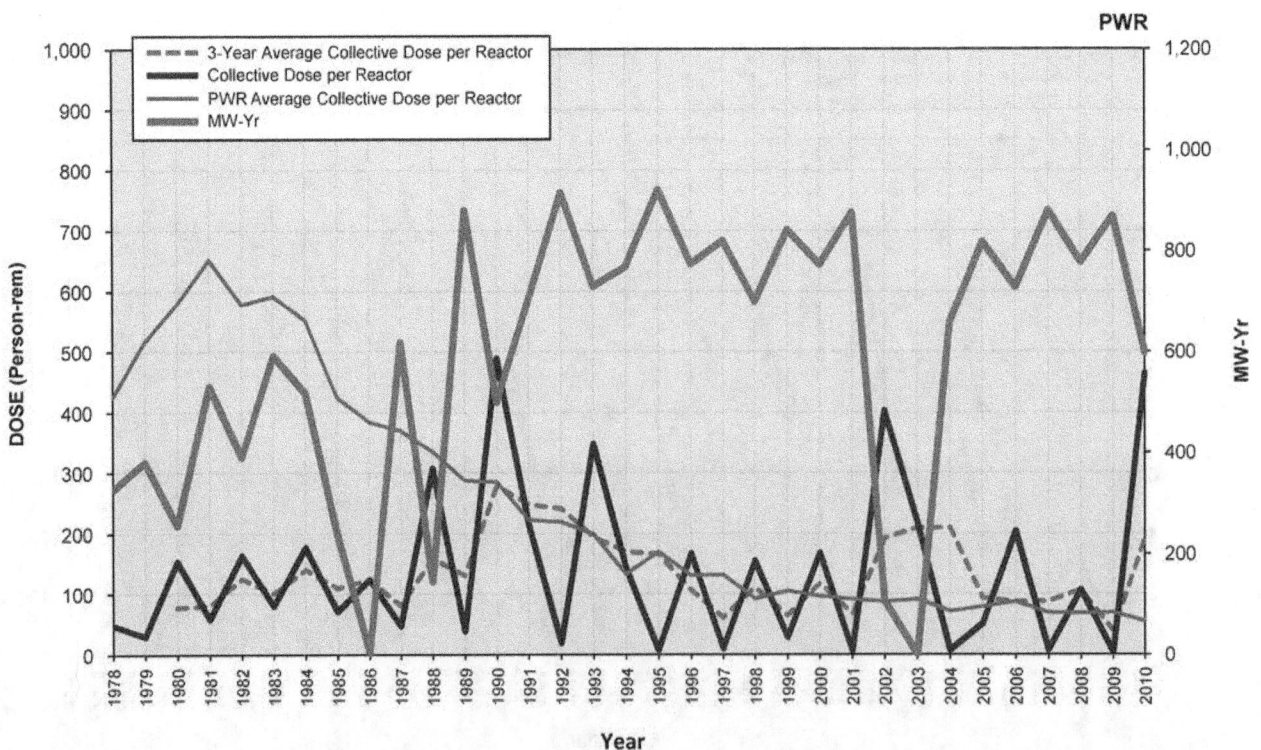

## DIABLO CANYON 1, 2
### Dose Performance Trends

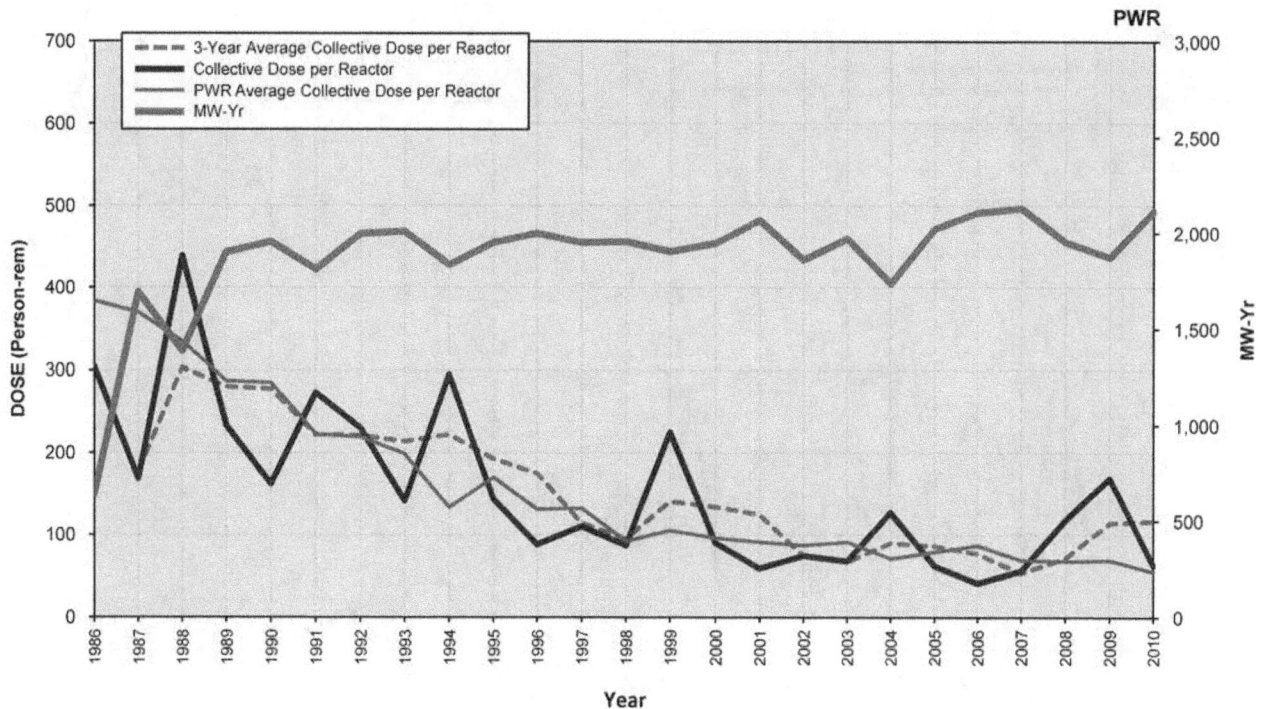

## DRESDEN 2, 3
### Dose Performance Trends

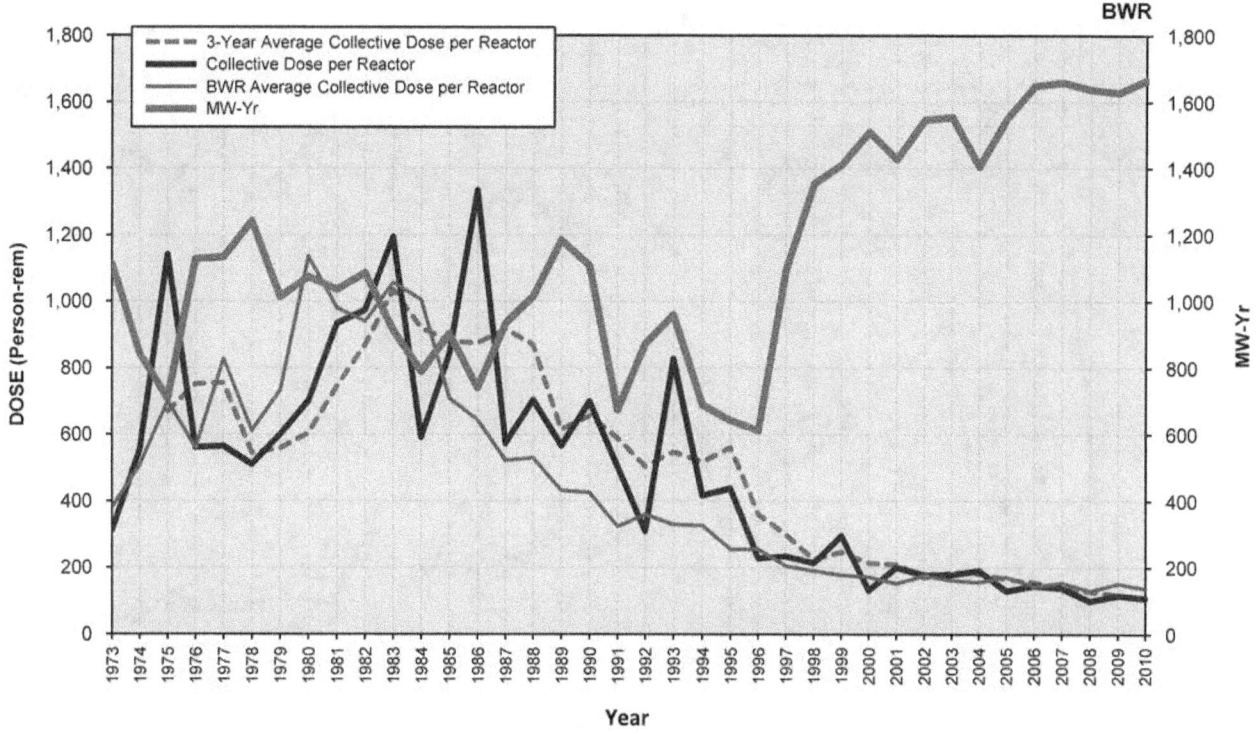

## DUANE ARNOLD
### Dose Performance Trends

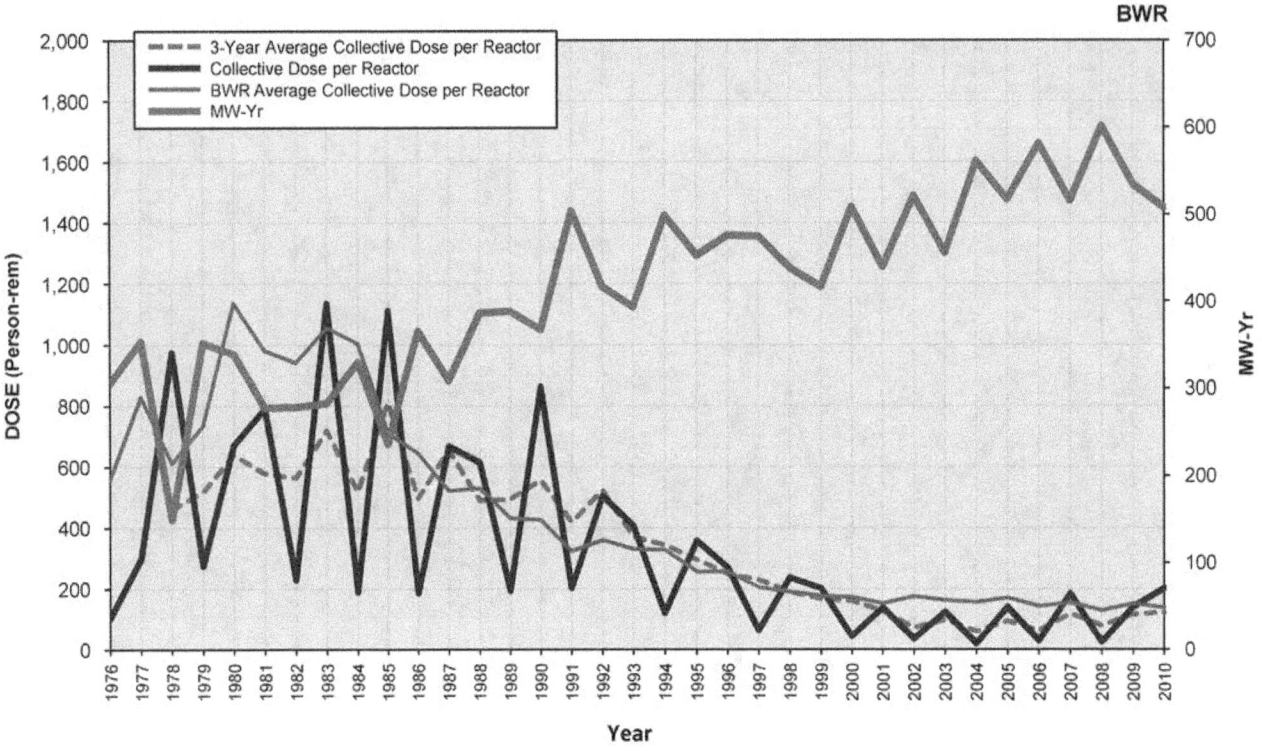

## FARLEY 1, 2
### Dose Performance Trends

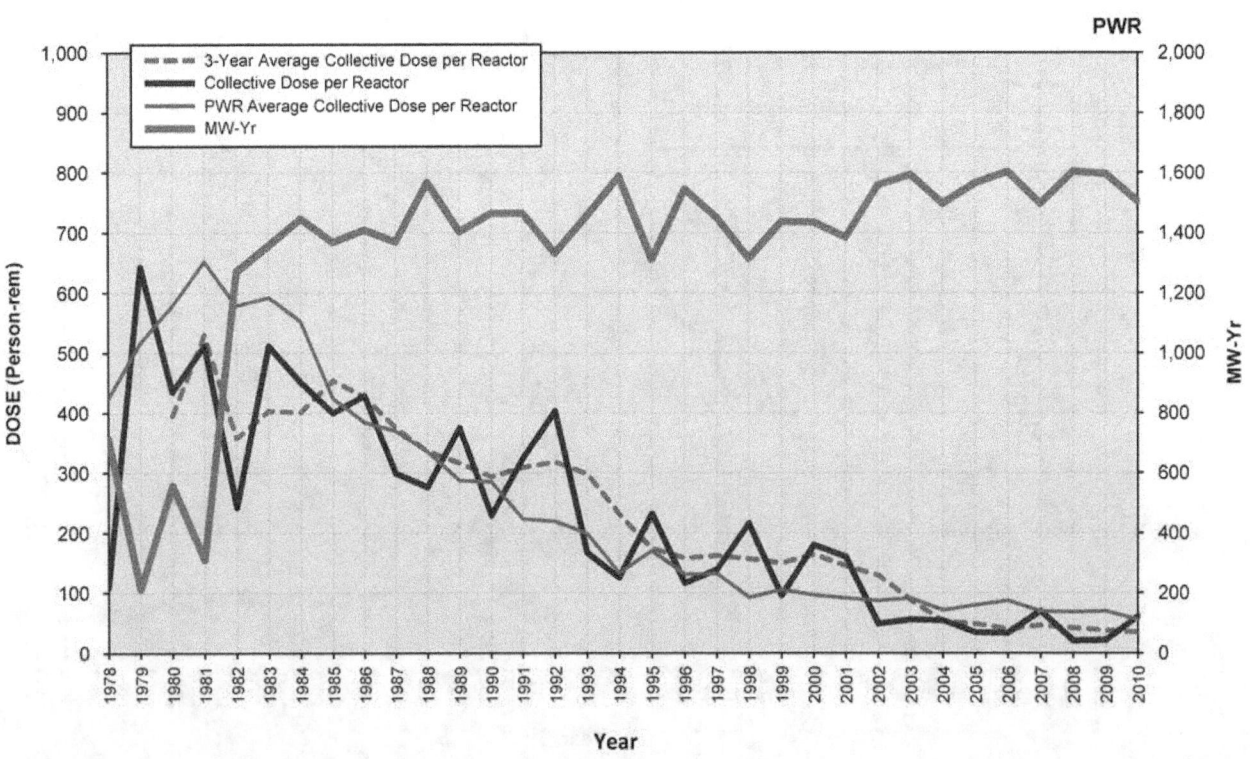

# FERMI 2
## Dose Performance Trends

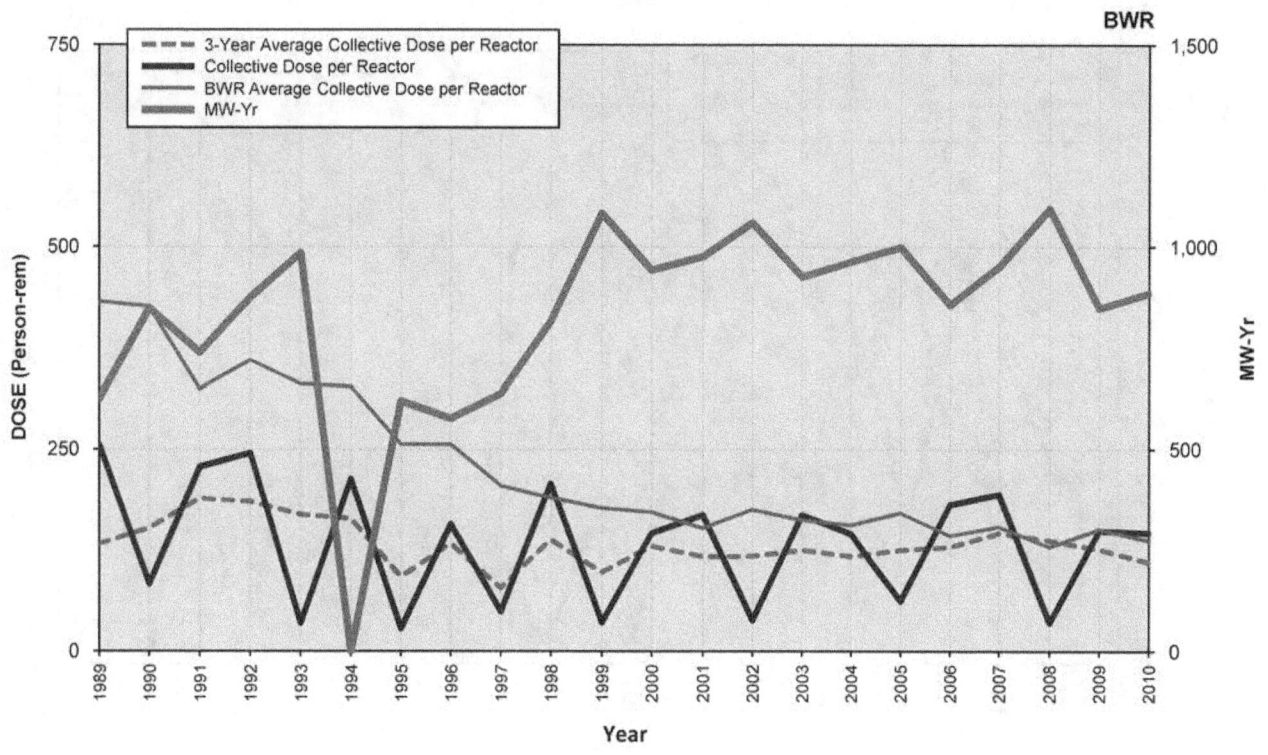

# FITZPATRICK
## Dose Performance Trends

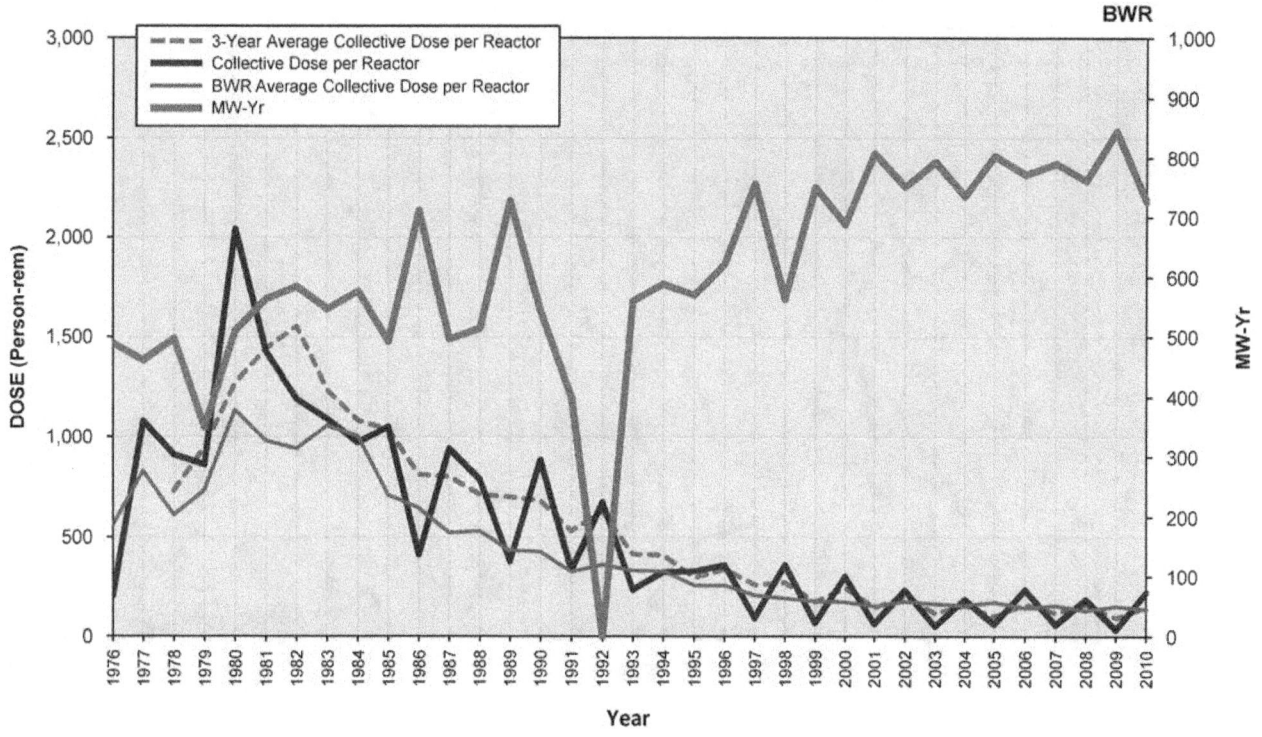

# FORT CALHOUN
## Dose Performance Trends

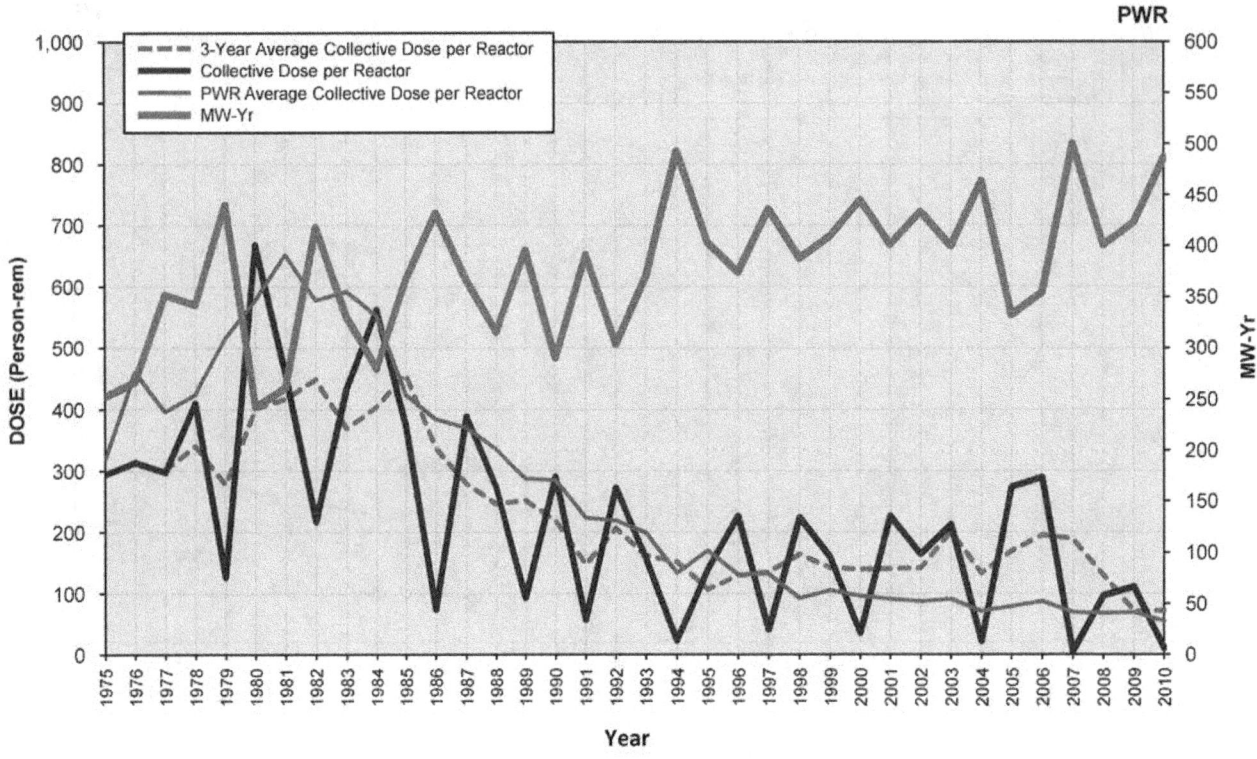

# GINNA
## Dose Performance Trends

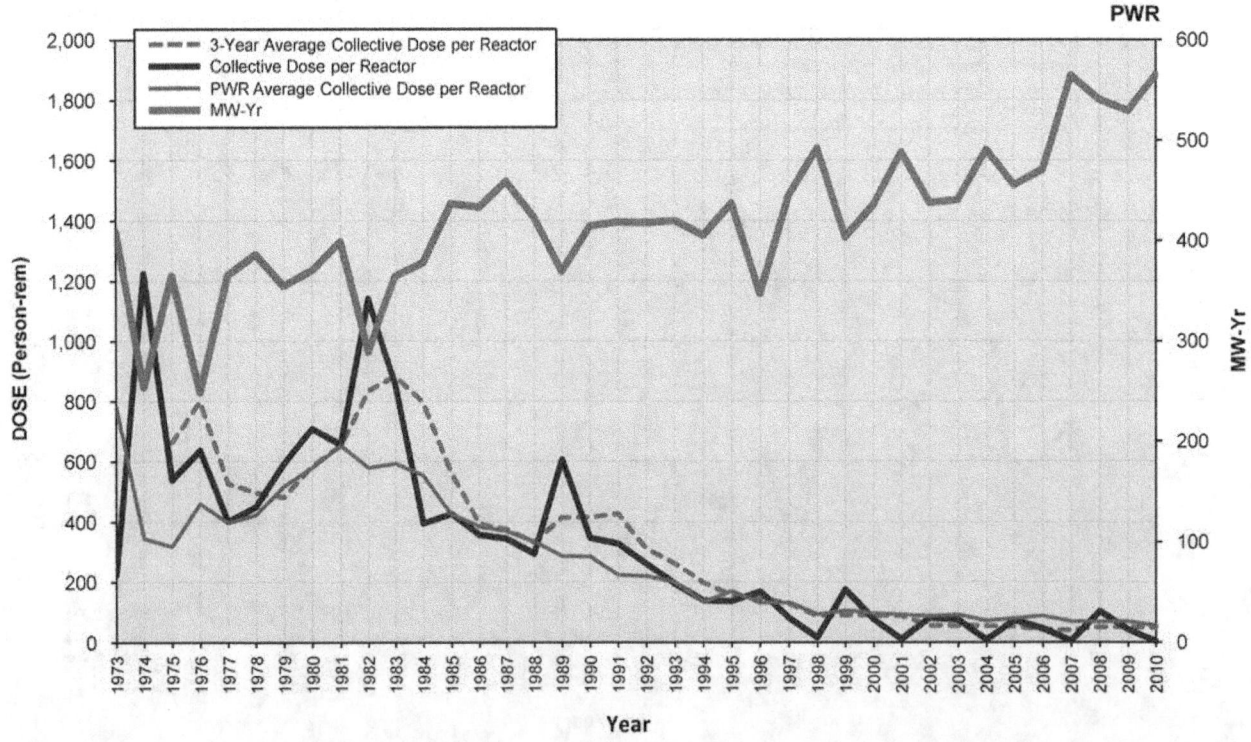

# GRAND GULF
## Dose Performance Trends

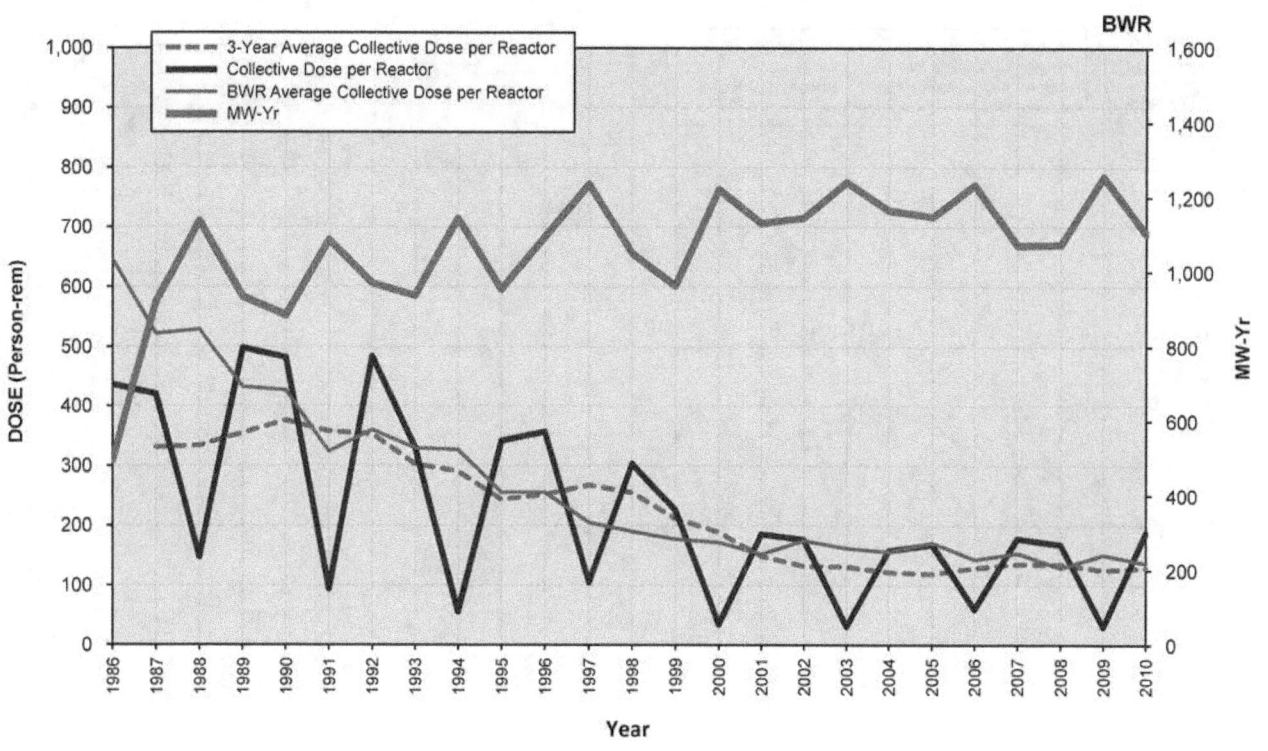

# HARRIS 1
## Dose Performance Trends

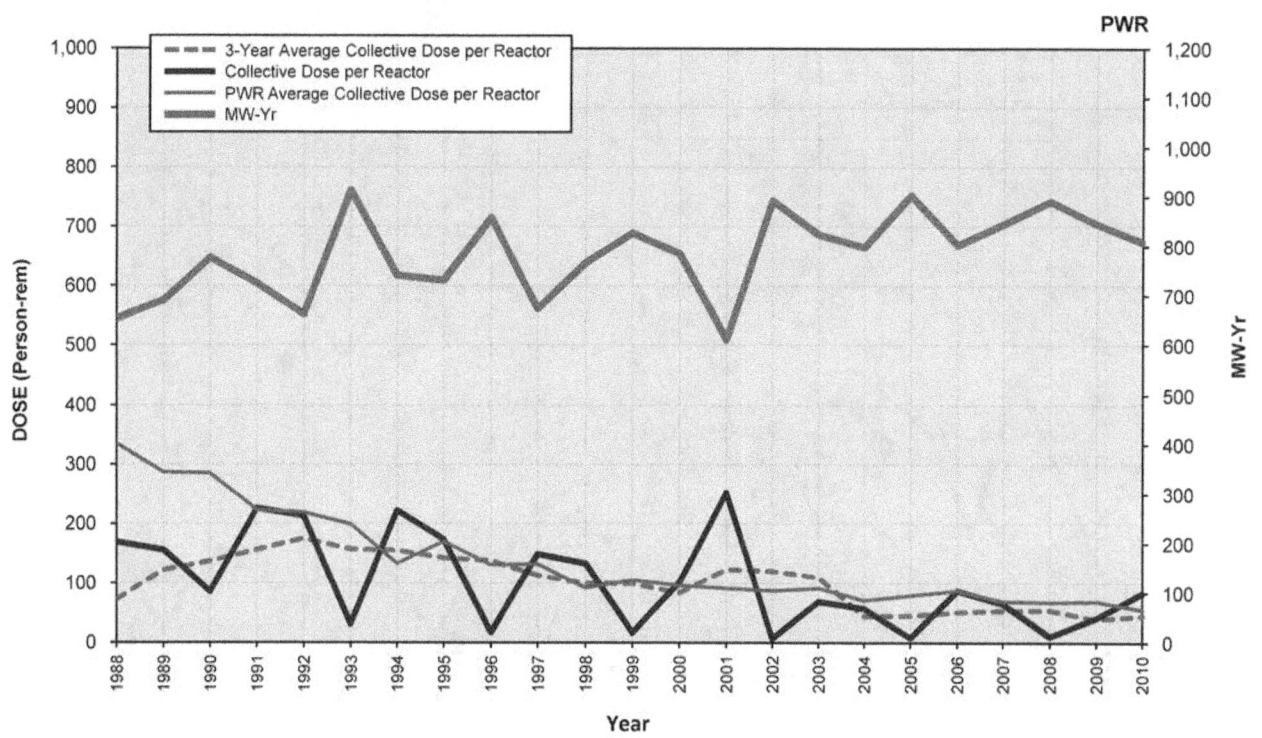

# HATCH 1, 2
## Dose Performance Trends

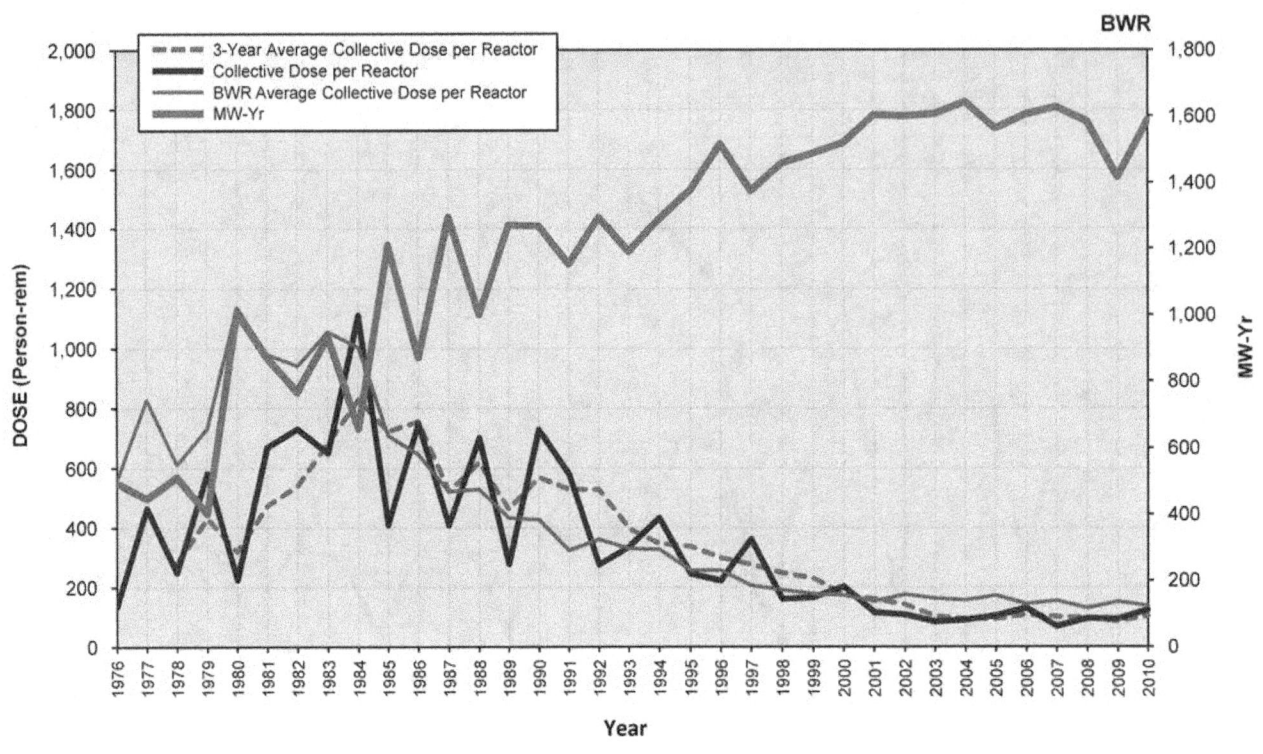

# HOPE CREEK 1
## Dose Performance Trends

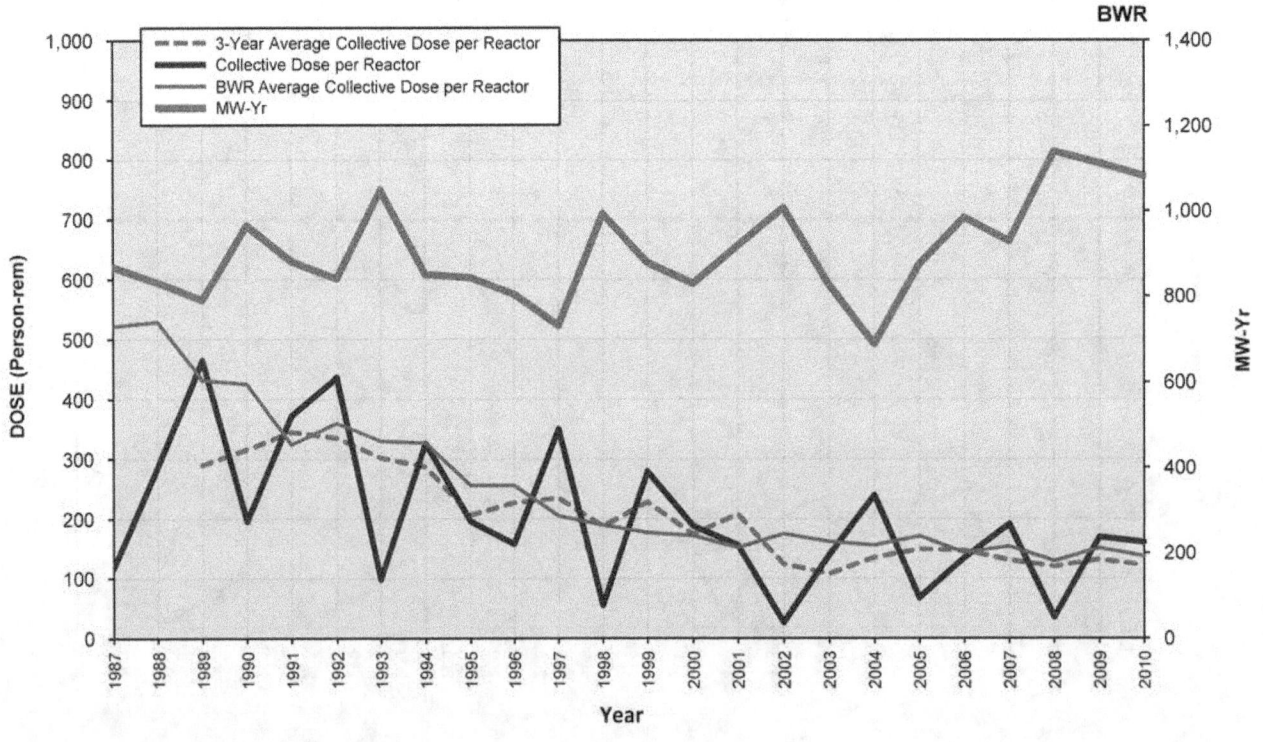

## INDIAN POINT 2
## Dose Performance Trends

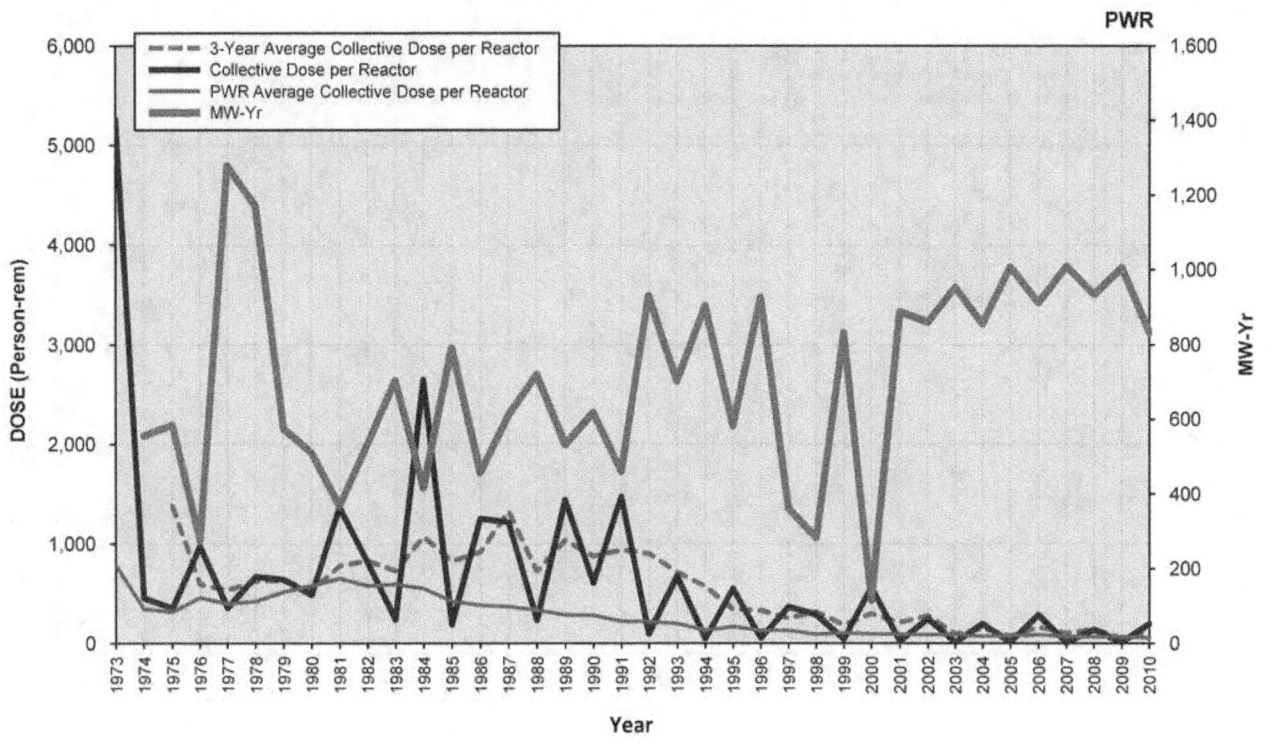

## INDIAN POINT 3
## Dose Performance Trends

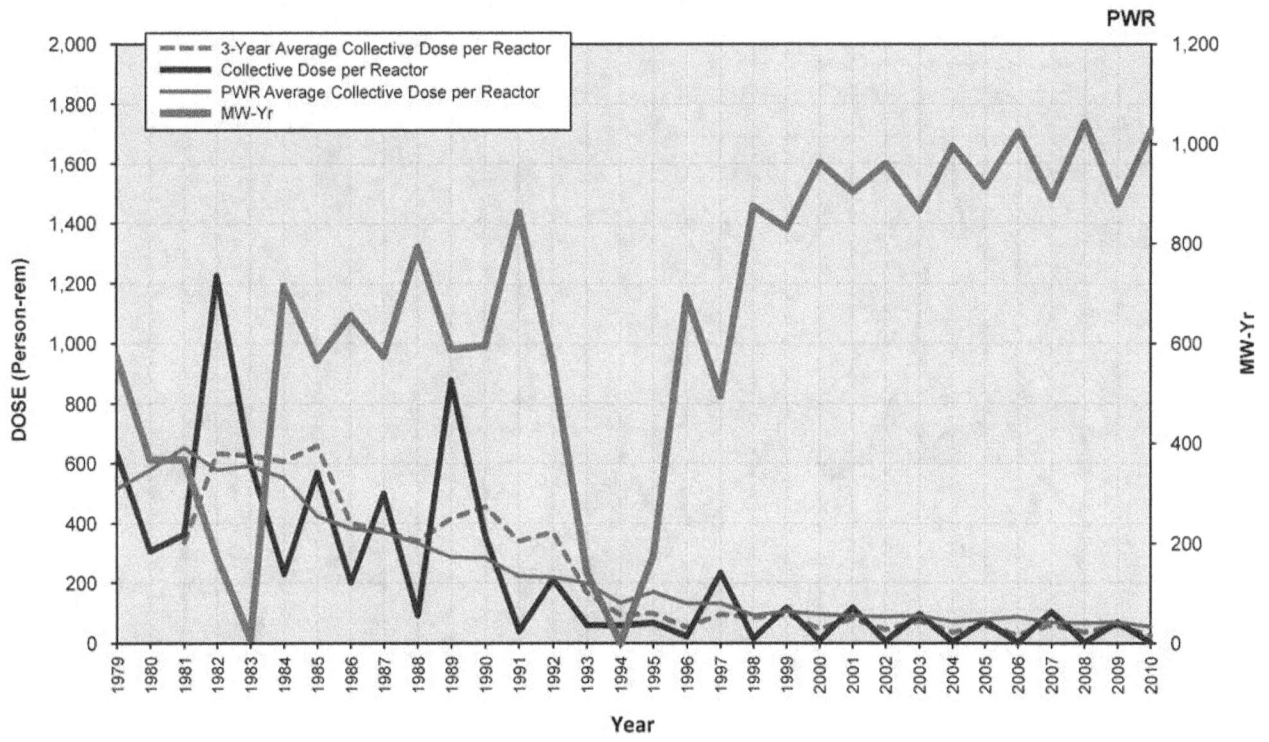

## KEWAUNEE
### Dose Performance Trends

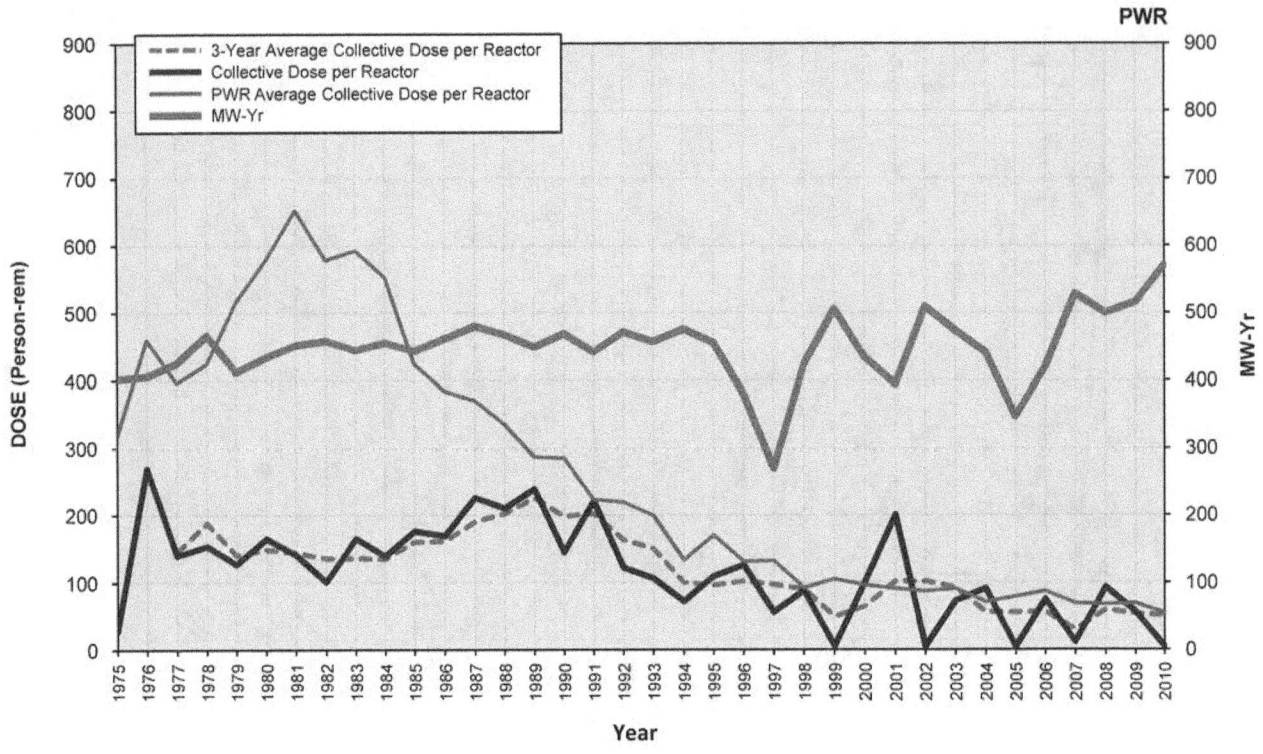

## LASALLE 1, 2
### Dose Performance Trends

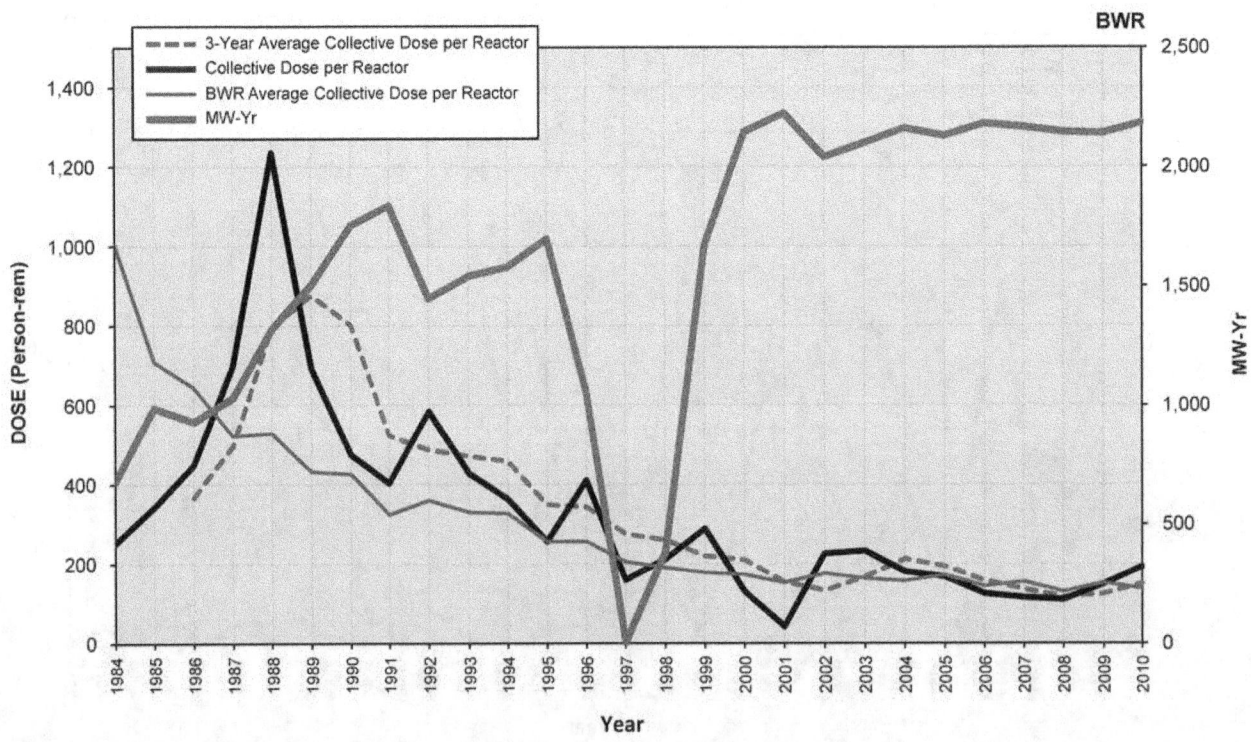

# LIMERICK 1, 2
## Dose Performance Trends

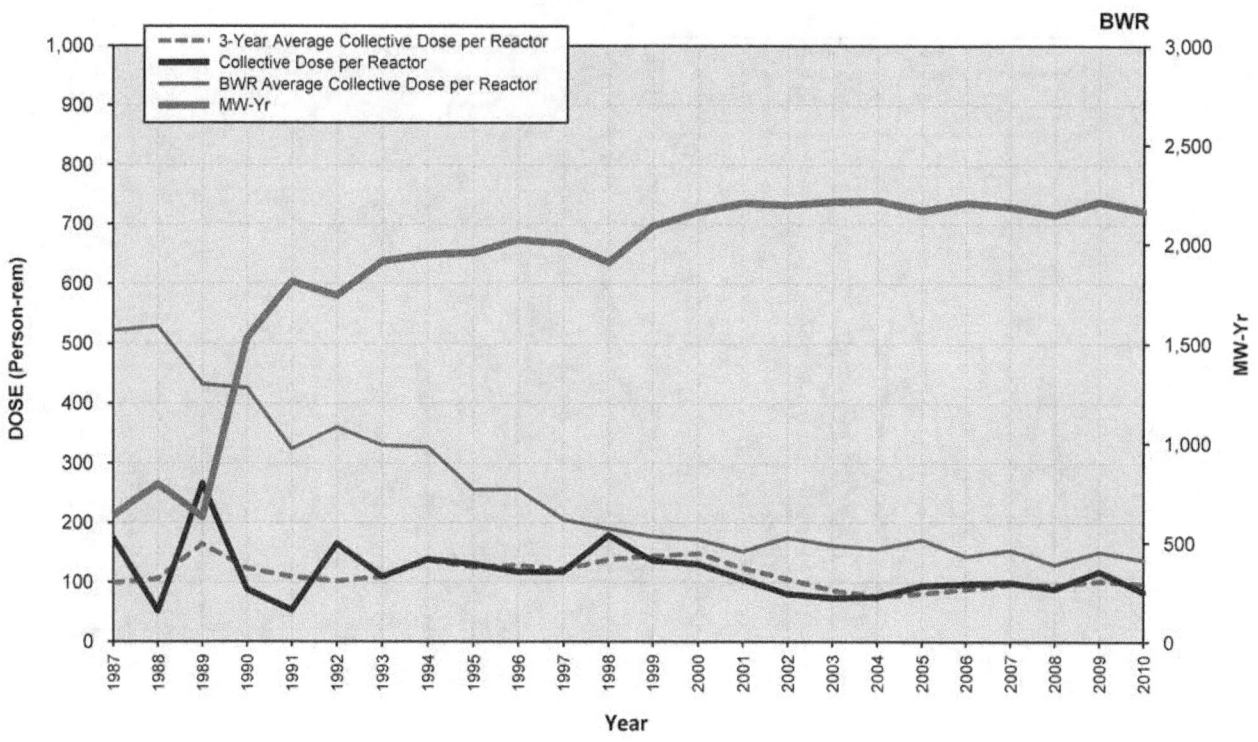

# MCGUIRE 1, 2
## Dose Performance Trends

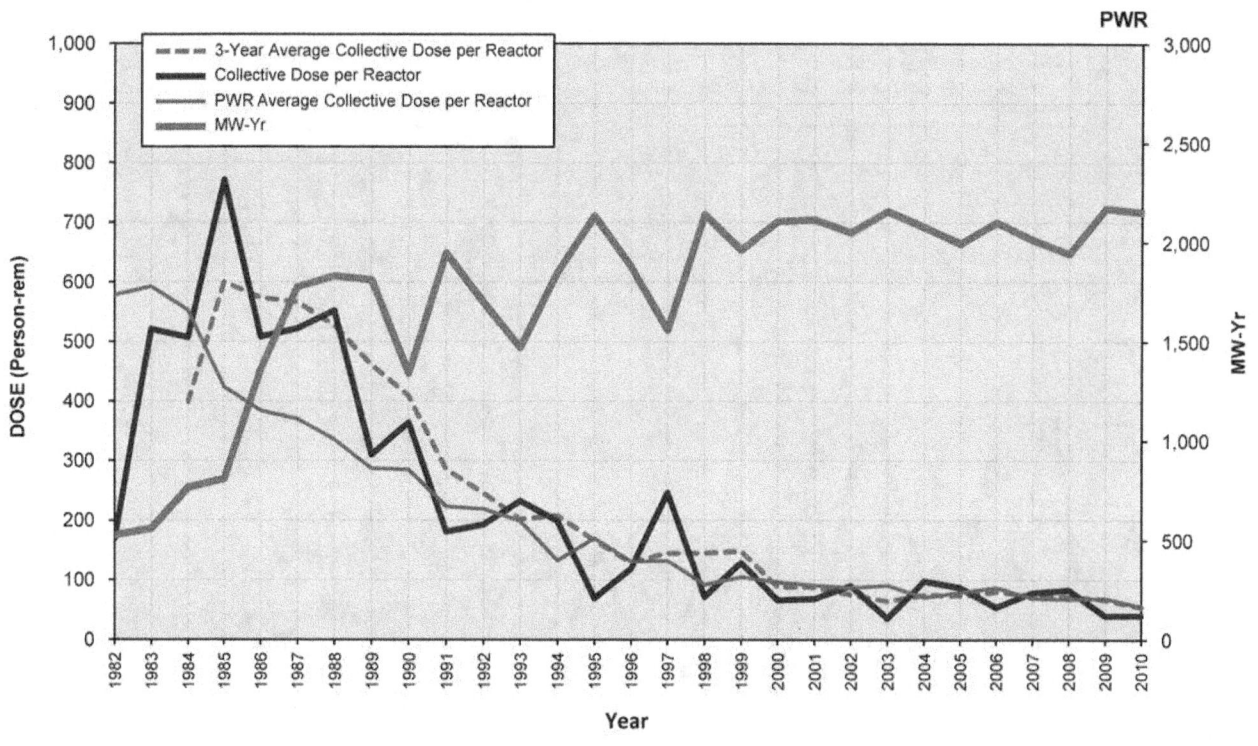

## MILLSTONE 2, 3
### Dose Performance Trends

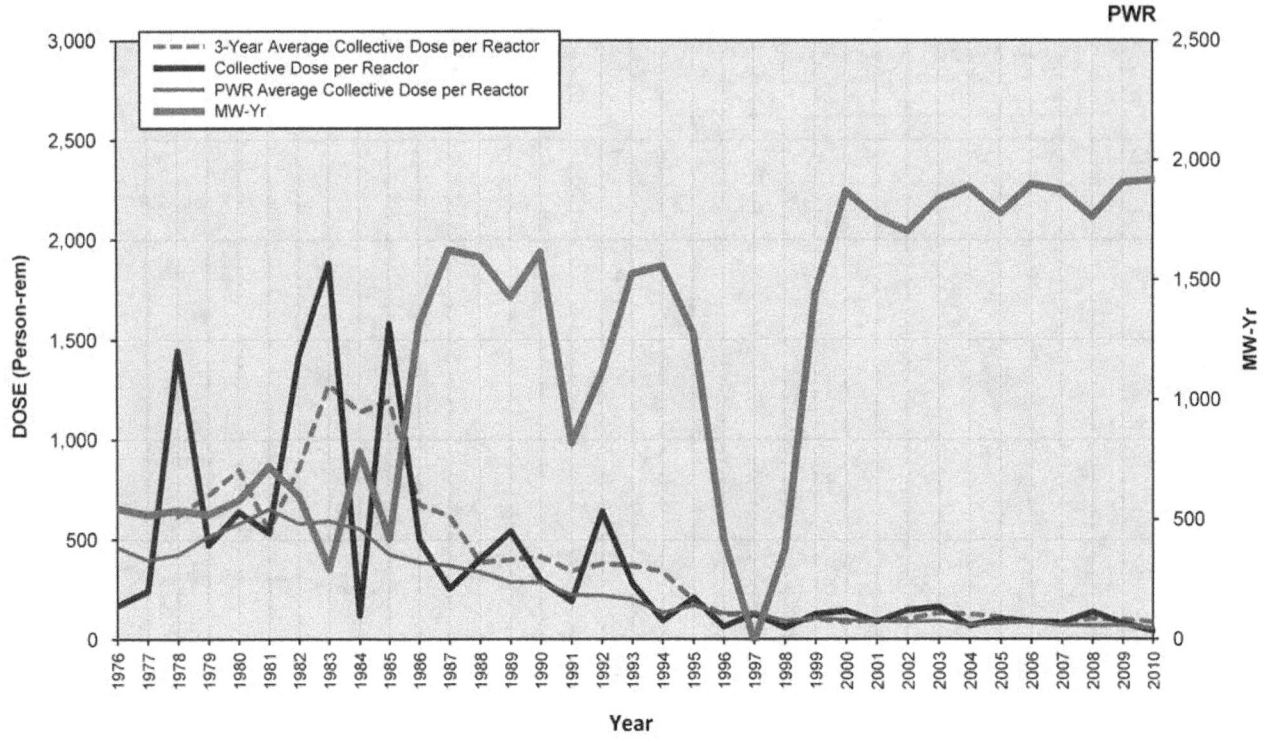

## MONTICELLO
### Dose Performance Trends

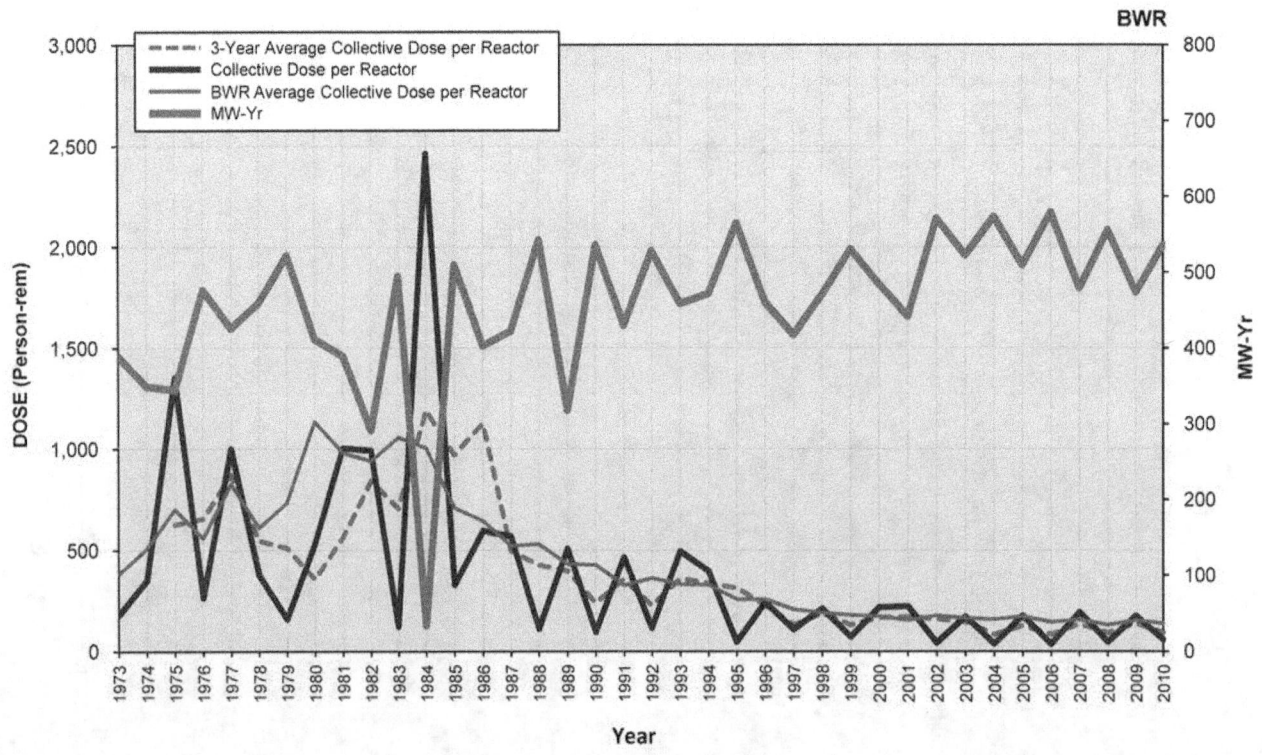

## NINE MILE POINT 1, 2
### Dose Performance Trends

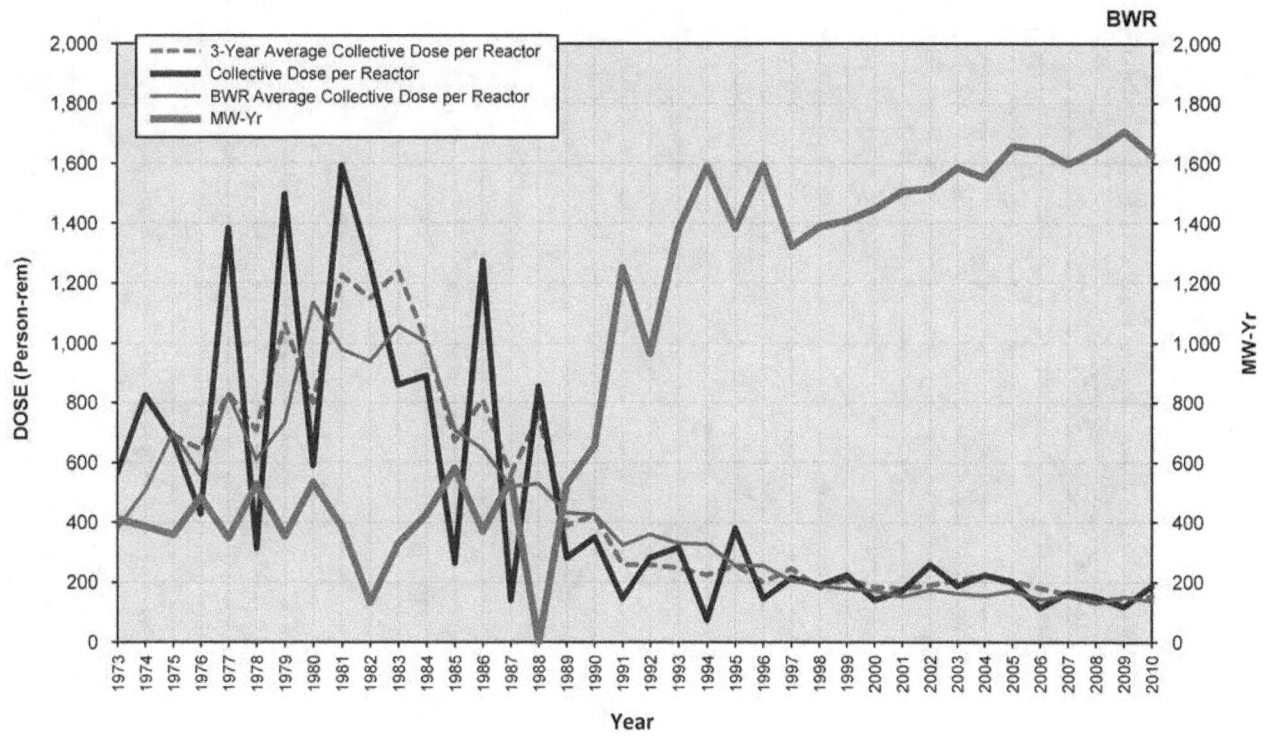

## NORTH ANNA 1, 2
### Dose Performance Trends

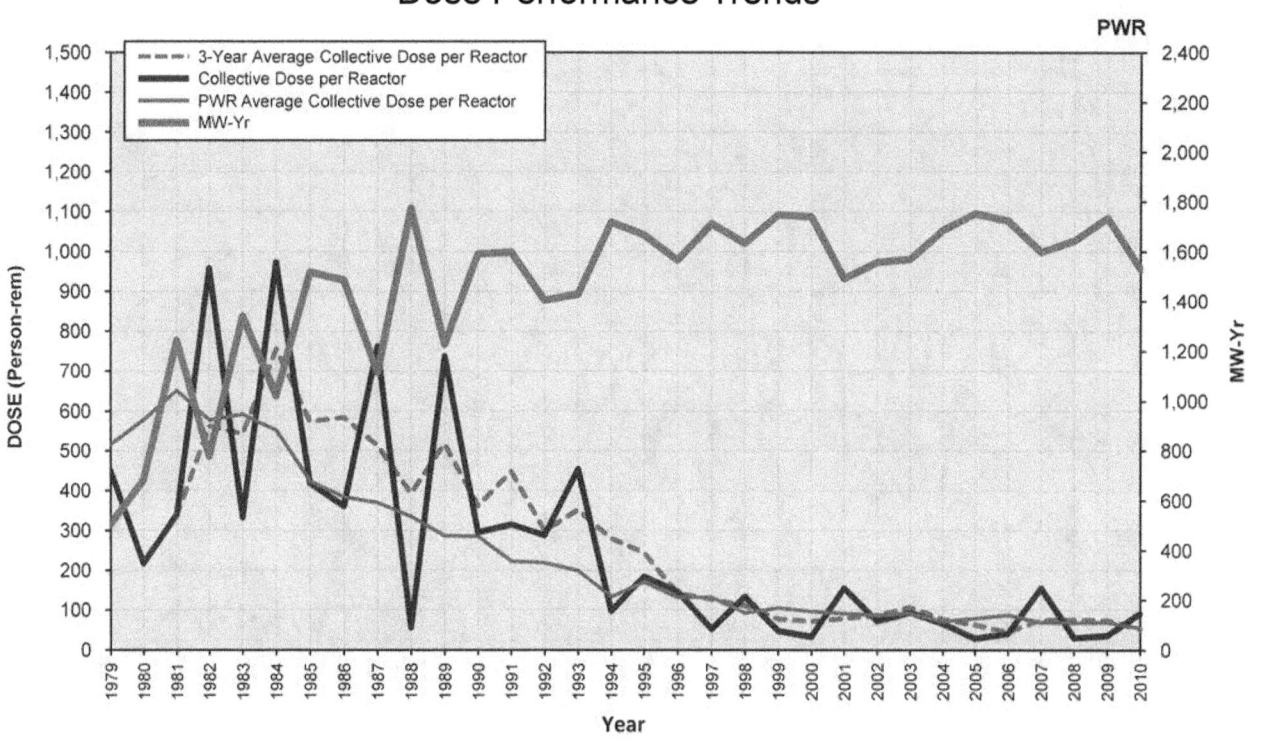

## OCONEE 1, 2, 3
### Dose Performance Trends

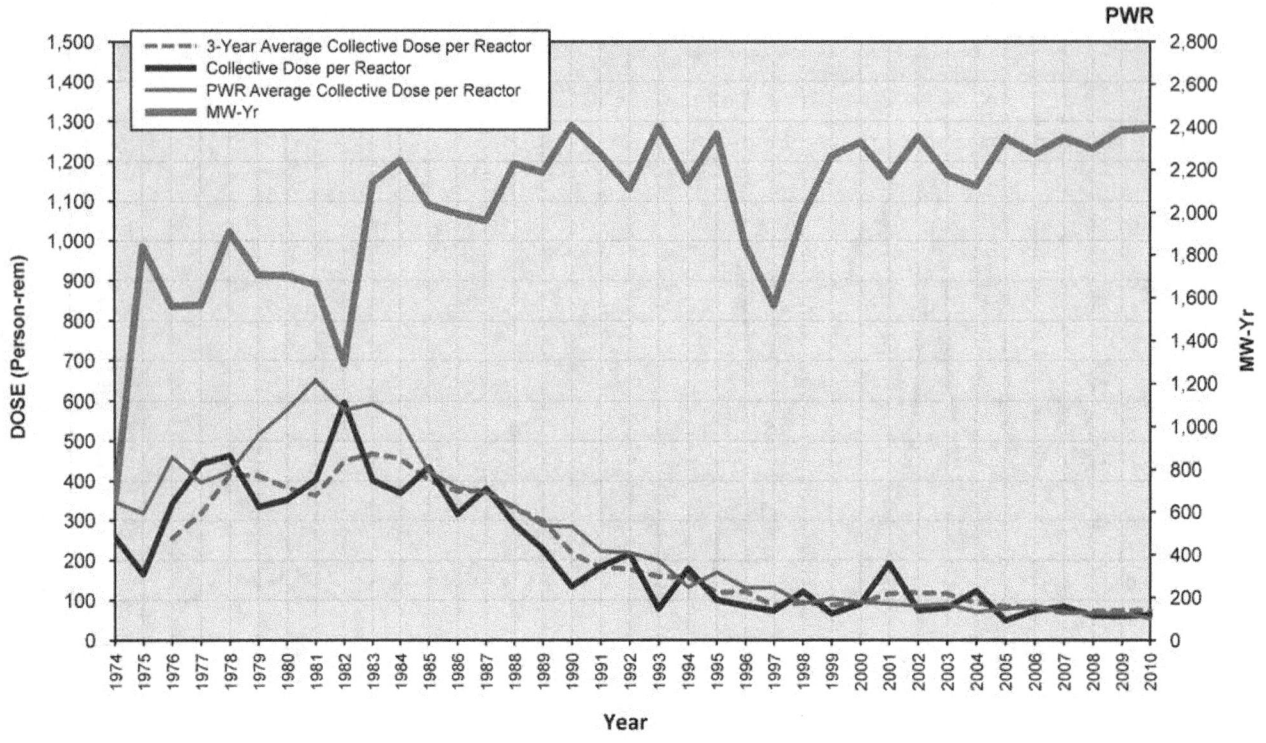

## OYSTER CREEK
### Dose Performance Trends

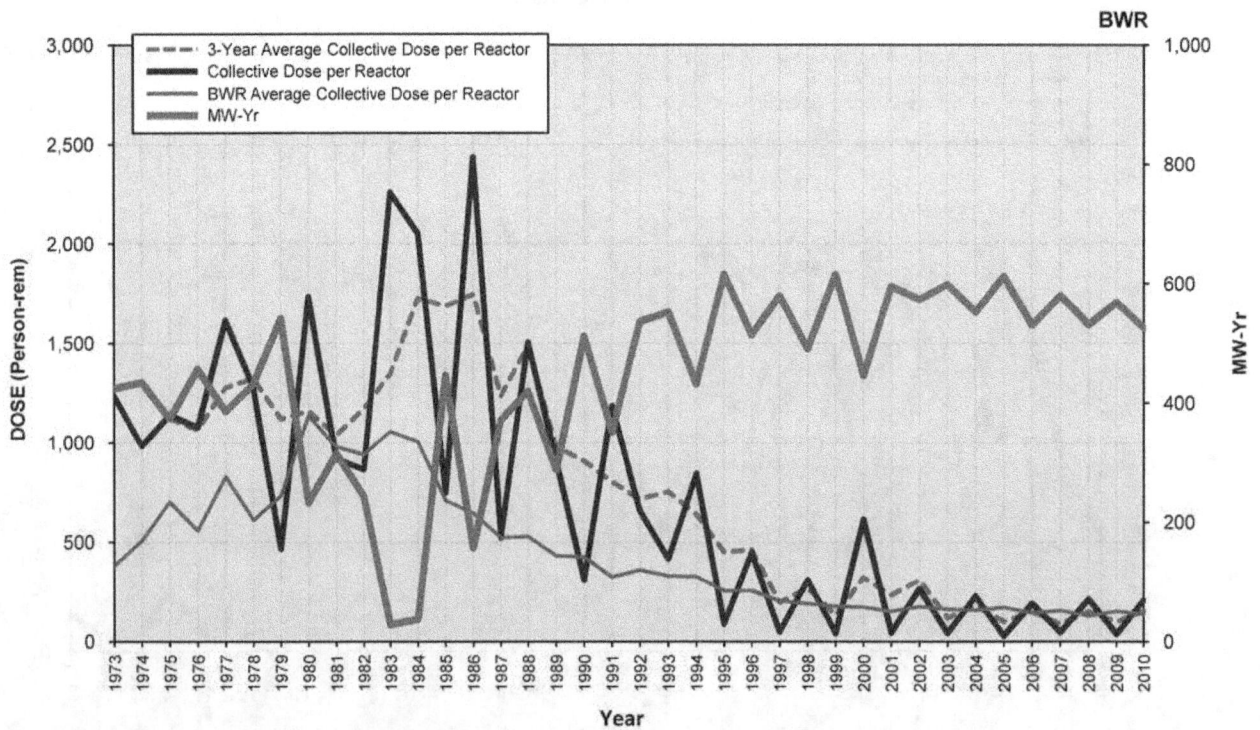

# PALISADES
## Dose Performance Trends

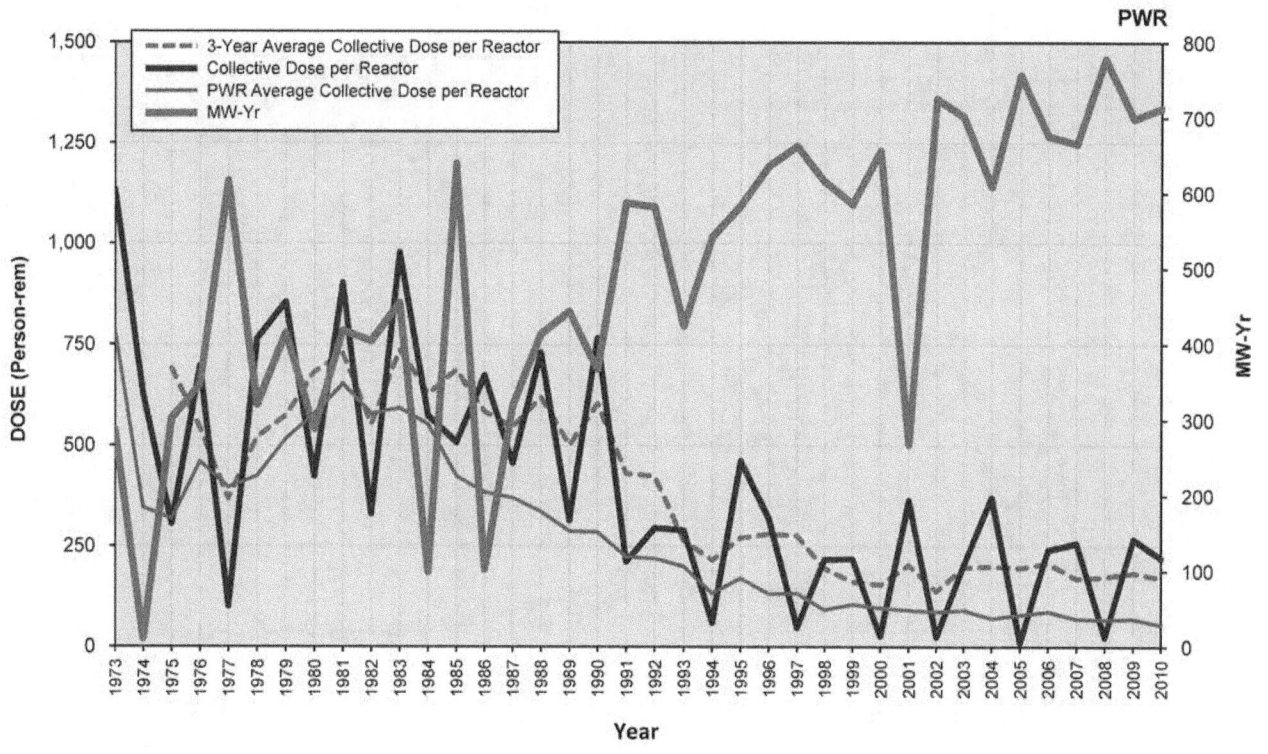

# PALO VERDE 1, 2, 3
## Dose Performance Trends

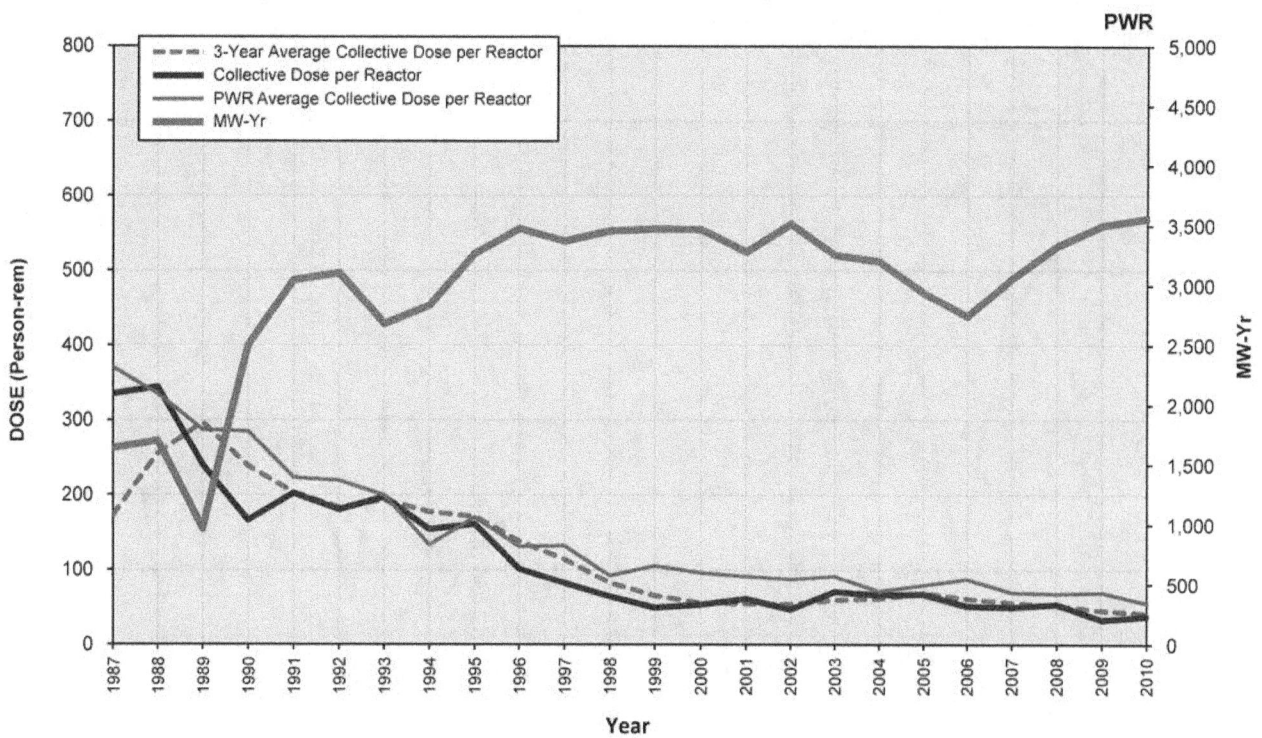

## PEACH BOTTOM 2, 3
### Dose Performance Trends

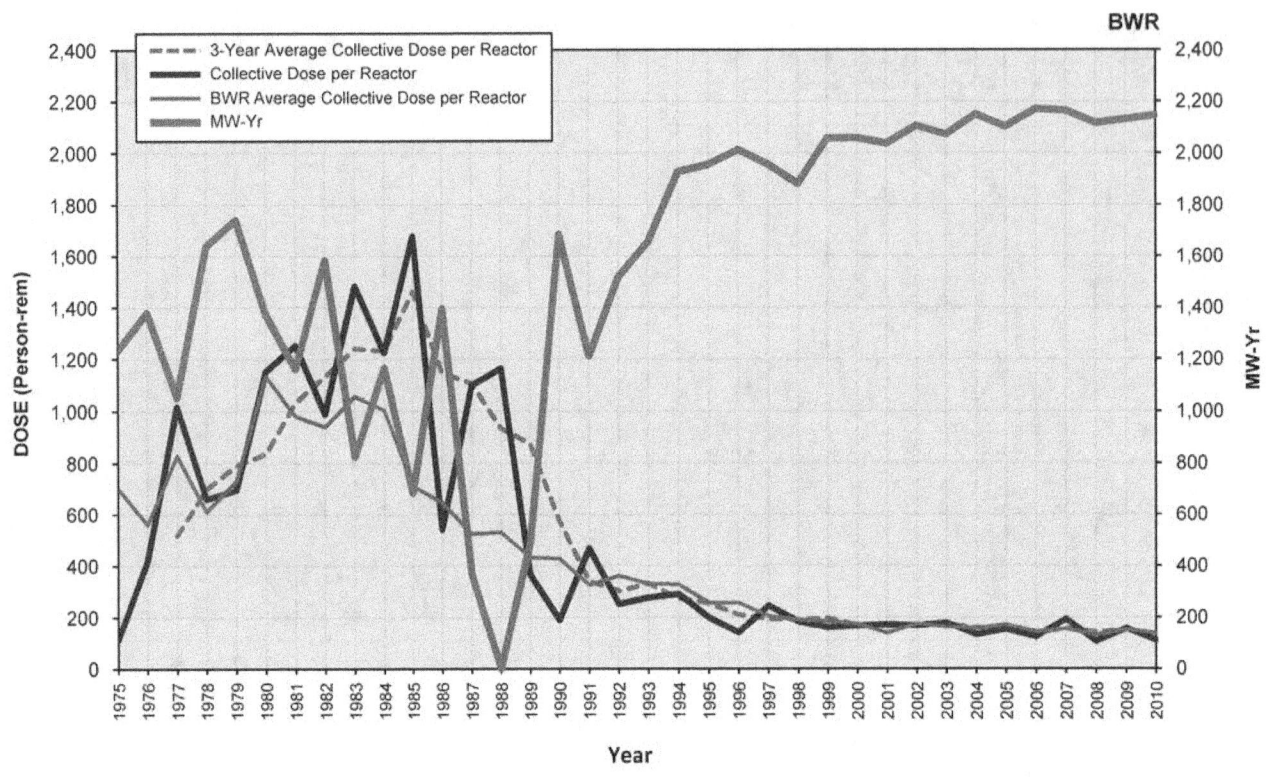

## PERRY 1
### Dose Performance Trends

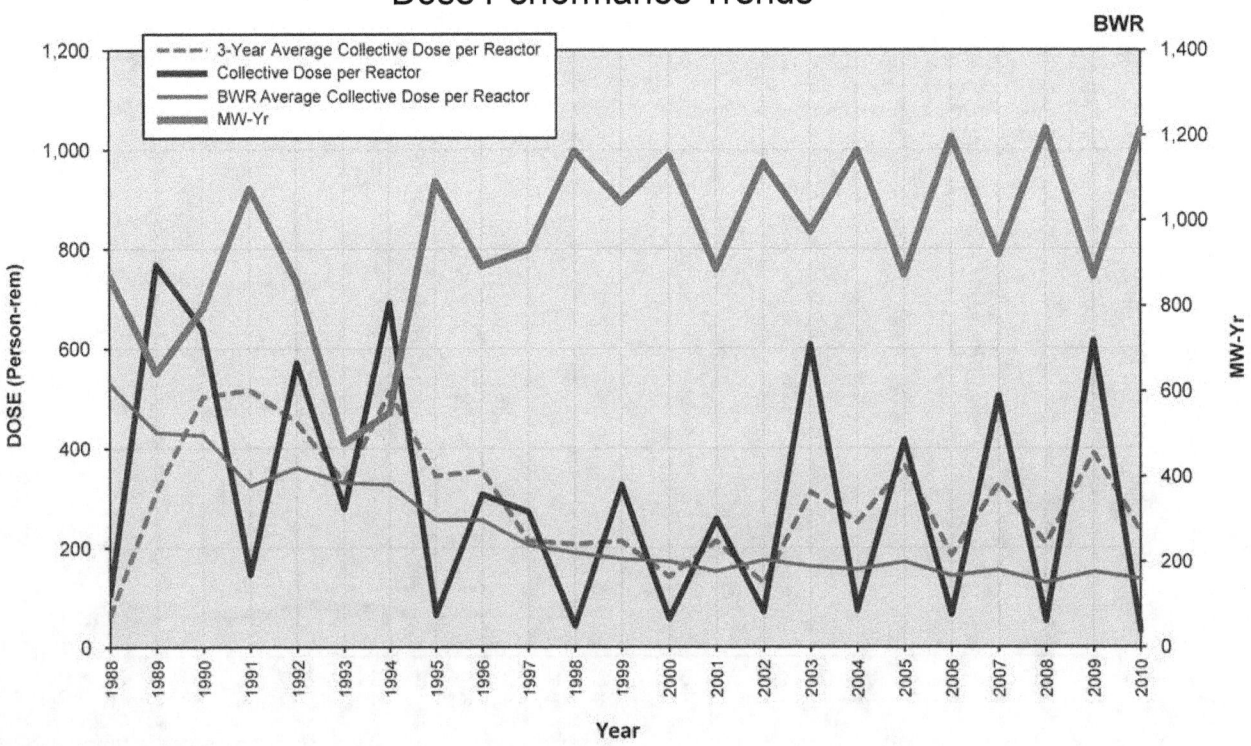

## PILGRIM 1
### Dose Performance Trends

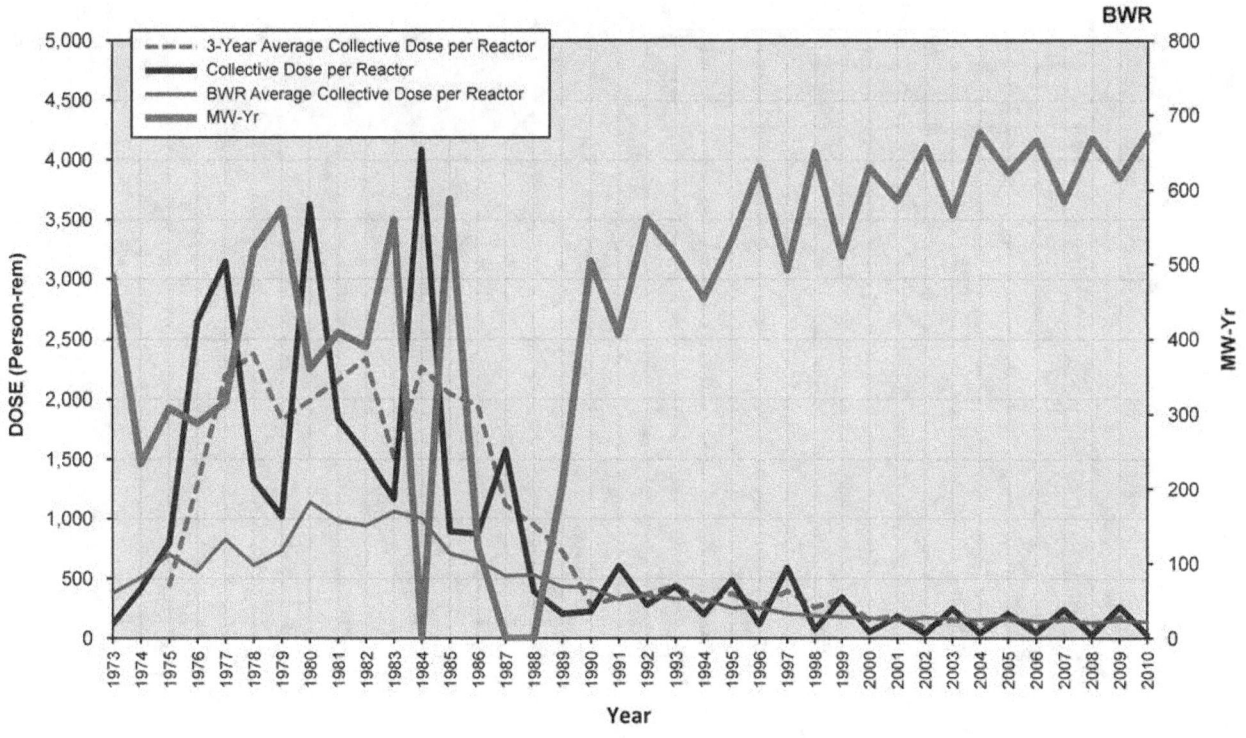

## POINT BEACH 1, 2
### Dose Performance Trends

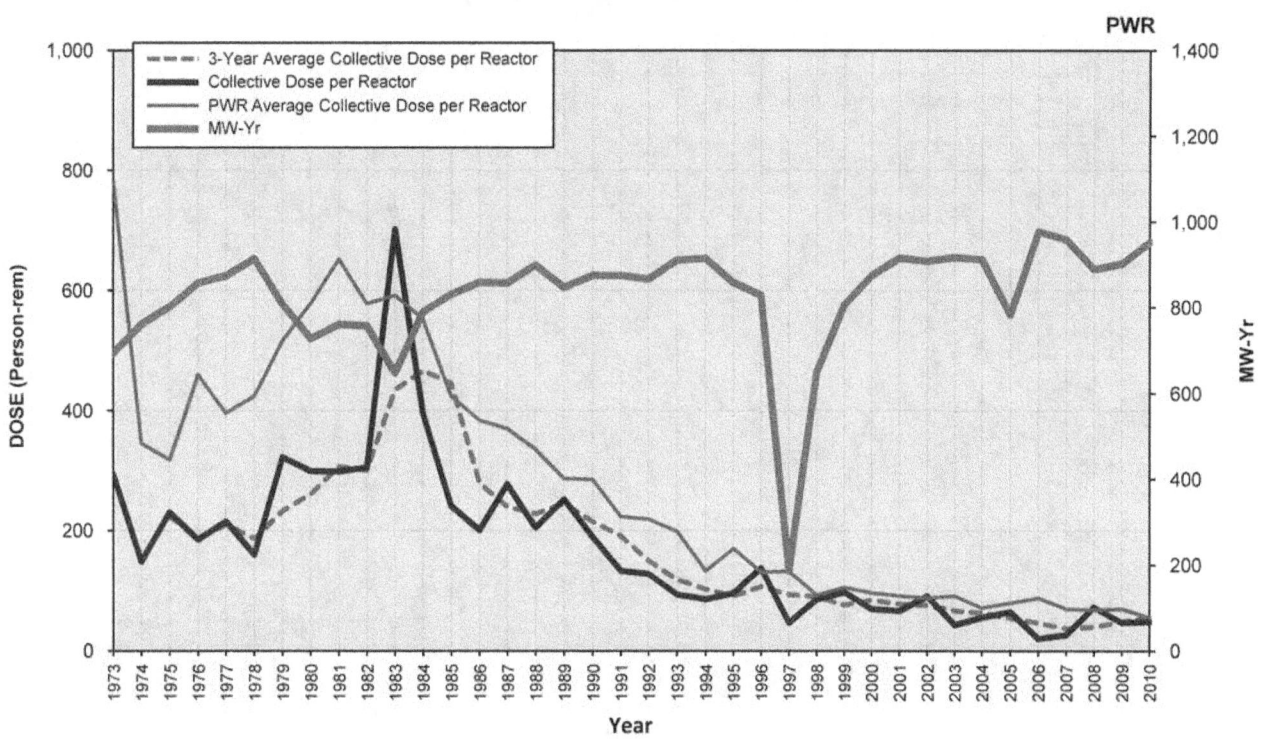

## PRAIRIE ISLAND 1, 2
### Dose Performance Trends

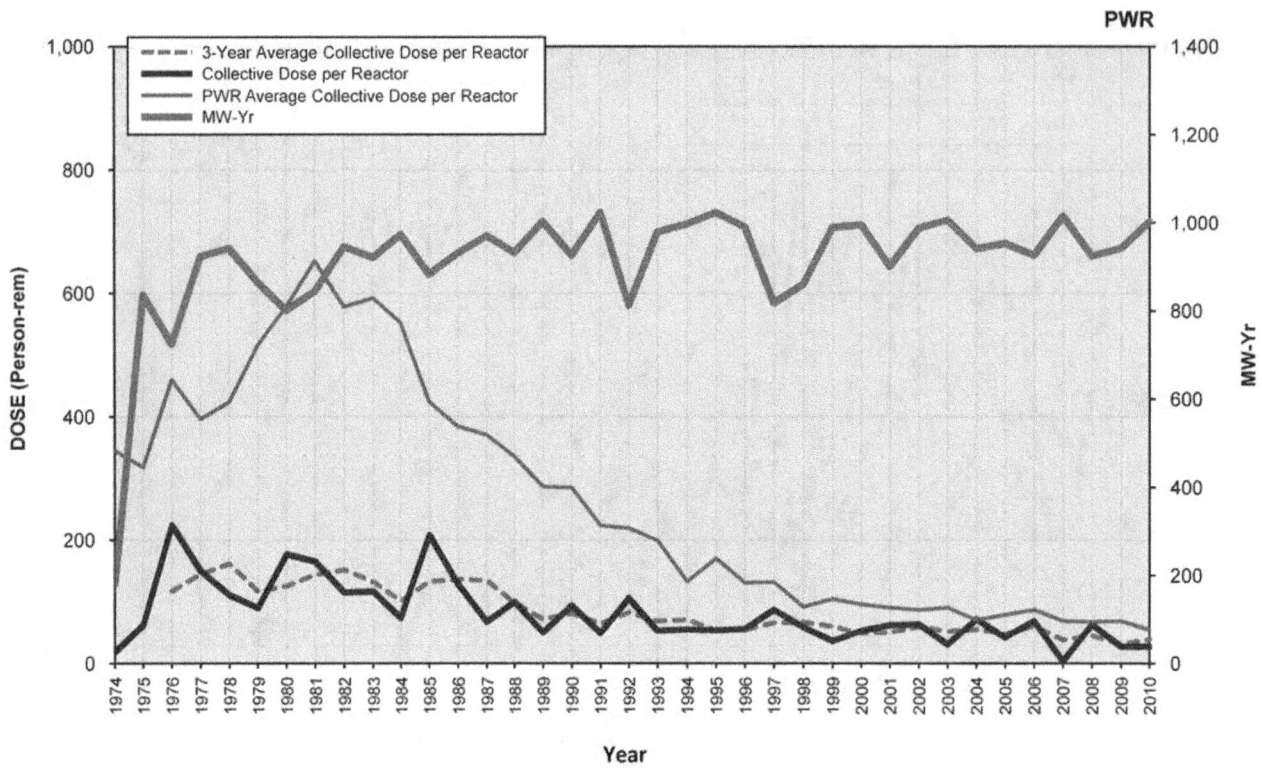

## QUAD CITIES 1, 2
### Dose Performance Trends

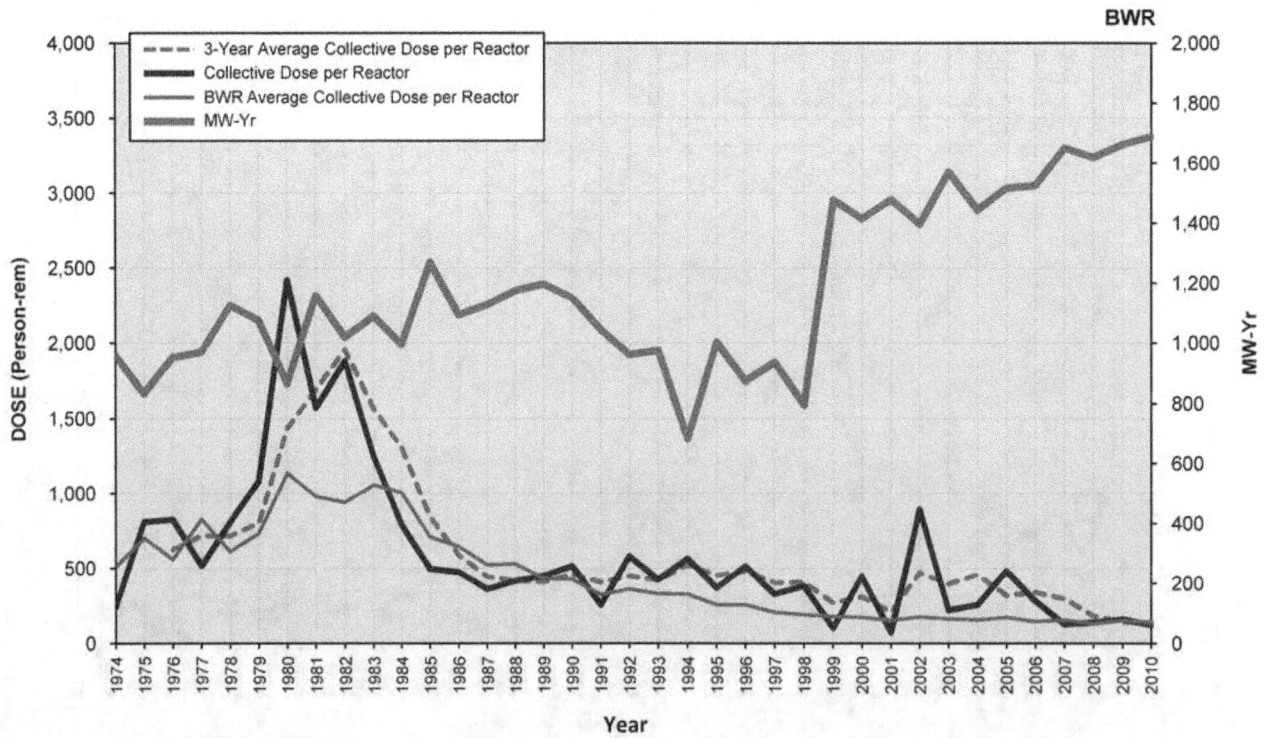

# RIVER BEND 1
## Dose Performance Trends

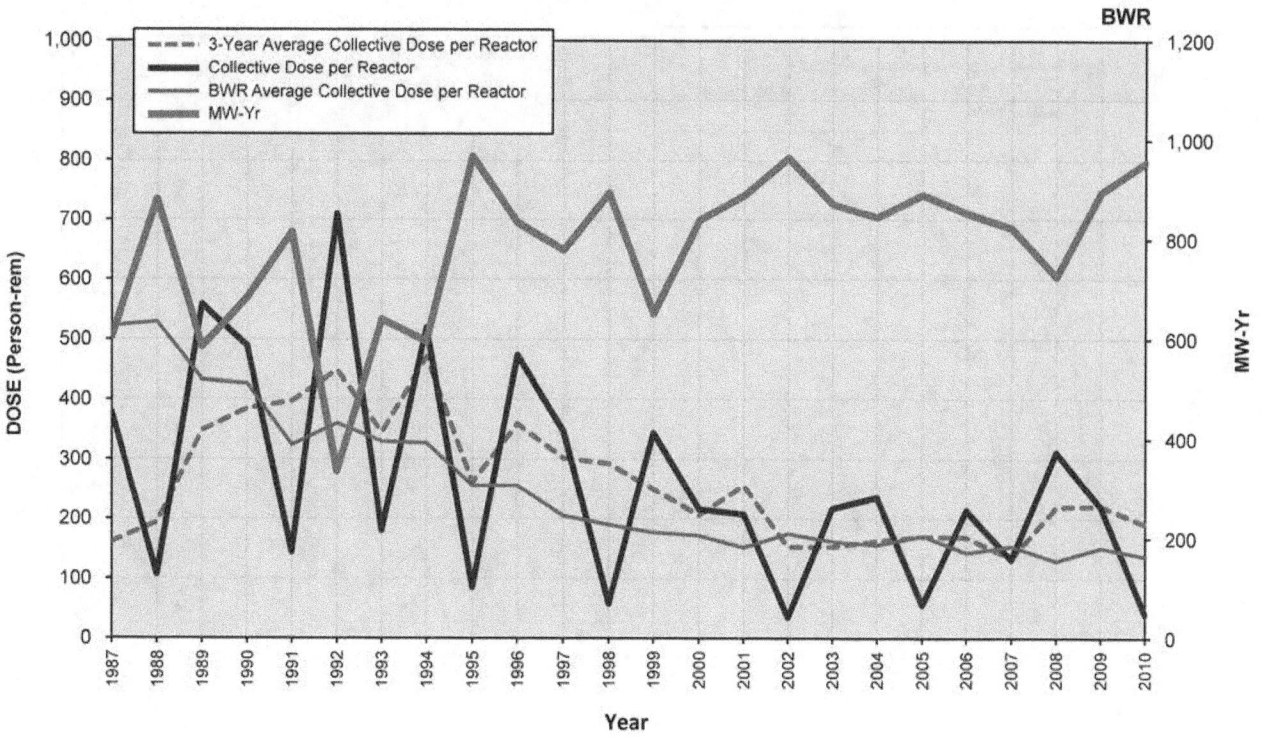

# ROBINSON 2
## Dose Performance Trends

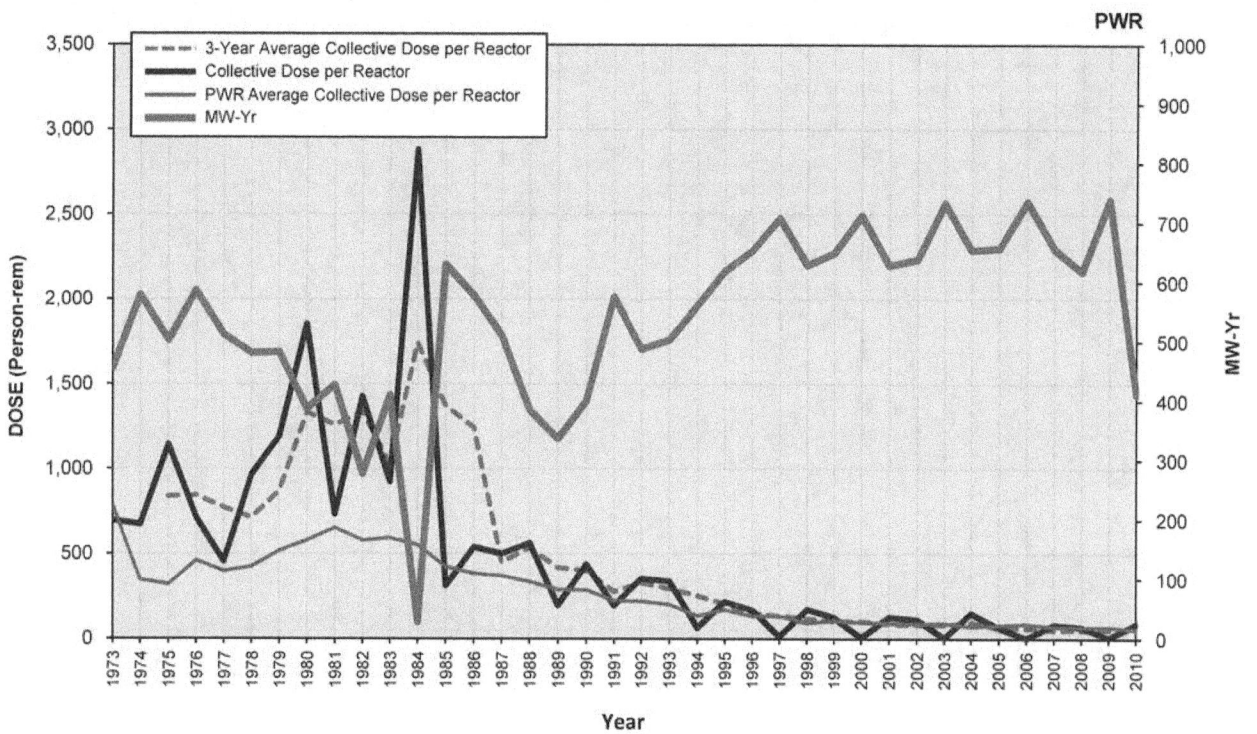

## SALEM 1, 2
## Dose Performance Trends

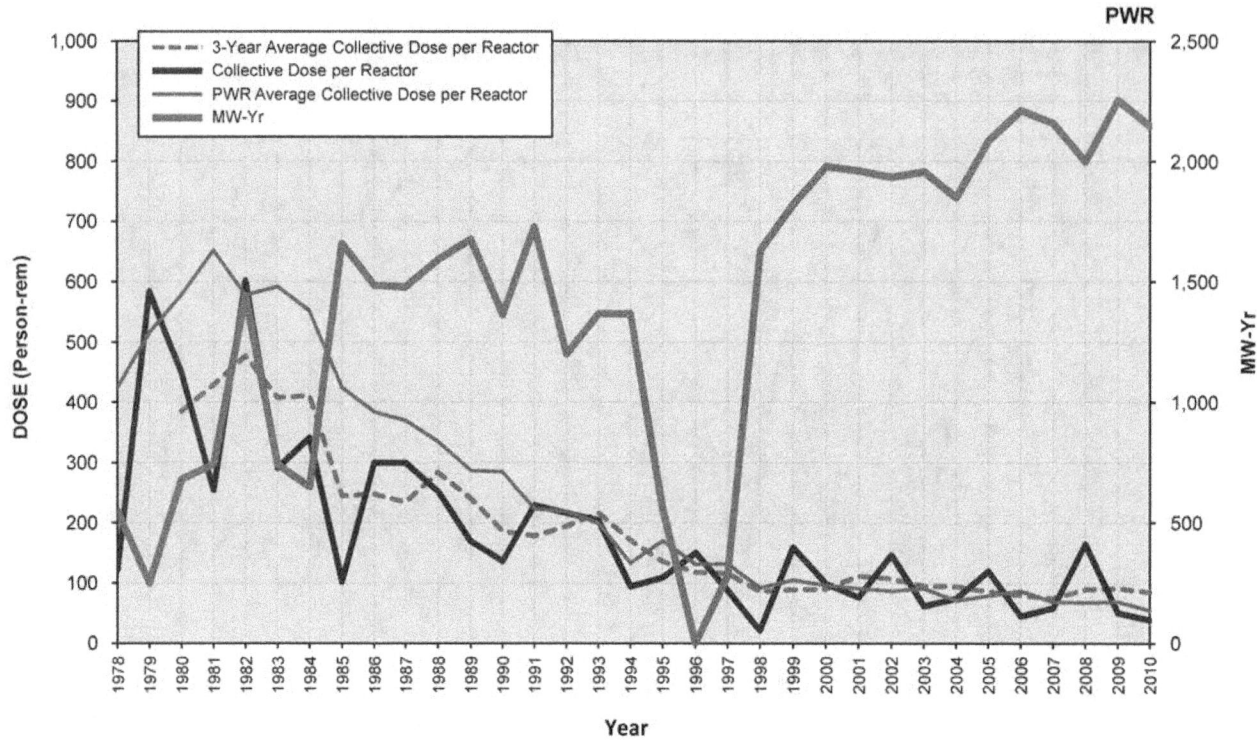

## SAN ONOFRE 1, 2, 3
## Dose Performance Trends

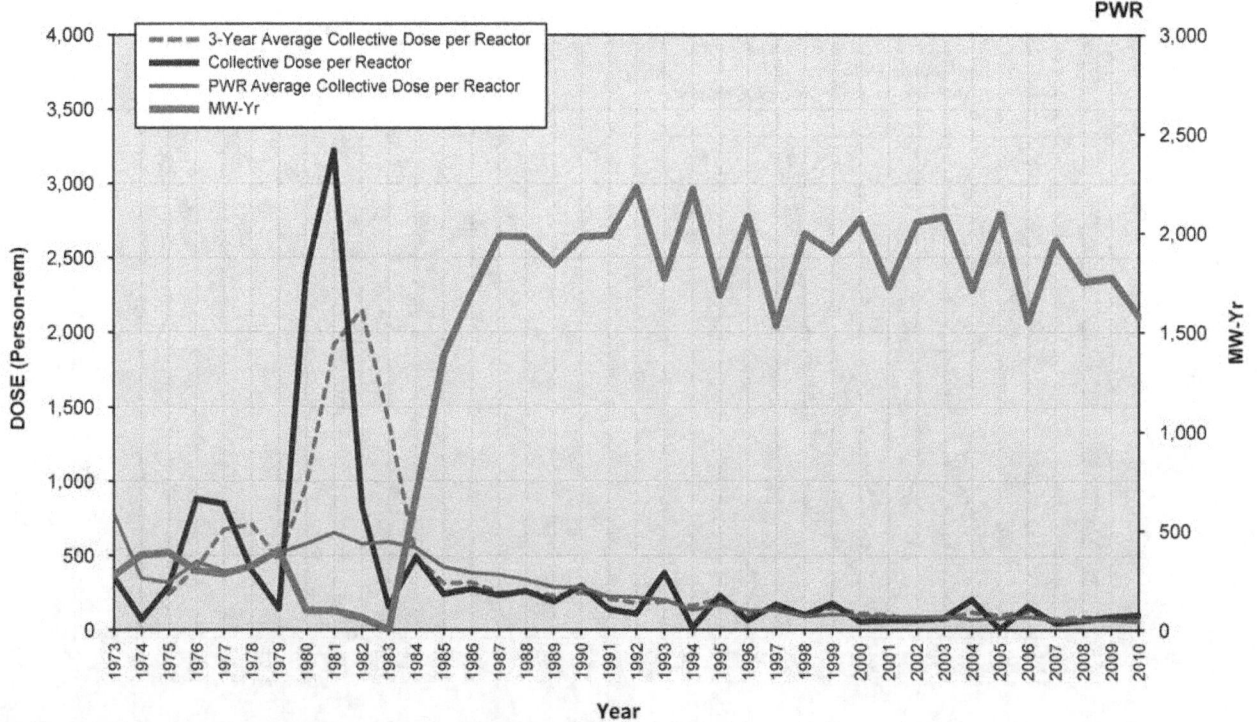

NOTE: Since 2001, data only includes San Onofre Units 2 and 3.

## SEABROOK
## Dose Performance Trends

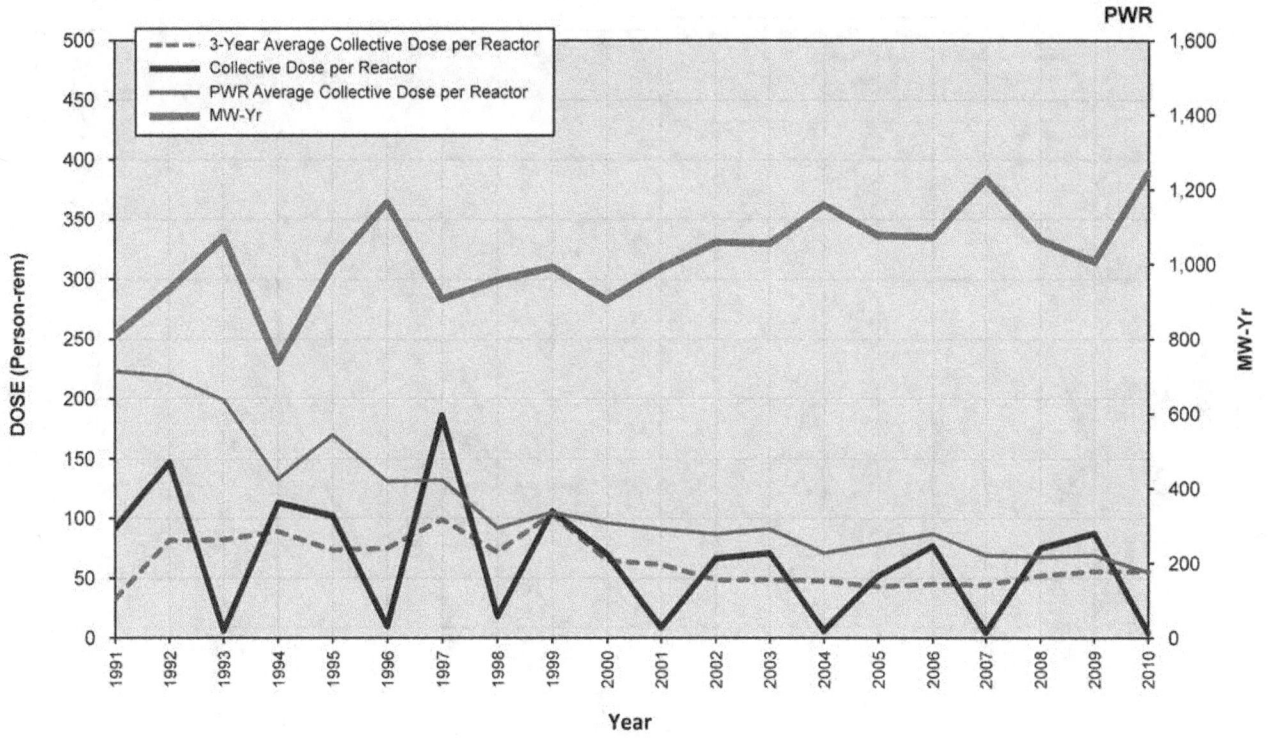

## SEQUOYAH 1, 2
## Dose Performance Trends

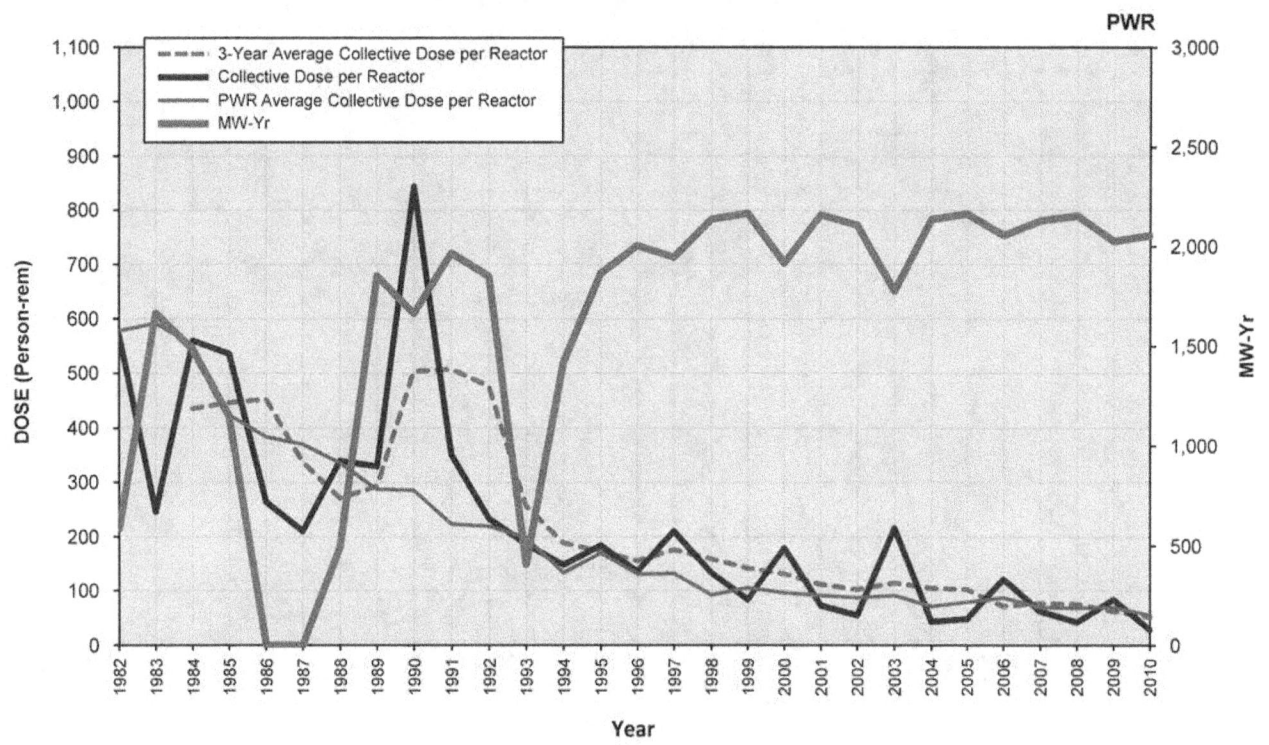

## SOUTH TEXAS 1, 2
### Dose Performance Trends

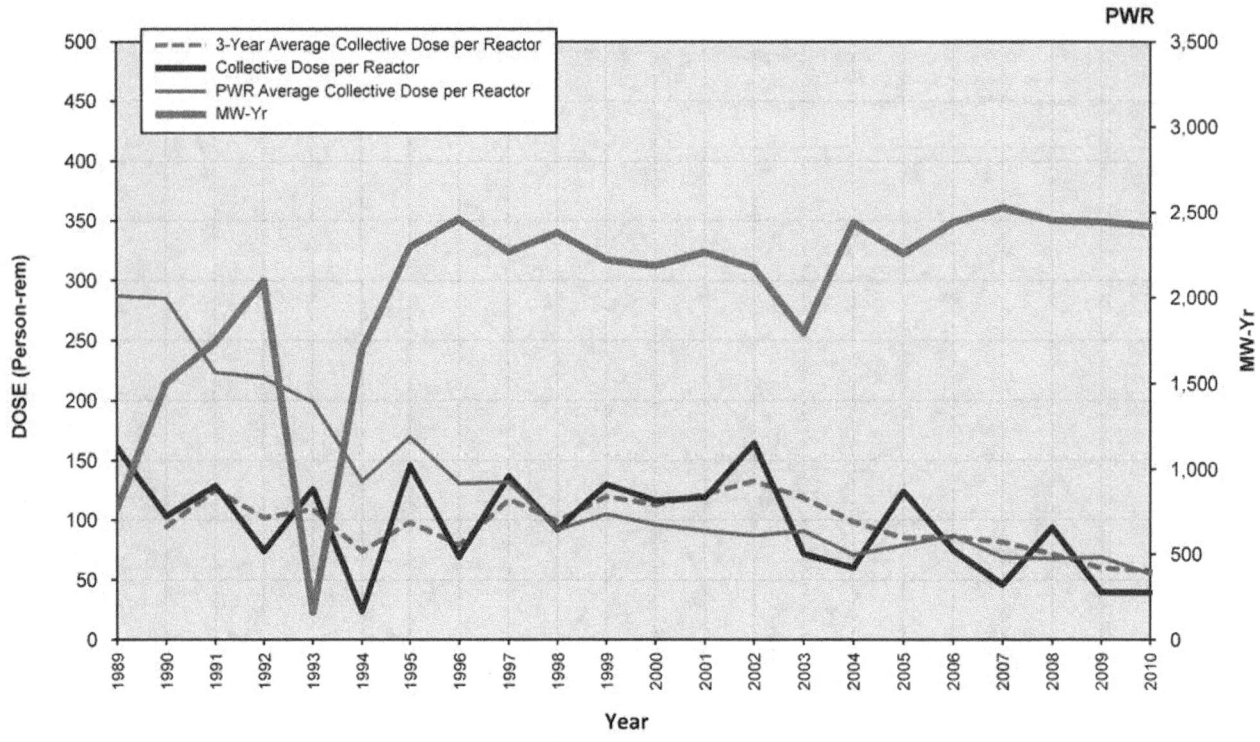

## ST. LUCIE 1, 2
### Dose Performance Trends

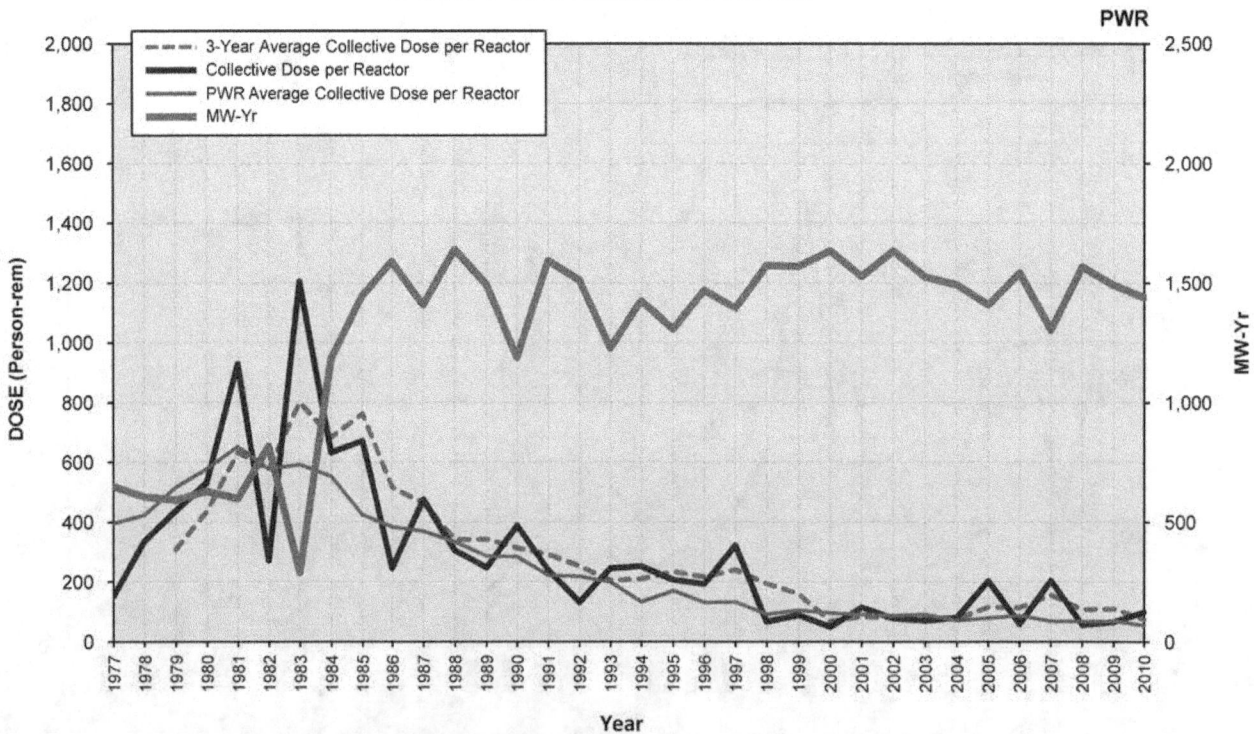

## SUMMER
### Dose Performance Trends

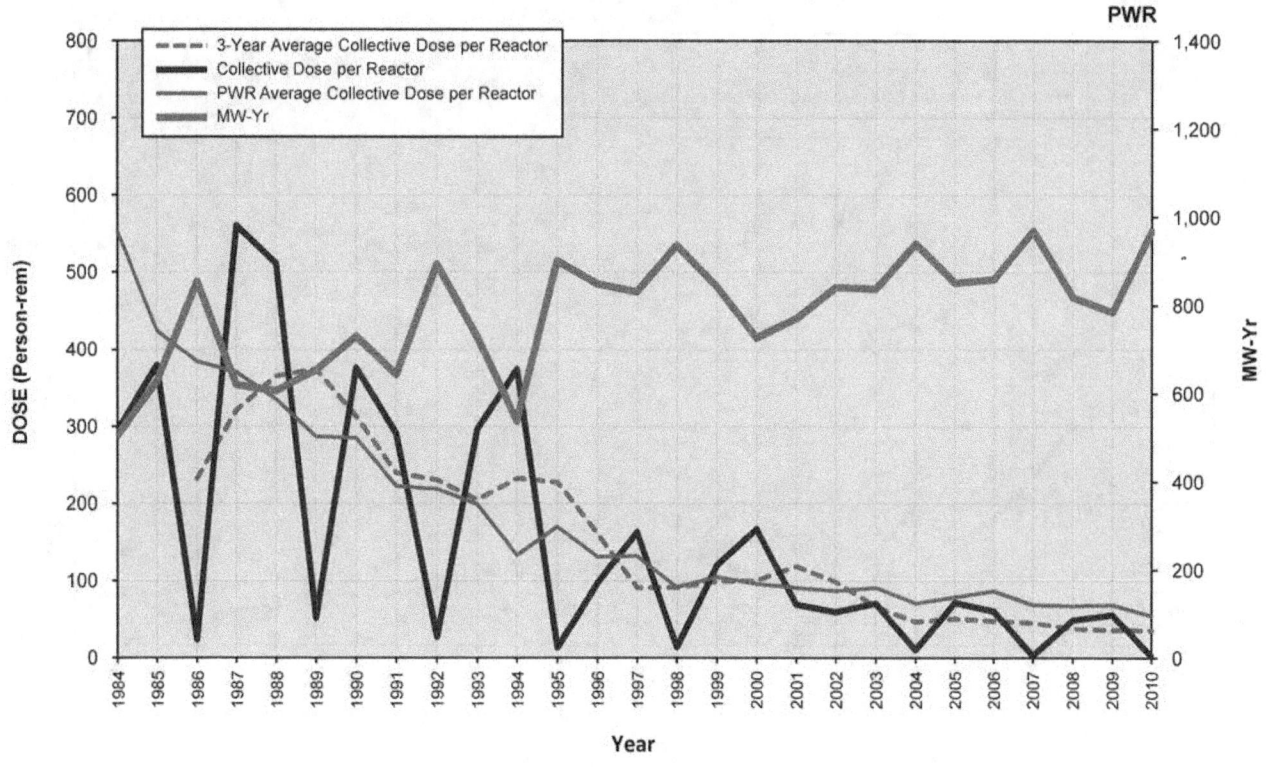

## SURRY 1, 2
### Dose Performance Trends

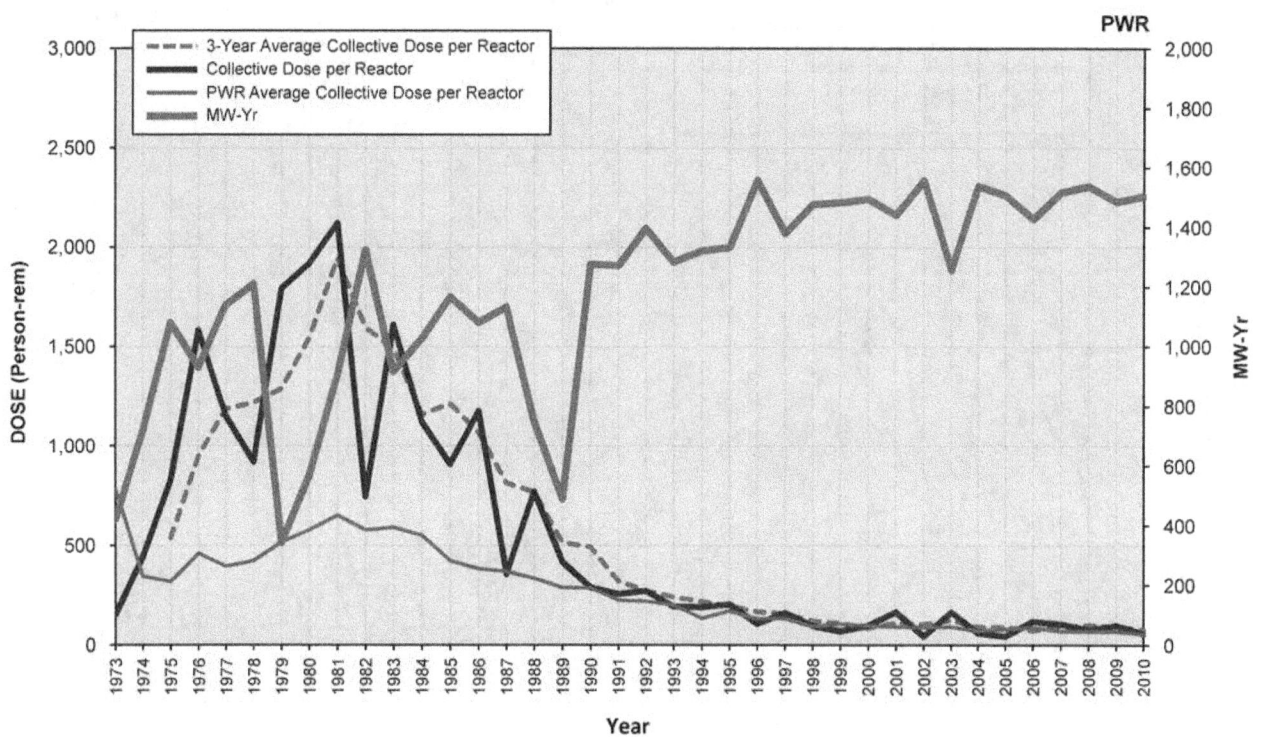

# SUSQUEHANNA 1, 2
## Dose Performance Trends

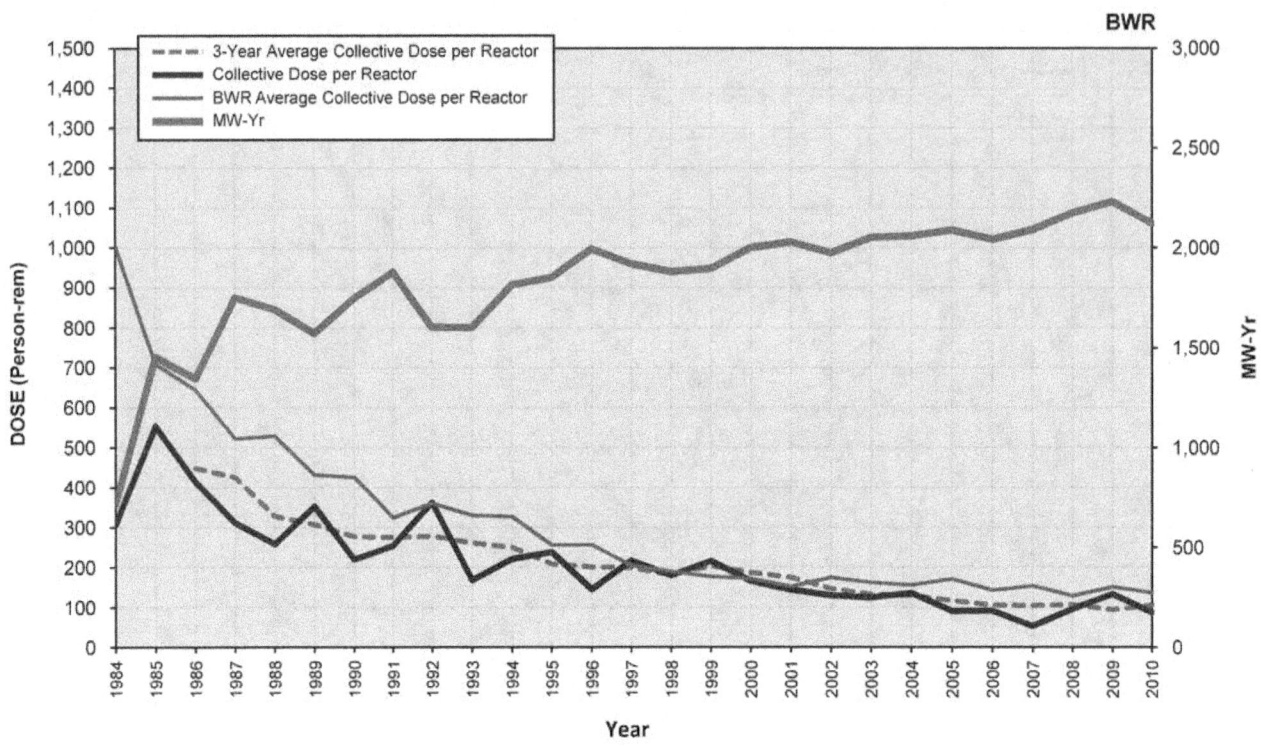

# THREE MILE ISLAND 1*
## Dose Performance Trends

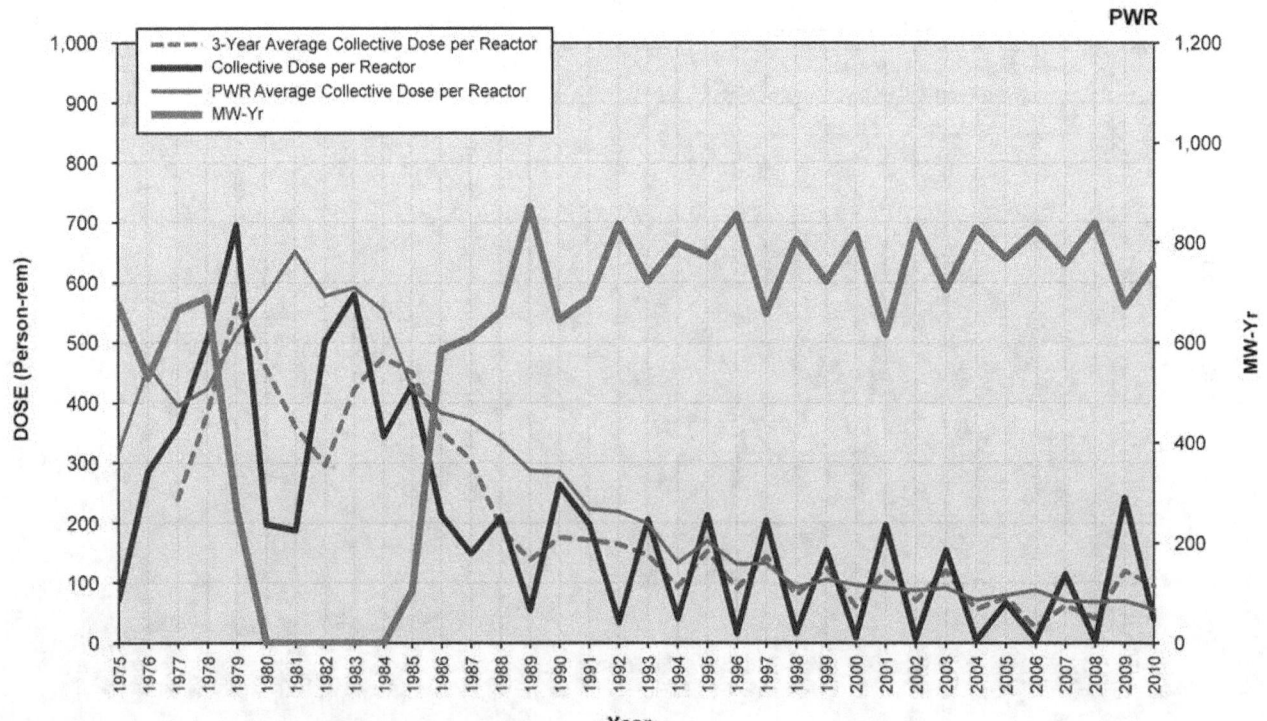

*Graph includes data for Three Mile Island 2 for the years 1975 – 1985.

## TURKEY POINT 3, 4
### Dose Performance Trends

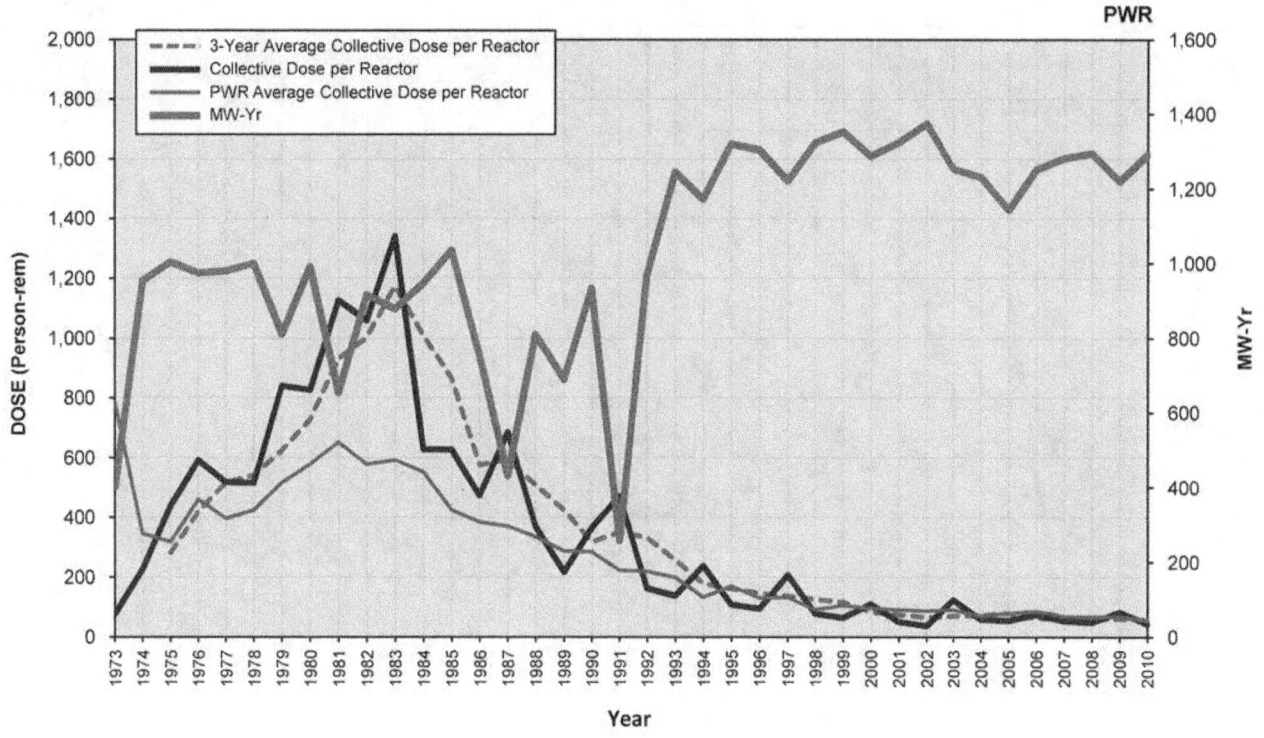

## VERMONT YANKEE
### Dose Performance Trends

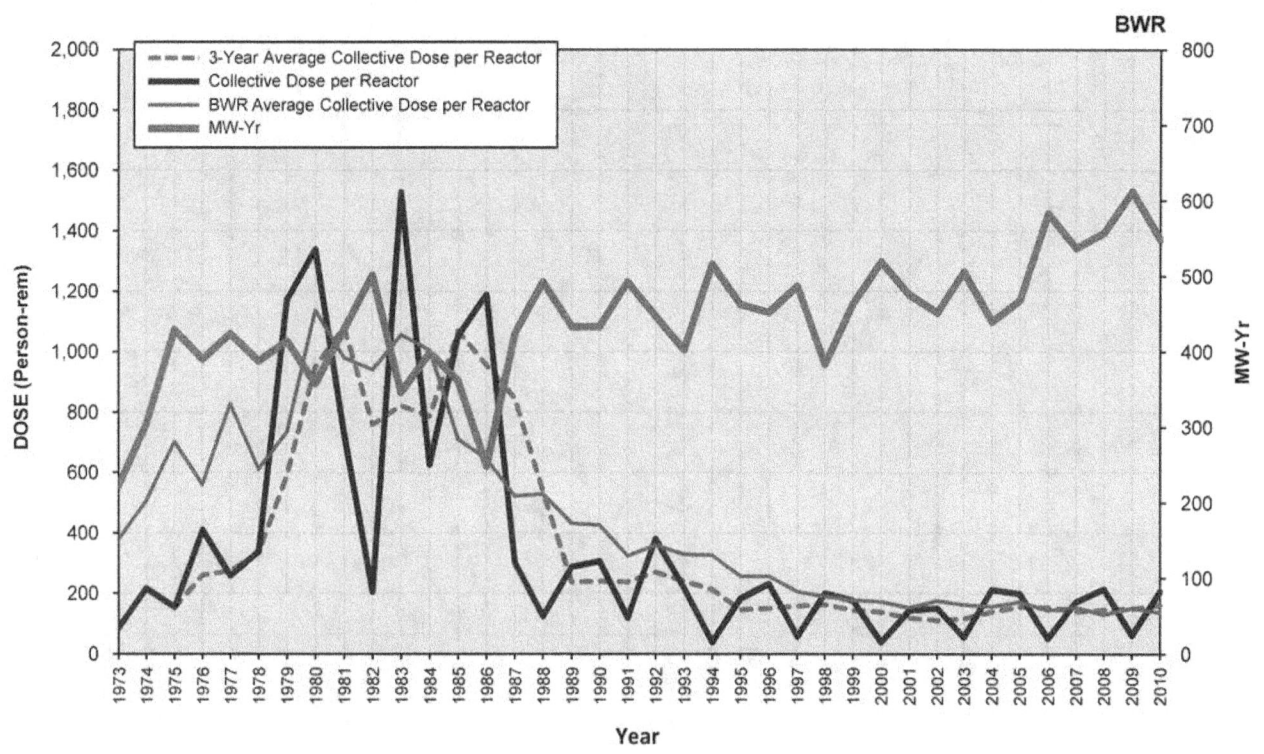

## VOGTLE 1, 2
### Dose Performance Trends

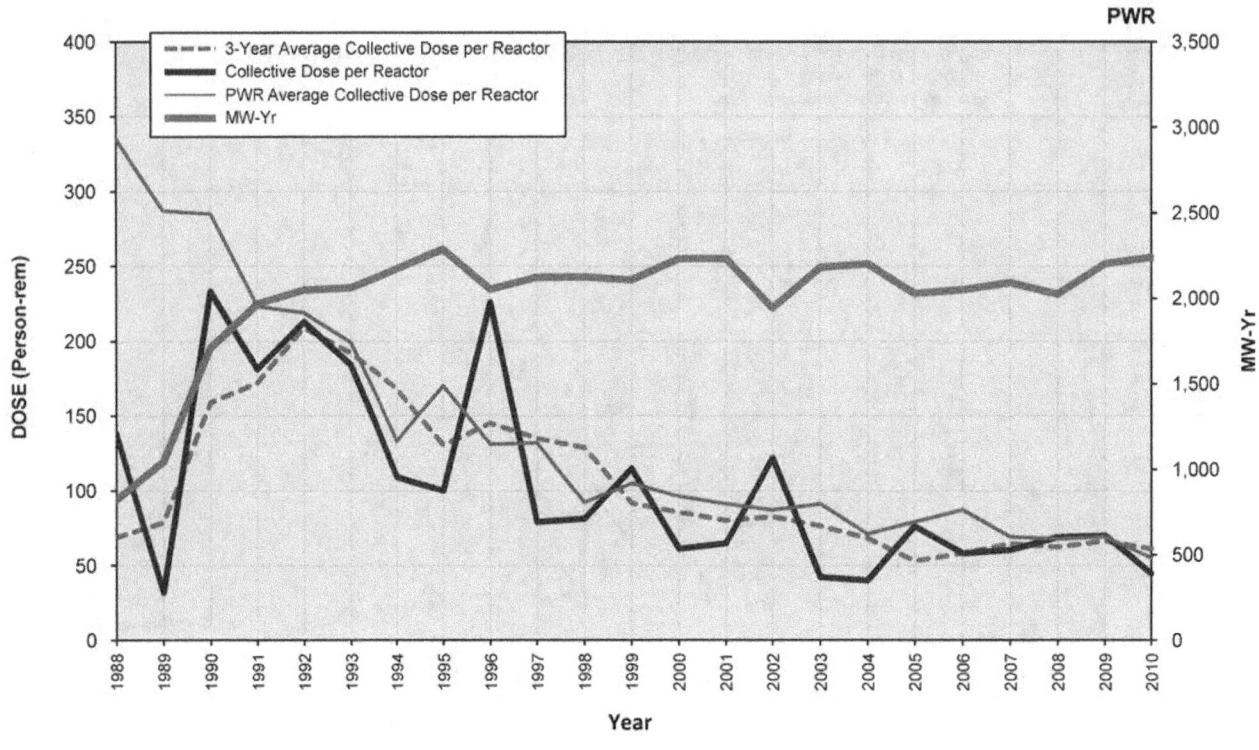

## WATERFORD 3
### Dose Performance Trends

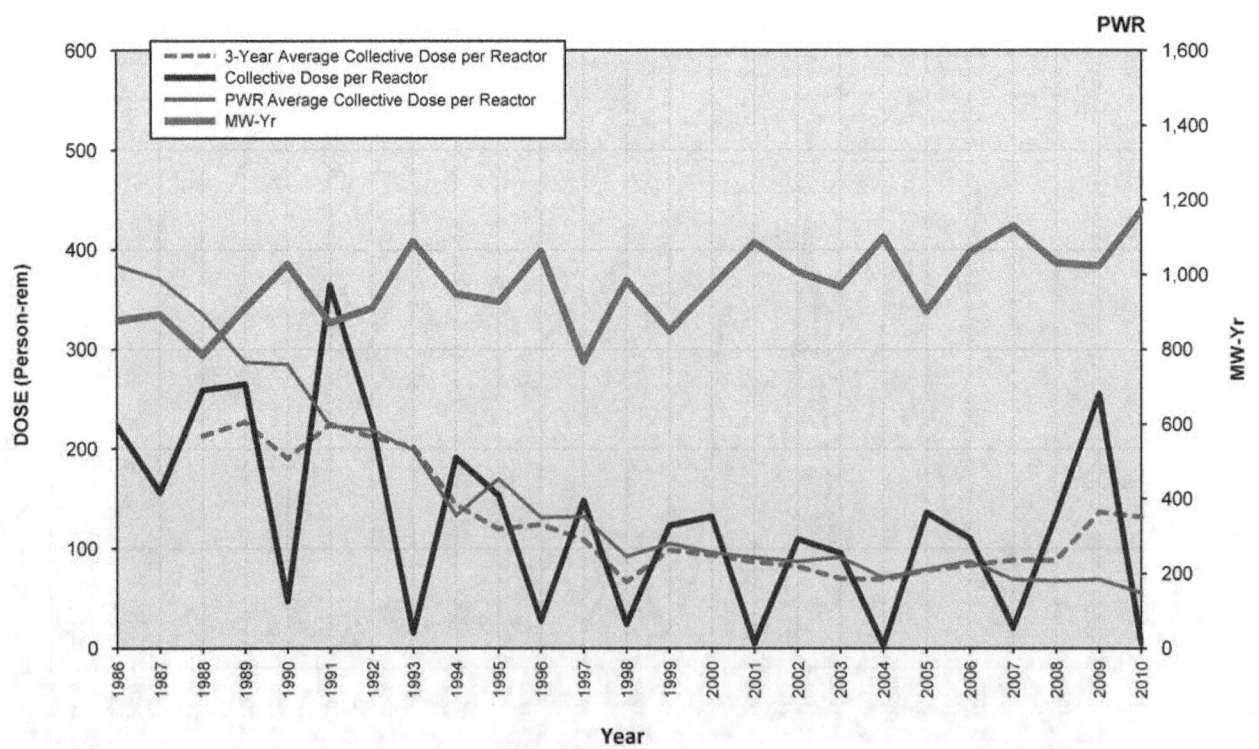

# WATTS BAR 1
## Dose Performance Trends

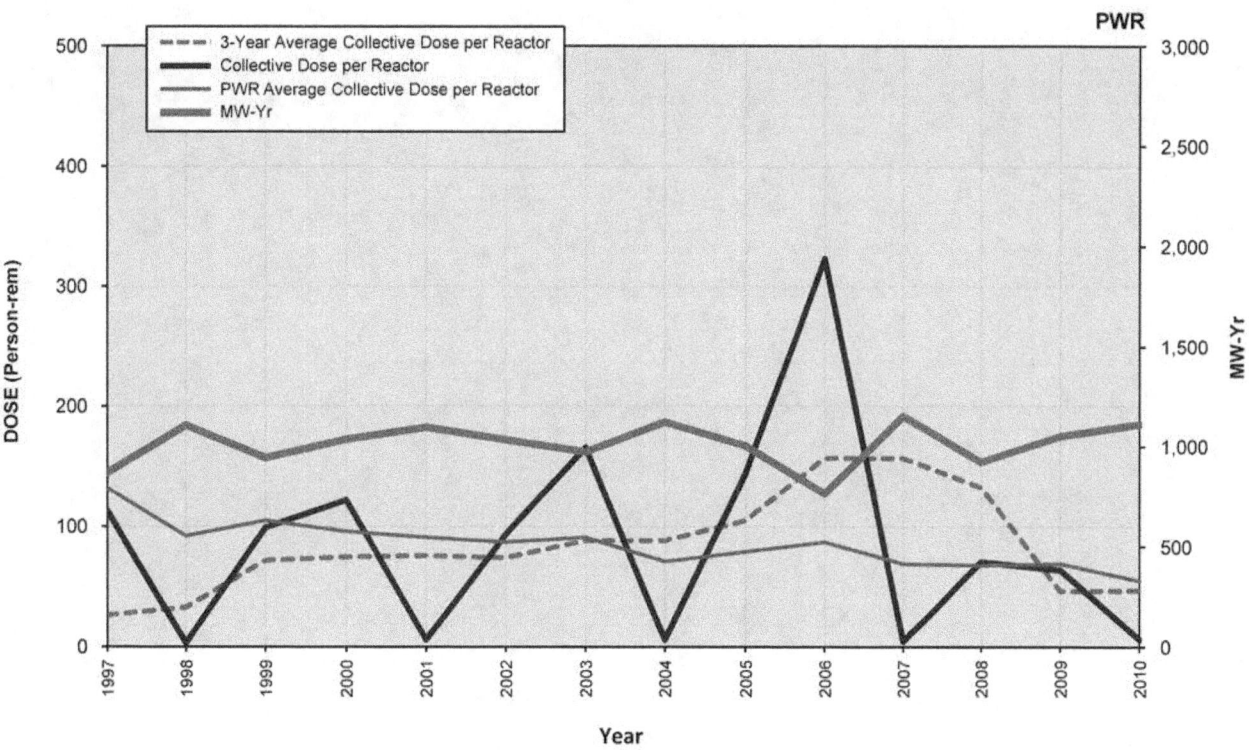

# WOLF CREEK 1
## Dose Performance Trends

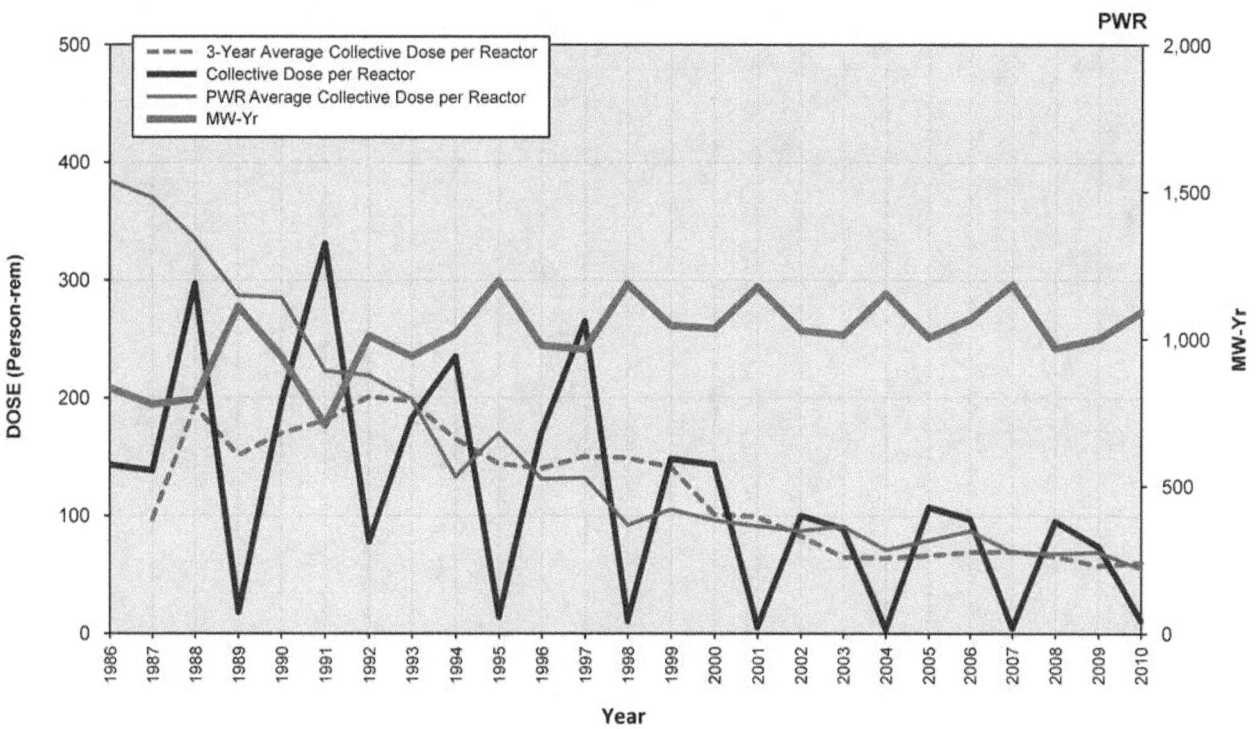

# Appendix E*

# PLANTS NO LONGER IN OPERATION

# 2010

* Information in this Appendix was obtained from Reference 18

## Big Rock Point

Big Rock Point (BRP) was a boiling water reactor rated at 75Mw electric, designed by General Electric Company and owned by Consumers Energy Company (CE). BRP permanently shut down on August 29, 1997 and fuel was transferred to the spent fuel pool by September 20, 1997. On March 26, 1998, CE submitted a revised PSDAR that showed conclusion of decommissioning about August 2005. Dry fuel storage will continue through 2012 or later, depending on when the U.S. Department of Energy (DOE) will accept spent fuel.

All systems and structures not needed for the independent spent fuel storage installation (ISFSI), except the intake piping and sanitary drainfield, have been removed. All remedial work has been completed and final status surveys were completed in 2006.

All fuel was transferred to the ISFSI by March, 2003. After fuel is removed from the site to a DOE facility, the ISFSI will be decommissioned and the license terminated.

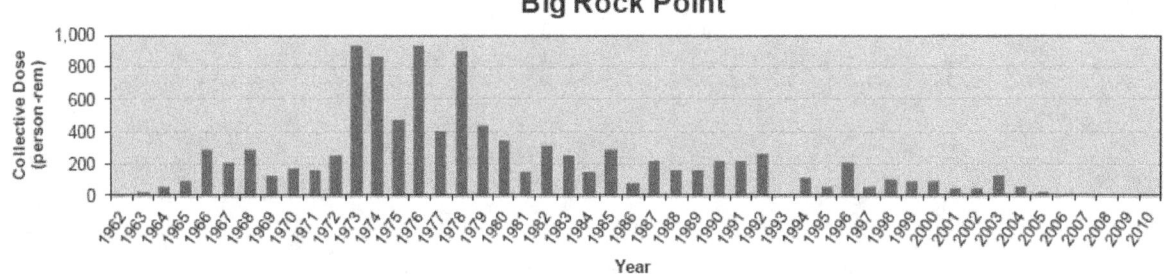

**Big Rock Point**

## Dresden Unit 1

Dresden Unit 1 produced power commercially from August 1, 1960 to October 31, 1978. Unit 1 was taken off-line on October 31, 1978 to backfit it with equipment to meet new federal regulations and to perform a chemical decontamination of major piping systems. While the unit was out of service for retrofitting, additional regulations were issued as a result of the March 1979 incident at Three Mile Island. The estimated cost to bring Unit 1 into compliance with these regulations was more than $300 million. Commonwealth Edison, the owner of the facility, concluded that the age of the unit and its relatively small size did not warrant the added investment and submitted a Decommissioning Plan to the NRC. NRC approved the Decommissioning Plan in September 1993. Dresden Unit 1 is currently in SAFSTOR.

During the SAFSTOR period, through 2027, the Unit 1 facility will be subjected to periodic inspection and monitoring. These activities will include condition monitoring of the ISFSI, ongoing environmental surveys, and maintenance of equipment required to support the SAFSTOR condition of the facility. The licensee plans that decontamination and dismantlement of Unit 1, including removal of any remaining spent fuel that is stored in the Unit 3 spent fuel pool, will take place from

2029 through 2031. In 2031, a comprehensive radiological survey will be initiated to demonstrate readiness for demolition of the Unit 1 portions of the facility. A four-year site restoration delay will follow the major decontamination and dismantlement of Unit 1 to allow for the decontamination and dismantlement of Units 2 and 3, with completion of these activities tentatively planned for 2035. Site restoration will be conducted in 2035 and 2036, concluding with a final site survey in late 2036. The licensee will monitor the ISFSI complex with site security and periodic inspections until final transfer of the spent fuel to DOE.

## Fermi Unit 1

The Enrico Fermi Atomic Power Plant, Unit 1 (Fermi 1) was a fast breeder reactor power plant cooled by sodium and operated at essentially atmospheric pressure. The reactor plant was designed for a maximum capacity of 430 Megawatt (MWt); however, the maximum reactor power was 200 MWt. The primary system was filled with sodium in December of 1960 and criticality was achieved in August 1963. The reactor was tested at low power in its first couple years of operation. Power ascension testing above 1 MWt commenced in December 1965, immediately following receipt of the high power operating license. In October 1966, during a power ascension, a zirconium plate at the bottom of the reactor vessel became loose and blocked sodium coolant flow to some fuel subassemblies. Two subassemblies started to melt. Radiation monitors alarmed and the operators manually shut down the reactor. No abnormal releases to the environment occurred. Three years and nine months later, the cause had been determined, cleanup completed, fuel replaced, and Fermi 1 was restarted. In 1972, the core was approaching the burnup limit. In November 1972, the Power Reactor Development Company made the decision to decommission Fermi 1.

The fuel and blanket subassemblies were shipped offsite in 1973. The non-radioactive secondary sodium system was drained and the sodium sent to Fike Chemical Company. The radioactive primary sodium was stored in storage tanks and in 55 gallon drums until the sodium was shipped offsite in 1984. Decommissioning of the Fermi 1 plant was originally completed in December 1975. The license for Fermi 1 expires in 2025. The licensee submitted a revised LTP in March 2010, and NRC staff completed an expanded acceptance review of the revised LTP for Fermi Unit 1.

## Haddam Neck – Connecticut Yankee

In 1996, Haddam Neck (a pressurized water reactor) ceased power operations. Steam generators, reactor coolant pumps, the pressurizer, the reactor vessel, and shield wall blocks from the Reactor Building have been disposed offsite and demolition of the administration and turbine building began in spring 2004. As of March 30, 2005, all spent fuel and greater than Class C waste have been transferred to the ISFSI which is currently operational.

Decommissioning at Haddam Neck was completed in 2007 and the Part 50 license requirements are in effect at the Haddam Neck ISFSI.

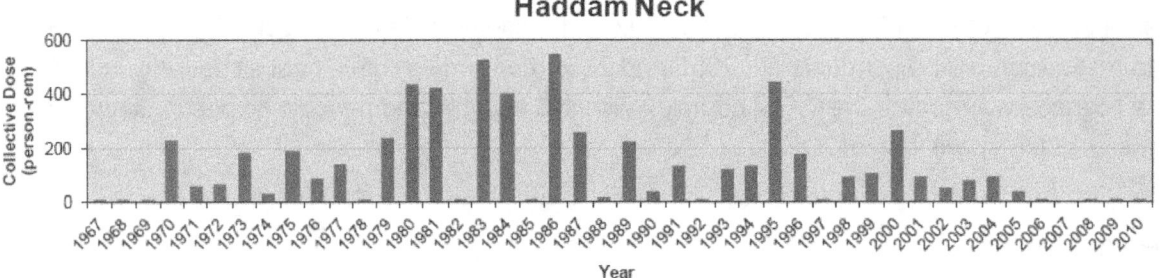

## Humboldt Bay Unit 3

Humboldt Bay Unit 3 produced power commercially from August 1, 1963 to July 1976. In July 1976, Unit 3 was shut down for seismic modifications. In 1983, with the plant still shutdown, Pacific Gas & Electric, the owner of the facility, determined that required seismic modifications and the requirements imposed as a result of the incident at Three Mile Island, made continued operations no longer economically feasible and decided to decommission the plant. The NRC approved the licensee's Decommissioning Plan in July 1988. Humboldt Bay Unit 3 has been in SAFSTOR since July 1976 until recently.

The licensee submitted a PSDAR in February 1998, and has begun incremental decommissioning activities. In December 2003, the licensee submitted an ISFSI application to the NRC. Humboldt Bay will have a unique ISFSI dry cask storage because of the short length of its fuel assemblies. Moreover, the casks will be stored below-grade to accommodate regional seismicity issues, security concerns, and site boundary dose limits. The NRC issued the ISFSI license on November 18, 2005, and the licensee began constructing the ISFSI in 2007. Following fuel loading into the ISFSI in 2008, the licensee began constructing two new units in 2008 and 2009 to replace Humboldt Bay Units 1 and 2. Decommissioning activities of the old Units 1 and 2 began in 2009 and 2010, respectively. During this period, only incremental decommissioning of Unit 3 has occurred. As decommissioning of Units 1 and 2 is completed, full decommissioning of Unit 3 will begin. It is estimated that all decommissioning activities will be completed in 2015.

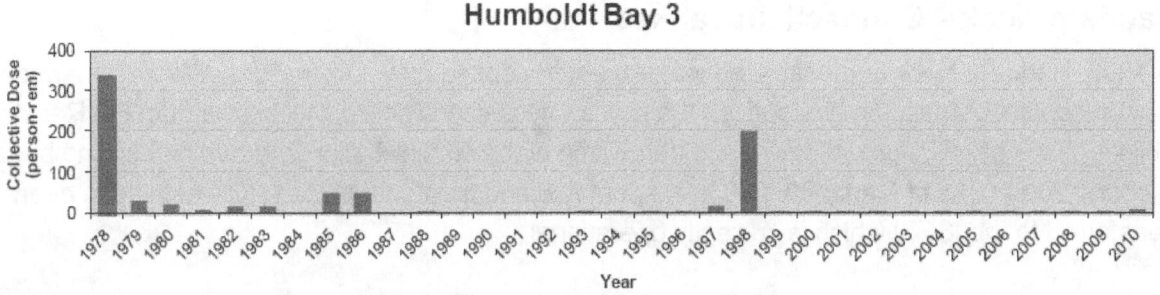

## Indian Point Unit 1

Indian Point Unit 1 (IP-1) produced power commercially from August 1962 to October 31, 1974. IP-1 was shutdown on October 31, 1974 because the emergency core cooling system did not meet regulatory requirements. Some decommissioning work associated with spent fuel storage was performed from 1974 through 1978. By January 1976, all spent fuel was removed from the reactor vessel. The NRC order approving SAFSTOR was issued in January 1996.

A PSDAR public meeting was held on January 20, 1999. The licensee plans to decommission IP-1 with Indian Point Unit 2 (IP-2), which is currently in operation. The licensee does not plan to begin active decontamination and decommissioning of IP-1 until the IP-2 license expires in September 2013. It is estimated that all decommissioning activities will be completed in 2026.

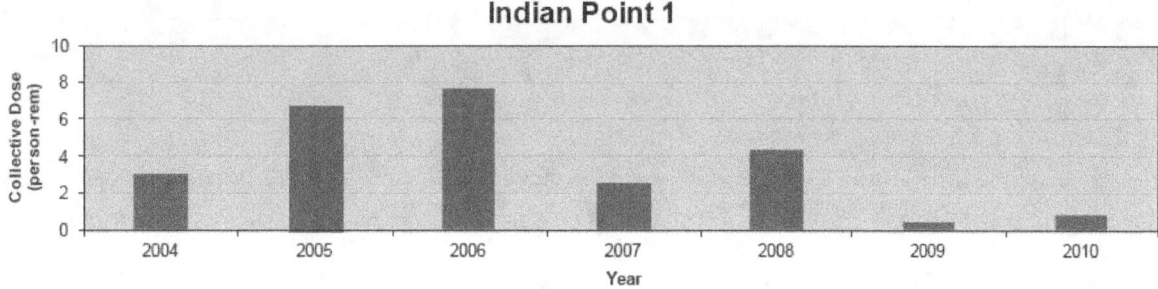

## La Crosse

The La Crosse Boiling Water Reactor (LACBWR) produced power commercially from November 1, 1969 to April 30, 1987. The plant was one of a series of demonstration plants funded, in part, by the U.S Atomic Energy Commission (AEC). The nuclear steam supply system and its auxiliaries were funded by the AEC, and the balance of the plant was funded by the Dairyland Power Cooperative (DPC). The AEC later sold the plant to DPC and provided them with a provisional operating license. LACBWR was shut down on April 30, 1987 and the NRC approved its Decommissioning Plan on August 7, 1991. LACBWR's Decommissioning Plan is also its PSDAR. LACBWR is currently in SAFSTOR.

NRC held a public meeting on LACBWR's PSDAR on May 13, 1998. DPC has been conducting dismantlement and decommissioning activities and is currently developing plans for an ISFSI. It is estimated that all decommissioning activities will be completed in 2026.

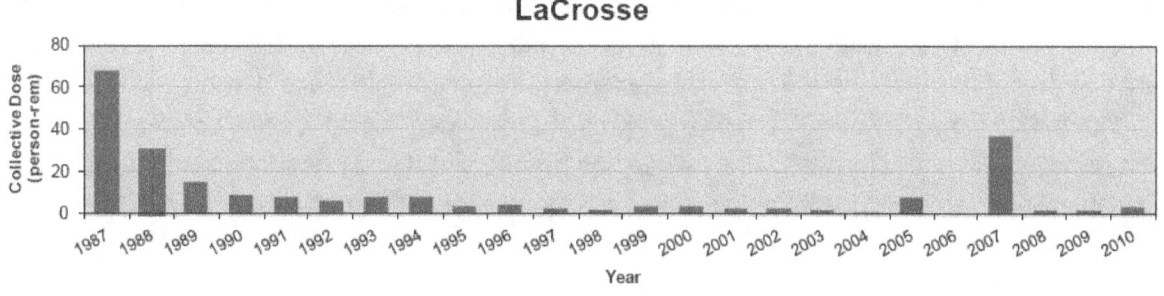

## Maine Yankee

Maine Yankee was a 900 Mw pressurized water reactor located on Bailey Point in Wiscasset. The Maine Yankee plant was shut down on December 6, 1996. Certification of permanent cessation of operations was submitted on August 7, 1997. The PSDAR was submitted on August 27, 1997 and the License Termination Plan (LTP) was approved on February 28, 2003.

In 2003, the reactor pressure vessel was shipped to Barnwell, South Carolina via barge. Spent nuclear fuel and greater than Class C Waste was transferred to the onsite ISFSI beginning in August 2002 and ending February 2004. Decommissioning was completed in June 2005 and Maine Yankee will retain its Part 50 license until the fuel is removed from the ISFSI.

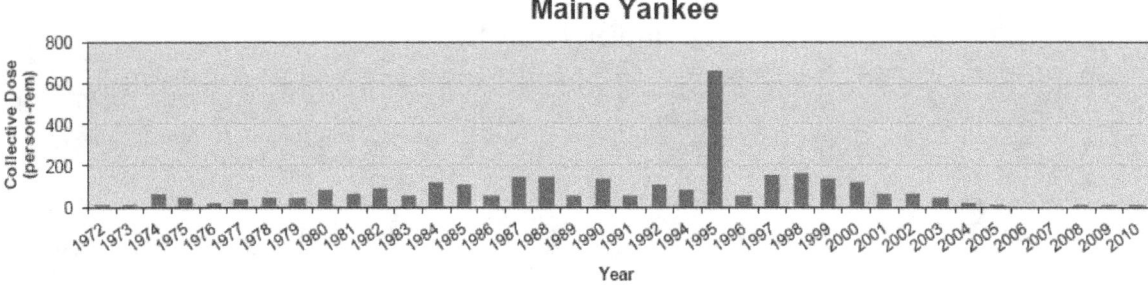

## Millstone Unit 1

Millstone Unit 1 produced power commercially from December 28, 1970 to November 4, 1995. On May 19, 1966, the AEC authorized a provisional construction permit for the construction of Millstone Unit 1. Construction of Millstone Unit 1 was completed and fuel loading began in October 1970. The plant went into commercial operation on December 28, 1970. Millstone Unit 1 was a single-cycle, boiling water reactor with a reactor thermal output of 2011 megawatts and a net electrical output of 652.1 megawatts. The unit was shut down on November 4, 1995. On July 21, 1998, pursuant to 10 CFR 50.82(a)(1)(i) and 10 CFR 50.82(a)(1)(ii), the licensee certified to the NRC that, as of July 17, 1998, Millstone Unit 1 had permanently ceased operations and that fuel had been permanently removed from the reactor vessel. Dominion Nuclear Connecticut, the owner of the facility, submitted its PSDAR to the NRC on June 14, 1999. Millstone Unit 1 is currently in SAFSTOR.

Safety related structures, systems, and components (SSCs) and SSCs important to safety remaining at Millstone Unit 1 are associated with the spent fuel pool island where the spent fuel is stored. Other than non-essential systems supporting the balance of plant facilities, the remaining plant equipment has been de-energized, disabled and abandoned in place or removed from the unit and can no longer be used for power generation. Irradiated reactor vessel components have been removed. The reactor cavity and vessel has been drained and abandoned with a radiation shield installed to limit occupational radiation doses to workers. Currently, the licensee has not provided an estimated date for completion of all decommissioning activities. However, the estimated closure date of this site has not been determined.

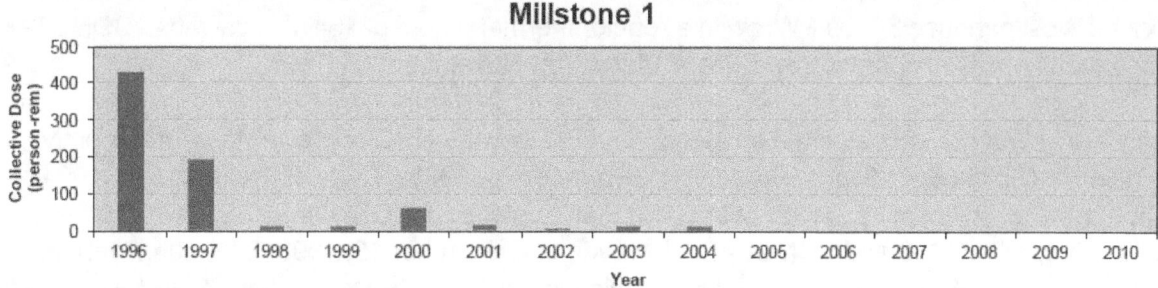

## Peach Bottom Unit 1

Peach Bottom Atomic Power Station, Unit 1 was a 200 MWt, high temperature, gas cooled reactor that was operated from June of 1967 to its final shutdown on October 31, 1974. All spent fuel has been removed from the site, and the spent fuel pool is drained and decontaminated. The reactor vessel, primary system piping, and steam generators remain in place.

The facility is currently in a SAFSTOR condition. The post-shutdown decommissioning activities report meeting was held on June 29, 1998. Final decommissioning is not expected until 2034 when Units 2 and 3 are scheduled to shut down.

## Rancho Seco

Rancho Seco Nuclear Generating Station was a 913 Mw pressurized water reactor owned by the Sacramento Municipal Utility District (SMUD). Rancho Seco permanently shut down in June 1989, after approximately 15 years of operation.

SMUD completed transfer of all the spent nuclear fuel to the Rancho Seco ISFSI in August 2002.

Rancho Seco completed decommissioning in 2009 and the site was released as greenfields, with the exception of a 6-acre ISFSI site.

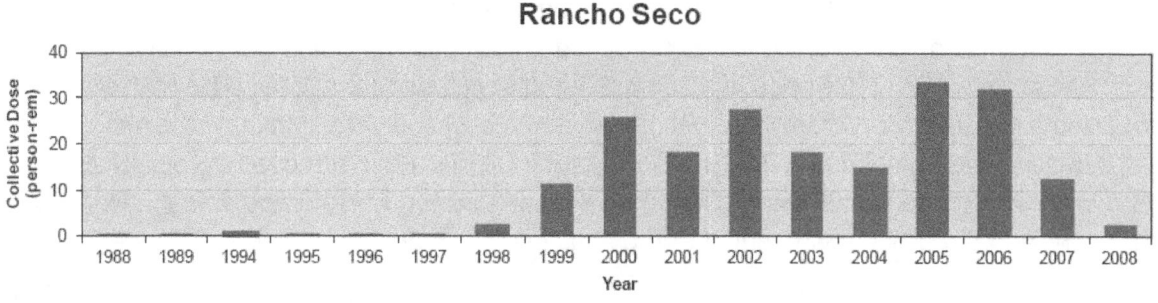

## San Onofre Unit 1

The San Onofre Nuclear Generating Station Unit 1 (SONGS-1), operated by Southern California Edison (SCE), produced power commercially from January 1, 1968 to November 30, 1992. Unit 1 was a Westinghouse 3-loop PWR with a reactor thermal output of 1347 megawatts. SONGS-1 subsequently ceased operation and was shutdown on November 30, 1992.

Defueling of SONGS-1 completed on March 6, 1993, and the NRC approved the Permanently Defueled Technical Specifications report on December 28, 1993. Then, on November 3, 1994, SCE submitted a Proposed Decommissioning Plan to place SONGS-1 in SAFSTOR until the shutdown of SONGS- 2 and SONGS- 3. However, on December 15, 1998, SCE submitted the PSDAR for SONGS-1, to commence decontamination in 2000. Since that time, SCE has been actively decommissioning the facility, which has since been almost entirely dismantled. Most of the structures and equipment have been removed and disposed. The SONGS-1 turbine building was removed and the licensee completed internal segmentation and cutup of the reactor pressure vessel. The licensee plans to store the vessel onsite for the foreseeable future, as long as licensed activities are ongoing. In addition, the licensee transferred SONGS-1 spent fuel to an onsite generally licensed ISFSI. The ISFSI will be expanded into the area previously occupied by SONGS- 1, as needed, in order to store all spent fuel from SONGS-2 and SONGS-3. SONGS-2 and SONGS-3 are expected to continue operating until 2022. In February 2010, NRC staff issued a license amendment to release off-shore portions of the San Onofre Unit 1 cooling intake and outlet pipes for unrestricted use. It is estimated that all decommissioning activities for SONGS-1 will be completed in 2030.

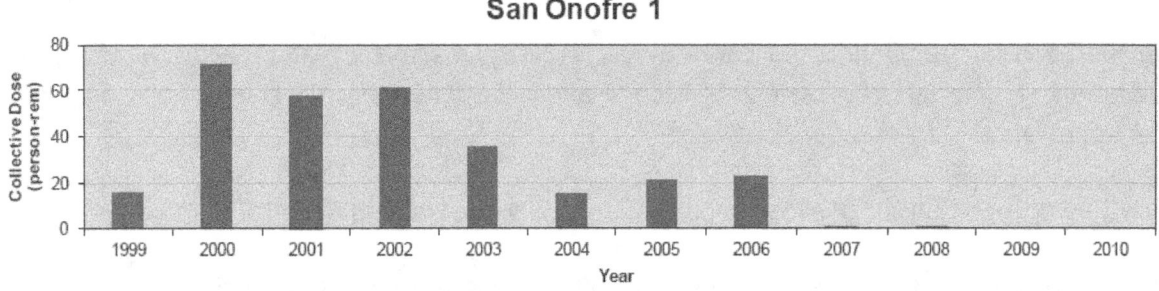

San Onofre 1

## Three Mile Island Unit 2

Three Mile Island Unit 2 (TMI-2) produced power commercially from December 30, 1978 to March 28, 1979. On March 28, 1979, the unit experienced an accident which resulted in severe damage to the reactor core. TMI-2 has been in a non-operating status since that time. The licensee conducted a substantial program to defuel the reactor vessel and decontaminate the facility. The plant defueling was completed in April 1990. All spent fuel has been removed except for some debris in the reactor coolant system. The removed fuel is currently in storage at Idaho National Laboratory, and the U.S. Department of Energy has taken title and possession of the fuel.

TMI-2 has been defueled and decontaminated to the extent the plant is in a safe, inherently stable condition suitable for long-term management. This long-term management condition is termed post-defueling monitored storage, which was approved in 1993. TMI-2 shares equipment with the operating TMI – Unit 1 (TMI-1). The licensee plans to actively decommission TMI-2 in parallel with the decommissioning of TMI-1. It is estimated that decommissioning activities for TMI-2 will be completed in 2036.

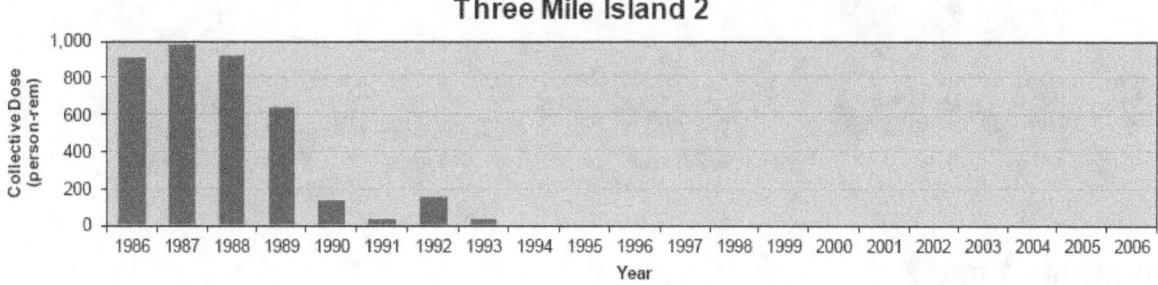

## Trojan

The Trojan plant was shut down in November 1992 and the steam generators and reactor vessel were shipped to the Hanford site. The licensee was granted a site-specific Part 72 license for an onsite ISFSI in March 1999 that is still in operation. The licensee began spent fuel transfer to the ISFSI in December 2002 and finished fuel transfer in August 2003.

In December 2004, the Trojan Nuclear Plant completed decommissioning activities. The NRC terminated Trojan's 10 CFR Part 50 operating license on May 23, 2005.

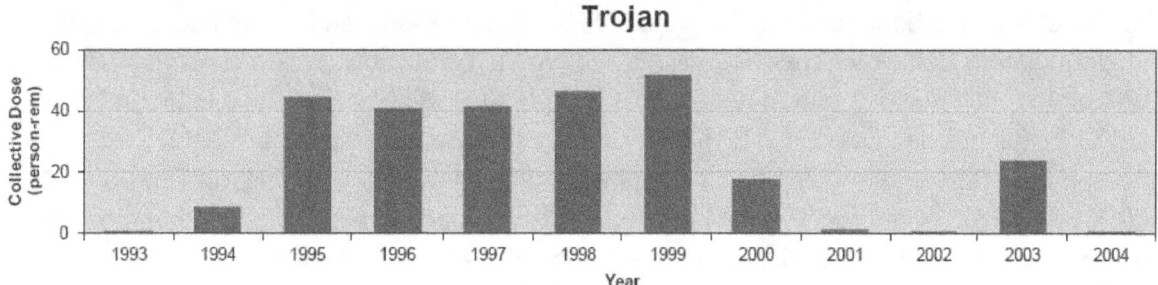

## Yankee Rowe

The plant was permanently shut down on October 1, 1991 and the steam generators were shipped to the Barnwell Low-Level Waste facility, in North Carolina, in November 1993. The reactor vessel was shipped to Barnwell in April 1997.

The owner completed construction of an onsite ISFSI and all the fuel from the spent fuel pool was transferred to the onsite ISFSI.

Yankee Rowe completed decommissioning in 2007. The license for the site was reduced to the two acres surrounding the ISFSI which is still in operation.

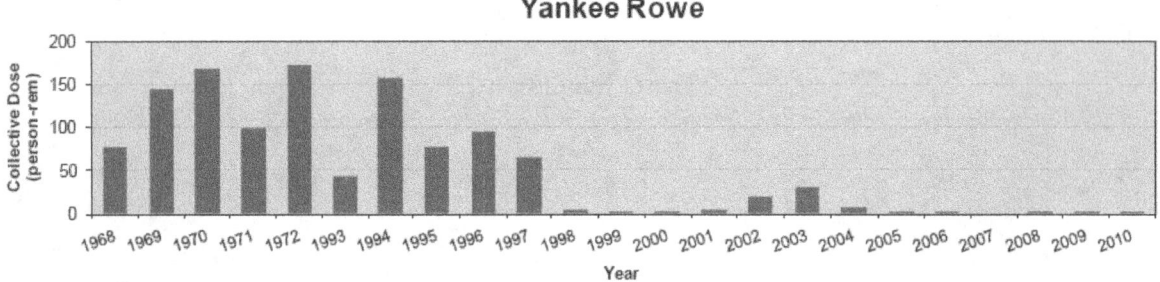

## Zion Units 1 and 2

Zion Nuclear Power Station (ZNPS) received a construction permit in December 1968 to begin building two nuclear power reactors. Unit 1 produced power commercially from December 31, 1973 to February 21, 1997 and Unit 2 produced power commercially from September 17, 1974 to September 19, 1996. On January 14, 1998, the Unicom Corporation and ComEd Boards of Directors, the joint owners of the facility, authorized the permanent cessation of operations at ZNPS for economic reasons. ComEd certified, in a letter dated February 13, 1998, to the NRC that operations had ceased at ZNPS.

On April 27, 1997, all fuel from Unit 1 was removed and on February 25, 1998 all fuel from Unit 2 was removed and placed in the spent fuel pool. On March 9, 1998, ComEd informed the NRC that all fuel had been removed from the ZNPS reactor vessels and committed to maintain them permanently defueled. The NRC acknowledged the certification of permanent cessation of power operation and permanent removal of fuel from the reactor vessels in a letter dated May 4, 1998. ZNPS has been placed in SAFSTOR, where it will remain until about 2013. The owner submitted the PSDAR, site-specific cost estimate, and fuel management plant on February 14, 2000. The SAFSTOR approach is the intended decommissioning method to be utilized for ZNPS which involves removal of all radioactive material from the site following a period of dormancy. In 2010 NRC staff finalized the transfer of the possession license for Zion Units 1 and 2 from Exelon Generating Company, LLC to Zion Solutions, LLC to facilitate decommissioning. Preparations for decontamination and dismantlement are scheduled to commence at the original license expiration date for ZNPS Unit 2 on November 14, 2013. It is estimated that all decommissioning activities will be completed at ZNPS in 2020.

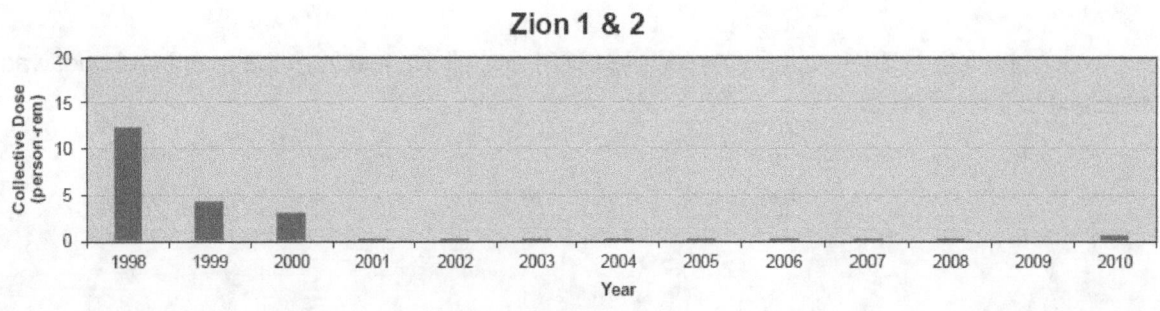

# Appendix F

# GLOSSARY

# 2010

*Agreement State:* as defined in 10 CFR 30.4, means any state with which the Atomic Energy Commission or the Nuclear Regulatory Commission has entered into an effective agreement under subsection 274b. of the [Atomic Energy] Act [of 1954, including any amendments thereto]. To simplify subsection 274b., an Agreement State is a state that has signed an agreement with the NRC under which the state regulates the use of certain byproduct, source, and small quantities of special nuclear material in that state.

*As low as is reasonably achievable (ALARA):* as defined in 10 CFR 20.1003, means making every reasonable effort to maintain exposures to radiation as far below the dose limits in 10 CFR 20 as is practical consistent with the purpose for which the licensed activity is undertaken, taking into account the state of technology, the economics of improvements in relation to the state of technology, the economics of improvements in relation to benefits to the public health and safety, and other societal and socioeconomic considerations, and in relation to utilization of nuclear energy and licensed materials in the public interest.

*Average measurable dose:* the dose obtained by dividing the collective dose by the number of individuals who received a measurable dose. This is the average most commonly used in this and other reports when examining trends and comparing doses received by workers, because it excludes those individuals receiving a less than measurable dose.

*Boiling water reactor (BWR):* reactor in which the water, used as both coolant and moderator, is allowed to boil in the core. The resulting steam can be used directly to drive a turbine and electrical generator, thereby producing electricity.

*Byproduct material:* as partially defined in 10 CFR 20.1003, means any radioactive material (except special nuclear material) yielded in, or made radioactive by, exposure to the radiation incident to the process of producing or using special nuclear material; and the tailings or wastes produced by the extraction or concentration of uranium or thorium from ore processed primarily for its source material content.

*Breeder:* a reactor that produces more nuclear fuel than it consumes. A fertile material, such as uranium-238, when bombarded by neutrons, is transformed into a fissile material, such as plutonium-239, which can be used as fuel. [Ref. 19]

*Class (or lung class or inhalation class):* as defined in 10 CFR 20.1003, means a classification scheme for inhaled material according to its rate of clearance from the pulmonary region of the lung. Materials are classified as D, W, or Y, which applies to a range of clearance half-times: for Class D (Days) of less than 10 days, for Class W (Weeks) from 10 to 100 days, and for Y (Years) of greater than 100 days.

*Collective dose:* as defined in 10 CFR 20.1003, is the sum of the individual doses received in a given period of time by a specified population from exposure to a specified source of radiation.

*Committed dose equivalent:* as defined in 10 CFR 20.1003, means the dose equivalent to organs or tissues of reference that will be received from an intake of radioactive material by an individual during the 50-year period following the intake. The acronym CDE is an NRC acronym used for this term.

*Committed effective dose equivalent:* as defined in 10 CFR 20.1003, is the sum of the products of the weighting factors applicable to each of the body organs or tissues that are irradiated and the committed dose equivalent to these organs or tissues. The acronym CEDE is an NRC acronym used for this term.

*Criticality:* the normal operating condition of a reactor, in which nuclear fuel sustains a fission chain reaction. A reactor achieves criticality (and is said to be critical) when each fission event releases a sufficient number of neutrons to sustain an ongoing series of reactions. [Ref. 19]

*DECON (immediate dismantlement):* soon after the nuclear facility closes, equipment, structures, and portions of the facility containing radioactive contaminants are removed or decontaminated to a level that permits release of the property and termination of the NRC license.

*ENTOMB:* radioactive contaminants that are permanently encased onsite in a structurally sound material such as concrete and appropriately maintained and monitored until the radioactivity decays to a level permitting restricted release of the property.

*Exposure:* as defined in 10 CFR 20.1003, means being exposed to ionizing radiation or to radioactive material.

*Independent Spent Fuel Storage Installation (ISFSI):* as defined in 10 CFR 72.3 means a complex designed and constructed for the interim storage of spent nuclear fuel, solid reactor-related GTCC waste, and other radioactive materials associated with spent fuel and reactor-related GTCC waste storage. An ISFSI which is located on the site of another facility licensed under 10 CFR 72 or a facility licensed under 10 CFR 50 of [Title 10 of the Code of Federal Regulations] and which shares common utilities and services with that facility or is physically connected with that other facility may still be considered independent.

*Lens dose equivalent (LDE):* as defined in 10 CFR 20.1003, applies to the external exposure of the lens of the eye and is taken as the dose equivalent at a tissue depth of 0.3 centimeter (300 mg/cm2).

*License:* as defined in 10 CFR 20.1003, means a license issued under the regulations in 10 CFR parts 30 through 36, 39, 40, 50, 60, 61, 63, 70, or 72 of [Title 10 of the Code of Federal Regulations].

*Licensee:* as defined in 10 CFR 20.1003, means the holder of the NRC license.

*Licensed material:* as defined in 10 CFR 20.1003, means source material, special nuclear material, or byproduct material received, possessed, used, transferred, or disposed of under a general or specific license issued by the [Nuclear Regulatory] Commission.

*Light water reactor (LWR):* the term used in this report to describe commercial nuclear reactors that use ordinary water as a coolant and are operated for the purposes of generating electricity. Light water reactors include boiling water reactors (BWRs) and pressurized water reactors (PWRs).

*Measurable dose:* a dose greater than zero rem (not including doses reported as "not detectable").

*Megawatt-year:* unit of electric energy, equal to the energy from a power of 1,000,000 watts over a period of one year.

*Mode of Intake:* the manner of intake into the body: inhalation (H), absorption through the skin (B), oral ingestion (G), and injection (J).

*Monitoring year:* interval during which the radiation exposure monitoring was performed.

*Non-reactor licensees:* NRC licensees that are not commercial nuclear power reactors. These licensees are industrial radiographers, fuel processors, fabricators, and reprocessors; manufacturers and distributors of byproduct material; independent spent fuel storage installations; facilities for land disposal of low-level waste; and geologic repositories for high-level waste.

*Number of individuals with measurable dose:* the count of unique individuals who received measurable dose during the monitoring year. In some instances in this report, the number of individuals with measurable dose may include individuals who are counted more than once since they may be monitored at more than one licensee during the year. (See Section 5 on the effect of transient individuals.) Tables that have been adjusted for transient workers are noted in the appropriate footnotes to the tables.

*Occupational dose:* as defined in 10 CFR 20.1003, means the dose received by an individual in the course of employment in which the individual's assigned duties involve exposure to radiation and to radioactive material from licensed and unlicensed sources of radiation, whether in the possession of the licensee or other person. Occupational dose does not include doses received from background radiation, from any medical administration the individual has received, from exposure to individuals administered radioactive material and released under [10 CFR] 35.75, from voluntary participation in medical research programs, or as a member of the public.

*Pressurized water reactor (PWR):* power reactor in which heat is transferred from the core to an exchanger by high temperature water kept under high pressure in the primary system. Steam used to turn a turbine and electrical generator is generated in a secondary circuit. The majority of reactors producing electric power in the United States are pressurized water reactors.

*Radionuclide:* a radioisotope. A radioisotope is an unstable isotope that undergoes spontaneous transformation, emitting radiation. [Ref. 20]

*REM:* as defined in 10 CFR 20.1004, is the special unit of any of the quantities expressed as dose equivalent. The dose equivalent in rems is equal to the absorbed dose in rads multiplied by the quality factor (1 rem = 0.01 sievert).

*SAFSTOR (often considered 'delayed DECON'):* a nuclear facility that is maintained and monitored in a condition that allows the radioactivity to decay; afterwards, it is dismantled.

*Shallow dose equivalent, maximum extremity (SDE-ME):* the external exposure of an extremity, taken as the dose equivalent at a tissue depth of 0.007 centimeter.

*Shallow dose equivalent, whole body (SDE-WB):* the external exposure of the skin, taken as the dose equivalent at a tissue depth of 0.007 centimeter.

*Sievert:* as defined in 10 CFR 20.1004, is the SI unit of any of the quantities expressed as dose equivalent. The dose equivalent in sieverts is equal to the absorbed dose in grays multiplied by the quality factor (1 Sv = 100 rems).

*Special nuclear material (SNM):* as defined in 10 CFR 20.1003, means plutonium, uranium-233, uranium enriched in the isotope 233 or in the isotope 235, and any other material that the [Nuclear Regulatory] Commission, pursuant to the provisions of section 51 of the [Atomic Energy] Act [of 1954, as amended], determines to be special nuclear material, but does not include source material. Any material artificially enriched by any of the foregoing but does not include source material.

*Total effective dose equivalent (TEDE):* as defined in 10 CFR 20.1003, means the sum of the effective dose equivalent (for external exposures) and the committed effective dose equivalent (for internal exposures).

*Transient individual:* one who is monitored at more than one licensed site during the calendar year.

*Unit availability factor:* the unit available hours (the total clock hours in the report period during which the unit operated online or was capable of such operation) times 100 divided by the period hours.

| NRC FORM 335<br>(12-2010)<br>NRCMD 3.7 | U.S. NUCLEAR REGULATORY COMMISSION<br><br>BIBLIOGRAPHIC DATA SHEET<br>*(See instructions on the reverse)* | 1. REPORT NUMBER<br>(Assigned by NRC, Add Vol., Supp., Rev., and Addendum Numbers, if any.)<br><br>NUREG-0713, Volume 32 |
|---|---|---|

| 2. TITLE AND SUBTITLE | 3. DATE REPORT PUBLISHED | |
|---|---|---|
| Occupational Radiation Exposure at Commercial Nuclear Power Reactors and Other Facilities, 2010 | MONTH<br>May | YEAR<br>2012 |
| | 4. FIN OR GRANT NUMBER | |

| 5. AUTHOR(S)<br><br>D.E. Lewis<br>* D.A. Hagemeyer | 6. TYPE OF REPORT<br><br>Annual |
|---|---|
| | 7. PERIOD COVERED (Inclusive Dates)<br><br>Calendar Year 2010 |

8. PERFORMING ORGANIZATION - NAME AND ADDRESS (If NRC, provide Division, Office or Region, U. S. Nuclear Regulatory Commission, and mailing address; if contractor, provide name and mailing address.)

Division of Systems Analysis                   *Oak Ridge Associated Universities (ORAU)
Office of Nuclear Regulatory Research          1299 Bethel Valley Road, SC-200, MS-21
US Nuclear Regulatory Commission               Oak Ridge, TN 37830
Washington, DC 20555-0001

9. SPONSORING ORGANIZATION - NAME AND ADDRESS (If NRC, type "Same as above", if contractor, provide NRC Division, Office or Region, U. S. Nuclear Regulatory Commission, and mailing address.)

Same as 8 above.

10. SUPPLEMENTARY NOTES

11. ABSTRACT (200 words or less)

This report summarizes the occupational radiation exposure data maintained in the U.S. Nuclear Regulatory Commission's (NRC's) Radiation Exposure Information and Reporting System (REIRS). The bulk of the information contained in this report was compiled from the 2010 annual reports submitted by five of the seven categories of NRC licensees subject to the reporting requirements in 10 CFR 20.2206. The annual reports submitted by these licensees consist of radiation exposure records for each monitored individual.

Annual reports were received from a total of 190 NRC licensees. Compilations of the reports submitted by the 190 licensees indicated that 192,424 individuals were monitored, 81,961 of whom received a measurable dose. The collective dose incurred by these individuals was 10,617 person-rem. In 2010, the average measurable dose per individual for all licensees calculated from reported data was 0.13 rem. Analysis of transient worker data indicate that 29,333 individuals completed work assignments at two or more licensees during the monitoring year. The corrected dose distribution resulted in an average measurable dose per individual for all licensees of 0.17 rem.

| 12. KEY WORDS/DESCRIPTORS (List words or phrases that will assist researchers in locating the report.)<br><br>occupational exposure<br>nuclear power reactor<br>fuel facility | 13. AVAILABILITY STATEMENT<br>unlimited |
|---|---|
| | 14. SECURITY CLASSIFICATION |
| | (This Page)<br>unclassified |
| | (This Report)<br>unclassified |
| | 15. NUMBER OF PAGES |
| | 16. PRICE |

NUREG-0713, Vol. 32

Occupational Radiation Exposure at Commercial Nuclear Power Reactors and Other Facilities 2010

May 2012

www.ingramcontent.com/pod-product-compliance
Lightning Source LLC
Chambersburg PA
CBHW080245180526

45167CB00006B/2423